Habib M. Ammari

# Challenges and Opportunities of Connected *k*-Covered Wireless Sensor Networks

# Studies in Computational Intelligence, Volume 215


**Editor-in-Chief**
Prof. Janusz Kacprzyk
Systems Research Institute
Polish Academy of Sciences
ul. Newelska 6
01-447 Warsaw
Poland
*E-mail:* kacprzyk@ibspan.waw.pl


Further volumes of this series can be found on our homepage: springer.com

Vol. 195. Vivek Bannore and Leszek Swierkowski
*Iterative-Interpolation Super-Resolution Image Reconstruction: A Computationally Efficient Technique,* 2009
ISBN 978-3-642-00384-4

Vol. 196. Valentina Emilia Balas, János Fodor and Annamária R. Várkonyi-Kóczy (Eds.)
*Soft Computing Based Modeling in Intelligent Systems,* 2009
ISBN 978-3-642-00447-6

Vol. 197. Mauro Birattari
*Tuning Metaheuristics,* 2009
ISBN 978-3-642-00482-7

Vol. 198. Efrén Mezura-Montes (Ed.)
*Constraint-Handling in Evolutionary Optimization,* 2009
ISBN 978-3-642-00618-0

Vol. 199. Kazumi Nakamatsu, Gloria Phillips-Wren, Lakhmi C. Jain, and Robert J. Howlett (Eds.)
*New Advances in Intelligent Decision Technologies,* 2009
ISBN 978-3-642-00908-2

Vol. 200. Dimitri Plemenos and Georgios Miaoulis *Visual Complexity and Intelligent Computer Graphics Techniques Enhancements,* 2009
ISBN 978-3-642-01258-7

Vol. 201. Aboul-Ella Hassanien, Ajith Abraham, Athanasios V. Vasilakos, and Witold Pedrycz (Eds.)
*Foundations of Computational Intelligence Volume 1,* 2009
ISBN 978-3-642-01081-1

Vol. 202. Aboul-Ella Hassanien, Ajith Abraham, and Francisco Herrera (Eds.)
*Foundations of Computational Intelligence Volume 2,* 2009
ISBN 978-3-642-01532-8

Vol. 203. Ajith Abraham, Aboul-Ella Hassanien, Patrick Siarry, and Andries Engelbrecht (Eds.)
*Foundations of Computational Intelligence Volume 3,* 2009
ISBN 978-3-642-01084-2

Vol. 204. Ajith Abraham, Aboul-Ella Hassanien, and André Ponce de Leon F. de Carvalho (Eds.)
*Foundations of Computational Intelligence Volume 4,* 2009
ISBN 978-3-642-01087-3

Vol. 205. Ajith Abraham, Aboul-Ella Hassanien, and Václav Snášel (Eds.)
*Foundations of Computational Intelligence Volume 5,* 2009
ISBN 978-3-642-01535-9

Vol. 206. Ajith Abraham, Aboul-Ella Hassanien, André Ponce de Leon F. de Carvalho, and Václav Snášel (Eds.)
*Foundations of Computational Intelligence Volume 6,* 2009
ISBN 978-3-642-01090-3

Vol. 207. Santo Fortunato, Giuseppe Mangioni, Ronaldo Menezes, and Vincenzo Nicosia (Eds.)
*Complex Networks,* 2009
ISBN 978-3-642-01205-1

Vol. 208. Roger Lee, Gongzu Hu, and Huaikou Miao (Eds.)
*Computer and Information Science 2009,* 2009
ISBN 978-3-642-01208-2

Vol. 209. Roger Lee and Naohiro Ishii (Eds.)
*Software Engineering, Artificial Intelligence, Networking and Parallel/Distributed Computing,* 2009
ISBN 978-3-642-01202-0

Vol. 210. Andrew Lewis, Sanaz Mostaghim, and Marcus Randall (Eds.)
*Biologically-Inspired Optimisation Methods,* 2009
ISBN 978-3-642-01261-7

Vol. 211. Godfrey C. Onwubolu (Ed.)
*Hybrid Self-Organizing Modeling Systems,* 2009
ISBN 978-3-642-01529-8

Vol. 212. Viktor M. Kureychik, Sergey P. Malyukov, Vladimir V. Kureychik, and Alexander S. Malyoukov
*Genetic Algorithms for Applied CAD Problems,* 2009
ISBN 978-3-540-85280-3

Vol. 213. Stefano Cagnoni (Ed.)
*Evolutionary Image Analysis and Signal Processing,* 2009
ISBN 978-3-642-01635-6

Vol. 214. Been-Chian Chien and Tzung-Pei Hong (Eds.)
*Opportunities and Challenges for Next-Generation Applied Intelligence,* 2009
ISBN 978-3-540-92813-3

Vol. 215. Habib M. Ammari
*Challenges and Opportunities of Connected k-Covered Wireless Sensor Networks,* 2009
ISBN 978-3-642-01876-3

Habib M. Ammari

# Challenges and Opportunities of Connected *k*-Covered Wireless Sensor Networks

## From Sensor Deployment to Data Gathering

Springer

Dr. Habib M. Ammari
Wireless Sensor and Mobile Ad-hoc Networks (WiSeMAN) Research Laboratory
Department of Computer Science
Hofstra College of Liberal Arts and Sciences
Hofstra University
Hempstead, NY 11549
USA

ISBN 978-3-642-26929-5 e-ISBN 978-3-642-01878-7

DOI 10.1007/978-3-642-01878-7

Studies in Computational Intelligence ISSN 1860949X


Softcover reprint of the hardcover 1st edition 2009



*Typeset & Cover Design:* Scientific Publishing Services Pvt. Ltd., Chennai, India.

Printed in acid-free paper

9 8 7 6 5 4 3 2 1

springer.com

This book is dedicated

To my first teachers: My mother, Mbarka, and my father, Mokhtar

To my very best friends: My wife, Fadhila, and my children,

Leena, Muath, Mohamed-Eyed, Lama, and Maitham

# Preface

*"The decomposition of the difficulties to be resolved, or the objects to be known, should be pushed up to the simplest elements ... Such elements are seized directly and completely by the intuition."*

René Descartes, Discours de la méthode (1637)

Wireless sensor networks have received significant attention because of their important role and many conveniences in our lives. Indeed, the recent and fast advances in inexpensive sensor technology and wireless communications have made the design and development of large-scale *wireless sensor networks* cost-effective and appealing to a wide range of mission-critical situations, including civilian, natural, industrial, and military applications, such as health and environmental monitoring, seism monitoring, industrial process automation, and battlefields surveillance, respectively. A *wireless sensor network* consists of a large number of tiny, low-powered devices, called *sensors*, which are randomly or deterministically deployed in a field of interest while collaborating and coordinating for the successful accomplishment of their mission. These sensors suffer from very scarce resources and capabilities, such as bandwidth, storage, CPU, battery power (or *energy*), sensing, and communication, to name a few, with energy being the most critical one. The major challenge in the design process of this type of network is mainly due to the limited capabilities of the sensors, and particularly, their energy, which makes them unreliable.

This book aims to develop a reader's thorough understanding of the opportunities and challenges of *k-covered wireless sensor networks*, where each point in a deployment field is covered (or *sensed*) by at least *k* sensors. Following René Descartes' most elegant methodology of dividing each difficulty into as many parts as might be possible and necessary to best solve it (*Discours de la Method*, 1637), this book presents a variety of theoretical studies based on percolation theory and computational geometry as well as protocols that lead to the design of a unified framework, where connected *k*-coverage, sensor scheduling, and data routing and dissemination are jointly considered. I have written this book given the tremendous interest of numerous researchers in *k*-covered wireless sensor networks, which has been expressed by their very active and productive research for the last 6 years and until now. Indeed, several protocols have been proposed to solve problems related to the design and implementation of energy-efficient *k*-covered wireless sensor networks that span a variety of topics, such as sensor deployment, network connectivity, sensing coverage, sensor scheduling (or duty-cycling), and data routing and dissemination.

This text is mainly based on my research work that has been focused so far on the study of *k*-covered wireless sensor networks, which has been the main contribution of my Ph.D. Dissertation. This is one of the major topics covered in both of the introductory and advanced courses on wireless sensor networks that I have taught at Hofstra University. This book will be useful to senior undergraduate and graduate students in computer science, computer engineering, and any related discipline. It will also be of interest to computer scientists, researchers, and practitioners in both academia and industry with interest in $k$-covered wireless sensor networks from their deployment until data gathering.

## Book Organization

The book is divided into four parts which are described as follows.

In Part 1, we introduce wireless sensor networks as a new emerging technology and the background necessary. Chapter 1 gives a brief introduction to wireless sensor networks and presents the major challenging problems in their design. Also, it describes a sample of their potential applications as well as a key set of design requirements of the protocols proposed in this book. Moreover, it states the problems being investigated in this book along with a brief description of their solutions. Chapter 2 introduces the background that is necessary for the description of all the protocols discussed in this book.

In Part 2, we address the problem of *almost sure* integrated coverage and connectivity in wireless sensor networks from the perspective of percolation theory. Specifically, we are interested in finding the critical sensor density above which the network is *almost surely* connected and the deployment field is *almost surely* covered. Chapter 3 proposes our solution to this problem in two-dimensional deployment fields using a probabilistic approach. Chapter 4 discusses our solution to the same problem in three-dimensional deployment fields using an approach that is not a generalization of the one proposed for a two-dimensional space.

In Part 3, we focus on the problem of connected $k$-coverage in densely deployed wireless sensor networks. Precisely, we are interested in achieving connected $k$-coverage with a minimum number of sensors. Chapter 5 describes our solution to this problem under the assumption of static and homogeneous sensors while considering a deterministic sensing model. Chapter 6 extends our above solutions to account for heterogeneous $k$-covered wireless sensor networks. Also, it introduces our solution to the same problem in the context of mission-oriented wireless sensor networks while considering sensor mobility. Chapter 7 addresses the problem of connected $k$-coverage using a more realistic, stochastic sensing model. Moreover, it investigates the problem of connected $k$-coverage in three-dimensional wireless sensor networks using a deterministic sensing model. Chapter 8 gives our measures of unconditional and conditional connectivity and fault tolerance of $k$-covered wireless sensor networks while considering two-dimensional deployment fields.

In Part 4, we are interested in the problem of energy-efficient data forwarding in wireless sensor networks. Chapter 9 describes our solution to this problem for always-on wireless sensor networks using both short-range and long-range data forwarding schemes. Chapter 10 presents our solution to the problem of finding a trade-off between energy and delay when forwarding data to a central gathering point, such as the *sink*. Chapter 11 proposes our solution to the energy sink-hole problem, which is inherent to static, always-on wireless sensor networks. Chapter 12 discusses our energy-efficient, unified framework for geographic forwarding in duty-cycled, $k$-covered wireless sensor networks while using deterministic sensing and communication models. Furthermore, it considers both two-dimensional and three-dimensional deployment fields.

In Part 5, we conclude and extend our measures of network and fault tolerance in Chap. 8 to account for three-dimensional settings. Chapter 13 focuses on the future trends in connected $k$-covered wireless sensor networks. Precisely, it concludes this book with both a summary and a discussion of potential future work and open problems that deserve more attention. Appendix provides measures of unconditional and conditional connectivity and fault tolerance of $k$-covered wireless sensor networks deployed in three-dimensional fields.

## Acknowledgements

Numerous family members and friends made this book a reality. And, it is time to thank them and acknowledge their excellent support and highly appreciated efforts, which result in writing this book.

First and foremost, I would very much like to express my sincere and permanent gratitude to Allah—the Most Gracious, the Most Merciful for the wonderful opportunity He has given me to put together my research work in this book and for His blessing by helping me finish this project—one of my precious dreams—and make it available to others who are interested in its topics. This modest book is dedicated to Him and I hope He would kindly accept it. And, I always remember His Saying "And of knowledge, you (mankind) have been given only a *little*".

Although writing a book is an individual process, it is in reality a shared experience with others who appreciate it and have tremendously helped me in order to make it a successful one. I have been fortunate to have shared this experience with my family closely or distantly. I would like to express my special thanks and deep appreciation to my beloved wife, Fadhila, for her beautiful mind and good humor, exceptional support, constant encouragements, and very constructive advices since the first time I thought about writing this book, and to my wonderful children, Leena, Muath, Mohamed-Eyed, Lama, and Maitham, for their very kind patience and understanding for the long time. I have spent away from them while setting in front of my laptop writing this book. Several times, Leena and Muath told me: "Dady, you are always spending too much time with that electronic rectangle!" – referring to my laptop. Both of my wife and children are providing me with the social and intellectual environment in which I thrive and survive. My hearty gratitude goes to my first teachers, my mother, Mbarka, and my father, Mokhtar, for teaching me the value of knowledge and the importance of family, for providing me with constant support and encouragements, for their wonderful prayers, and more importantly, for believing in me. Also, I would like to thank all my sisters, sisters in law, brothers, brothers in law, nieces, and nephews for their endless support and encouragement.

This book is dedicated to them and to the fond memory of my grand-parents, Fatma and Abdelkarim, and my uncle Mahfoudh, who always wished me to be very successful and well received.

Writing a book is a great pleasure yet a challenging project that could not have been accomplished without the support of the people around me who made this experience more joyful. I would like to thank my colleagues and friends at Hofstra University and the Department of Computer Science for the environment they created for me to finish this book. In particular, I am very thankful to Dr. Bernard Firestone, Professor of Political Science and Dean of the Hofstra College of Liberal Arts and Sciences (HCLAS), for supporting my research with a generous new faculty start-up research grant and his great support to WiSeMAN Research Lab; Mr. Richard Apollo, Senior Assistant Dean for HCLAS Planning and Budget, for his wonderful and continuous support to WiSeMAN Research Lab; Dr. John Impagliazzo, Professor Emeritus of Computer Science, Dr. Gretchen Ostheimer, Associate Professor of Computer Science, Dr. Krishnan Pillaipakkamnatt, Associate Professor of Computer Science, and Ms. Zsa Zsa Tucker, Senior Executive Secretary, for their kind support and encouragement; and Mrs. Sofia Kakoulidis, Associate Provost for Research and Sponsored Programs, for her kind assistance with my research grant proposals. Their support has been very valuable since I joined the Department of Computer Science at Hofstra University on September 1, 2008.

I always remember the wonderful support and encouragements of my Professors and friends in my Department of Computer Science and Engineering at The University of Texas at Arlington—my *Alma Mater Studiorum*. Indeed, one of the steps towards my goals and success is the exciting and unforgettable years I have spent in the Department of Computer Science and Engineering completing my Ph.D. studies in Computer Science and Engineering. Over those years, I learned a lot from them and acquired solid background and knowledge in computer science, and particularly, the field of wireless sensor networks. Especially, I am very grateful to the following faculty members, Dr. Sajal K. Das (Ph.D. Advisor), Dr. Gautam Das (Ph.D. Committee Member), Dr. Bahram Khalili (Graduate Advisor), Dr. Yonghe Liu (Ph.D. Committee Member), Dr. Fillia Makedon (Chairperson), Dr. Lynn Peterson (Senior Associate Dean of the College of Engineering), Dr. Bob Weems (Ph.D. Committee Member, Associate Chair), and Dr. Gergely Zaruba, and the staff members, Camille Costabile, Pamela Mcbride, and Sherri Warwick for their kindness and for making things easier for me during my doctoral studies. I would like to sincerely thank all of them for helping me achieve my goals and objectives in my career. I am very proud to have graduated with my Ph.D. degree in Computer Science and Engineering from my Department of Computer Science and Engineering at The University of Texas at Arlington, and with two prestigious awards: *John Steven Schuchman Award for 2006–2007 Outstanding Research by a PhD Student* in February 2008 and *Nortel Outstanding CSE Doctoral Dissertation Award* in February 2009! I received both awards at the Engineering Week Banquet at my *Alma Mater*, The University of Texas at Arlington.

I would like to thank numerous experts and researchers for contributing to the growth and development of the field of wireless sensor networks, and particularly the topics of this book. Especially, I gratefully thank Drs. X. Wang, G. Xing, Y. Zhang, C. Lu, R. Pless, and C. D. Gill, who introduced the concept of *k*-coverage as well as integrated coverage and connectivity in wireless sensor networks in their prestigious SenSys'03 conference and ACM TOSN 2005 journal papers.

I would like to give my sincere thanks to my professional editors, Dr. Janusz Kacprzyk, Editor-in-Chief, and Dr. Thomas Ditzinger, Senior Editor, of the book series, Studies in Computational Intelligence, Springer-Verlag, Heidelberg (Germany) and New York (USA), for their kind invitation, which stimulated me to write this book and devote considerable time to finish it, their enthusiasm, and their generous assistance throughout the entire writing process. It was a great pleasure to work with both of them. The other thing that most attracted me to have Drs. Kacprzyk and Ditzinger as my preferred editors for this book is the red-and-blue cover of their prestigious Studies in Computational Intelligence book series! I would like to acknowledge the publisher, Springer, who made this book possible. Last but not least, my special thanks go to SCI Data Processing Team – SPS for their professionalism and hard work.

January 2009 Habib M. Ammari

# Contents

## Part 3: Summary and Further Extensions

# List of Figures

# List of Tables

# Chapter 1
# Overview of Wireless Sensor Networks

This chapter gives a brief introduction to wireless sensor networks (WSNs) and presents the major challenging problems in their design. Moreover, it describes a sample of their potential applications as well as a key set of design requirements of the protocols proposed in this book. Furthermore, it states the problems being investigated in this book along with a brief description of their solutions.

## 1.1 Introduction

Recent advances in miniaturization, low-cost and low-power circuit design, and wireless communications have led to the development of low-cost, low-power, and tiny communication devices, called *sensors*. Like nodes (or computers, laptops, etc.) in traditional wireless networks, such as mobile ad hoc networks, the sensors have data storage, processing, and communication capabilities. Unlike those nodes, the sensors have an extra functionality related to their *sensing* capability. However, the sensors suffer from severe limitations of several resources and capabilities, such as battery power (or *energy*), storage, processing, sensing, and communication, compared to personal computers, for instance, with energy being the most crucial one. For an excellent survey on sensor networks, the interested reader is referred to [6, 7].

A *wireless sensor network* is composed of a large number of sensors that are densely deployed in a field of interest to monitor specific phenomena. The sensors can be engaged in a variety of sensing tasks, such as temperature, sound, vibration, light, humidity, etc. These sensors sense specific environment phenomenon and perform in-network processing on the sensed data before sending their results to a central gathering node, called the *sink*. In this type of network, sensors communicate with each other (possibly) through multi-hop wireless communication links and forward sensed data on behalf of others so the sink can receive them on-time for further processing and analysis. WSNs can be used for a wide variety of applications dealing with monitoring (health environments monitoring, seism monitoring, etc), control (object detection and tracking), and surveillance (battlefields surveillance).

Compared to traditional wireless networks, such as mobile ad hoc wireless networks, WSNs have several inherent characteristics. First, the sensors are very tiny and hence more susceptible to hardware failure. It is worth mentioning that *battery power* (or *energy*) is the most crucial resource, and hence sensors can also fail due to low energy. Second, the sensors are deployed in a field with high density to extend the network lifetime. Indeed, using a large number of sensors facilitates multi-hop communication between them, and hence the sensors can save their energy by transmitting or forwarding their sensed data through short distances. Third, the network topology may change very frequently as sensors join and/or leave the network. Thus, protocols designed for WSNs should account for all these features, which are inherent to these types of networks so they remain operational as longer as possible.

The remainder of this chapter is organized as follows. Section 1.2 gives the challenges that face the design of WSNs while Sect. 1.3 describes a sample of their potential applications. Section 1.4 presents the motivations of the work in this book. Section 1.5 describes the major requirements driving the design of the protocols that we propose in this book. Section 1.6 discusses the major contributions of this book by stating the main problems that are addressed in this book along with a brief overview of the proposed solutions. Section 1.7 summarizes this chapter.

## 1.2 Major Challenges

The design of network protocols for WSNs, including those for coverage configuration and data dissemination, is a challenging problem due to several constraints. Next, we describe these constraints that are imposed not only by the characteristics of the individual sensors, the behaviour of the network, and the nature of physical environments (or *deployment fields*) but also by the requirements of the sensing applications in terms of some desirable metrics.

### *1.2.1 Limited Resources and Capabilities*

Because of their inherent characteristics, the design of WSNs for different applications running in different deployment fields is facing several challenges. First of all, energy efficiency is the primary concern in the design of WSNs. Indeed, the sensors forming a network suffer from the limitations of several resources, such as storage, CPU, bandwidth, communication, sensing, and battery power (or *energy*). In particular, energy is the most crucial resource as it determines the lifetime of the sensors and hence the lifetime of the entire network. Energy poses a serious problem for network designers especially in hostile environments, such as battlefields, where it is difficult or even impossible to access the sensors and recharge or renew their batteries. Furthermore, when the energy of the sensors reaches a certain threshold, they become unreliable (or *faulty*) and would not be able to function properly. As a consequence, the behaviour of those faulty sensors will have a major impact on the network performance. Thus, network protocols and algorithms designed to be run by the sensors should be as energy efficient as possible to extend their lifetime and hence prolong the network lifetime while guaranteeing good performance overall.

### 1.2.2 Location Management

Sensor location management is another major challenge in the design of deployment strategies to achieve a certain degree of coverage. In most of the protocols designed for WSNs, the sensors are aware of their locations through either the use of *global positioning system* (GPS) receivers or some localization technique, such as the ones proposed by Bulusu et al. [52] and Ji and Zha [114]. On one hand, a GPS receiver-based solution provides the sensors with highly accurate locations but is not cost-effective for densely deployed sensors given that each sensor should be equipped with a GPS receiver. On the other hand, the use of a localization technique does not require any additional cost but may not guarantee high sensor location accuracy.

### 1.2.3 Sensor Deployment

As mentioned earlier, a deployment field may cause a problem not only to access the sensors for replacing and/or recharging their batteries but also for their deployment. Thus, a deterministic sensor deployment strategy is not always possible. Such a strategy would help cover the field appropriately and minimize the total number of sensors required to achieve the specific needs of sensing applications in terms of their expected type of coverage. Indeed, an application may demand partial coverage where only a certain percentage of the field is covered; full coverage, where the entire field is covered; or redundant coverage, where every location in the field is covered by multiple sensors simultaneously. In the case where the sensors cannot be deployed deterministically because of the field nature, random deployment is the only remaining strategy. However, there is no guarantee that the coverage required by the application would be satisfied. There might be some areas that are not covered well or even not covered at all and this would lead to a problem, known as *coverage hole*. Moreover, all the deployed sensors are not guaranteed to be connected to each other or to the sink. This would lead to another problem, known as *connectivity hole*. These are two of the reasons why most of time WSNs are designed with densely deployed sensors. Thus, the nature of the field has an influence on the network and this is a challenge for the designer and the investing party at least cost-wise. As will be discussed later, one of the most widely used assumption in the design of routing and data dissemination protocol is highly sensor density. Although highly dense deployed WSNs involve more than necessary sensors, they help guarantee network connectivity and achieve the coverage demanded by the application.

### 1.2.4 Time-Varying Network Characteristics

The topology of a wireless sensor network, which is defined by the sensors and communication links between them, changes frequently due to sensor addition and deletion. When new sensors decide to join the network, the neighbour sets of some sensors have to be updated. Indeed, it may seem necessary to add more sensors to maintain certain properties of coverage of the deployment field and network connectivity. Similarly, when the sensors deplete all their energy, they are considered faulty and no longer belong to the network. Thus, the neighbour sets of the fault sensors' neighbours should be updated. Also, in mobile WSNs, the network topology gets updated as the sensors move in the deployment field. Consequently, any topology change in the network will have an impact on the communication paths (or routes) between the sensors in the network. Therefore, routing and data dissemination paths should consider network topology dynamics due to limited energy and mobility of the sensors as well as increasing the size of the network to maintain specific application needs in terms of coverage and connectivity. It is worth noting that connectivity to the sink is very important. In fact, coverage would be meaningless if the sensed data cannot reach the sink, i.e., there is no communication paths between the source sensors (or data generators) and the sink. Thus, connectivity between all source sensors and the sink either directly or indirectly should be guaranteed for the correct operation of the network.

### 1.2.5 Network Scalability, Heterogeneity, and Mobility

The design of protocols for connected $k$-covered WSNs should consider network scalability. In other words, these protocols should scale with the number of sensors in the network, sensor mobility, and the size of the deployment field. For large-size networks, the number of sensors could be on the order of thousands. Also, the sensors may not necessarily be static given that sensor mobility helps achieve better quality of coverage [76, 144]. Furthermore, the sensors could be heterogeneous with regard to their storage, processing, communication, sensing, and energy capabilities and resources. In real-world applications, WSNs are composed of heterogeneous sensors that have a potential to increase the network lifetime and reliability without causing significant increase in its cost [209]. Indeed, deploying heterogeneous sensors reduces the probability of simultaneous failure of the entire neighbour set of a sensor [13].

### 1.2.6 Sensing Application Requirements

In most sensing applications, the sensed data should be as accurate as possible to assure better decision making by the sink. Moreover, the sensed data needs to reach the sink in a timely manner. Thus, the *delay* metric should also be considered in the design process of WSNs; otherwise, the underlying network may not be useful. Also, for several sensing applications, data redundancy is desirable in that it increases data accuracy. For instance, in an intruder detection and tracking application, multiple sensors should be active at the same time to gather enough information about the intruder and track its motion accurately. Therefore, the design of coverage configuration and routing and data dissemination protocols should guarantee data delivery and accuracy so the sink can gather the required knowledge about the physical phenomenon on time. Furthermore, the sensors may deplete their energy before expected and become faulty. As discussed earlier, a deployment field may not be accessible and thus replacing those faulty sensors would be impossible. Hence, a wireless sensor network should tolerate the presence of faulty sensors and remain functional in spite of those failures. The degree of fault tolerance of a network depends on the underlying sensing application. Thus, coverage configuration and routing and data dissemination protocols for WSNs should be fault tolerant for this type of sensor failure. It is worth noting that the link and sensing unit failures may also occur during the operation of a wireless sensor network. While sensing unit failure are due to imperfections in manufacturing or aging, link failures are caused by sensor failures and sensor mobility. In this book, however, we only consider sensor failures due to low battery.

## 1.3 Sample Sensing Applications

The design of WSNs should also be guided by the very specific requirements of the target applications. The knowledge gathered about the underlying application would help a network designer deploy more appropriate types of sensors and develop algorithms and protocols that meet the needs of the application. In this section, we describe some potential applications of WSNs spanning health, home, environmental, and military areas [6, 7].

- *Tracking and monitoring a hospital:* Sensors may be attached to patients and doctors. For a patient, specific sensors are used to perform a particular task. For instance, to detect the heart rate, a special sensor needs to be used. Also, to detect the blood pressure, another specific type of sensor has to be used. For a doctor, sensors may be used to track their locations in the hospital to facilitate their mission.
- *Smart environment:* One of the home applications is the design and development of a smart home (or environment), where a wireless sensor network can be deployed to satisfy the specific needs of habitants. The sensors could be embedded anywhere in a room (or apartment) and communicate with each other to offer services desired by habitants. For instance, for saving energy, the light and temperature in a room could be controlled by the sensors. In this case, the light is on only when the habitants are in the room and the temperature should be set to appropriate value depending on the time and season, for instance. The goal of this type of network is to provide habitants with the level of comfort they wish to have without any human intervention.
- *Forest fire detection:* The sensors could be randomly and densely deployed in a forest to detect the origin of a fire and report this information in a timely fashion to the end users to act accordingly before the fire spreads. This helps avoid catastrophic situations that may result. In this type of application, the sensors may be used for a long period of time, and hence have to be equipped with continuous source of energy, such as solar cells. Furthermore, the sensors need to collaborate with each other in their

sensing activity to overcome several problems, such as obstacles. Also, the sensors should be densely deployed for a quick and accurate detection.

- *Intruder detection and tracking:* Business stores, for instance, could be covered with special sensors to detect and track the motion of intruders. To achieve high accuracy of detection and tracking of an intruder, sensor redundancy is desirable and hence a dense network should be deployed. When an intruder is detected by some sensors, several other sensors become awake to cover the trajectories of the intruder. The collected information about an intruder is reported to end users for analysis and processing.
- *Battlefield surveillance:* A wireless sensor network can be deployed in a battlefield for performing detection and tracking of target objects, such as tanks and vehicles, and sending real-time information about the enemy mobility to a central control unit. Precisely, a network should be able to detect and classify multiple targets, such as vehicles and troop movements, using sensors that are capable of sensing acoustic and magnetic signals generated by different target objects.

## 1.4 Motivations of This Book

There are several critical applications, such as intruder detection and tracking, where WSNs need to be deployed in a field in such a way that every point is sensed (or covered) by at least one sensor. In particular, it is sometimes desirable to deploy a dense wireless sensor network to achieve redundant coverage of a deployment field, where every point in the field is guaranteed to be covered by at least $k$ sensors simultaneously and we say that the network is configured to provide *$k$-coverage*.

Our main interest in writing this book on $k$-covered WSNs is motivated by at least the following three applications, which require a degree of coverage that is at least equal to three, i.e., $k \geq 3$. First, to cope with the problem of sensor failures due their fragility, the design of sensor networks for planet exploration [184] should be as reliable as possible since failed sensors in space cannot be easily diagnosed and replaced. Sun et al. [184] simulated a Confidence Weighted Voting technique on top of a $k$-cover deployment strategy. They showed that $k$-cover deployment with $k \geq 3$ is necessary to guarantee data redundancy, which improves data reliability and fault tolerance of sensing applications. Indeed, high coverage degree helps achieve higher sensing accuracy and stronger robustness against sensor failures. Second, multiple-sensor data fusion [121] was found to be useful for at least a three-sensor system, i.e., system whose degree of coverage is $k \geq 3$. Klein [121] discussed how this type of system helps detect, classify, and track the target objects. Since at least all the three sensors participate in the decision, it is unlikely that a false target would be detected as a true target. Third, the design of triangulation-based positioning systems [162] requires that each point in a target field be covered by at least three sensors. Nicules and Nath [162] showed that this type of positioning system helps increase the accuracy of the positions of the sensors.

The design of network configuration protocols for WSNs faces a challenging problem, namely *energy conservation*, due to the constrained battery power of the sensors. Several energy conservation protocols for WSNs networks have been proposed at the MAC and network layers, such as the ones suggested by Biswas and Morris [43], Casari et al. [56], and Zorzi and Rao [226, 227]. Moreover, a variety of energy-efficient coverage configuration schemes have been suggested, such as the ones proposed by Wang et al. [197], Xing et al. [205], and Zhang and Hou [218]. In general, the sensors are deployed with high density, and hence the design of network configuration protocols should benefit from this fact to provide $k$-coverage. It is well known that the best approach to save the energy of the sensors is *duty-cycling* so the sensors remain operational for as long as possible. Using a duty-cycling approach, the sensors can be turned *on* (i.e., active) or *off* (i.e., inactive) according to some sleep–wakeup scheduling protocol while guaranteeing $k$-coverage all the time. Achieving $k$-coverage becomes difficult especially in hostile environments, such as battlefields, where access to the sensors is not feasible or even impossible. This implies that the sensors cannot be always on but rather duty-cycled. Otherwise, they will deplete their energy and die quickly. Also, $k$-coverage of a field should use as minimum number of active sensors as possible to extend the network lifetime.

Before all, the main goal of sensor deployment is to monitor a field and report data to the sink for further analysis and processing. Hence, the sensors should also be able to forward data on behalf of each other. More importantly, the load of data forwarding should be evenly distributed amongst all the active sensors, which currently $k$-cover the field, so all the sensors have the same chance to relay data for others. This implies that the network of active sensors should be connected. Otherwise, the sensed data will not reach the sink. Indeed, network connectivity is required for data routing and information dissemination. Therefore, it is important that the network provide $k$-coverage while maintaining connectivity between *all* active sensors. In addition, it is well known that geographic forwarding is an energy-efficient and practical scheme for WSNs. Indeed, the sensors are not required to maintain global and detailed information on the topology of the entire network. The sensors need to only maintain local knowledge on their one-hop neighbours. Thus, for more effective sensor deployment, the load of $k$-coverage and data forwarding should be evenly distributed amongst all the sensors to maximize the network lifetime.

## 1.5 Design Requirements

In this section, we summarize the major requirements driving the design, analysis, and development of protocols for $k$-covered WSNs. Particularly, we identify those requirements that are necessary for building our unified framework, where coverage, duty-cycling, and geographic forwarding are jointly considered. These requirements are summarized as follows:

- *Energy Awareness:* The sensors have severe limitations in terms of storage, computational, communication, and sensing capabilities. In particular, the sensors have scarce battery power (or *energy*). Thus, prolonging the network lifetime is the major challenge in the design and development of WSNs. With this in mind, it is essential that we design energy-aware protocols in all facets of the network operations, including sensor deployment, coverage and connectivity, scheduling, and data routing and dissemination.

- *On-Demand-Connected k-Coverage:* Several sensing applications prefer collecting redundant data to guarantee the most accurate decision-making process. Indeed, intruder detection and tracking applications may require more than one sensor to be active to collect information about the *intruder* (or *malicious node*) and track its motion accurately. Also, the sensors may die quickly because of their low energy, and thus, the network should be *fault tolerant*. Unlike most related work that focused on connected $k$-coverage with static sensors and fixed degree $k$ of coverage, where each point in a field is *covered* (or *sensed*) by at least $k$ sensors, our proposed framework supports mobile connected $k$-coverage, where a region of interest is $k$-covered using mobile sensors and $k$ may change over time. Also, the degree $k$ of coverage of a region of interest may not be the same for another region in the field. We call this *on-demand connected k-coverage*. Our framework adopts centralized and distributed strategies to ensure mobile connected $k$-coverage of a region of interest in a deployment field while maintaining network connectivity.
- *Autonomous and Purposeful Mobility:* In addition to their mobility, the sensors are autonomous, and thus are able to make their own decision based on the information they receive from the rest of the network. Particularly, these sensors should be able to move to designated locations in a region of interest whenever necessary to ensure its $k$-coverage and accomplish the target mission that has been determined by the sink. To account for mission-oriented WSNs, where the mobility of the sensors is controlled by the underlying mission, our framework enables *purposeful mobility* of the sensors and this mobility must be traded-off against the goals of the missions.
- *Situation Awareness and Intelligent Collaboration:* Situation awareness is an essential and critical foundation for successfully accomplishing all missions, and this requires intelligent collaboration between the sensors. Thus, all the sensors should be aware of any mission that has to be accomplished. This situation awareness assumes that there is a central entity, such as the *sink*, which decides the type of mission that has to be accomplished and in which region in the deployment field. This information should be propagated within the network while minimizing the total energy consumption needed to advertise this information about the mission.
- *Self-Organization:* The sensor mobility and scheduling may affect the topology of the network, which could possibly result in a *connectivity-hole* problem and/or a *coverage-hole* problem. One of the goals of our proposed framework is to make the sensors *self-organizing* and *adaptive* so they guarantee network connectivity, which is required for communication between the sensors, data forwarding from the sensors to the sink, and data

dissemination from the sink to the sensors. Also, it enables the sensors to achieve any degree of coverage needed by an application for a mission.

- *Fault Tolerance:* In most sensing applications, the sensed data should be accurate to ensure better decision making by the sink. Also, data redundancy is desirable in that it increases data accuracy. For instance, in the intruder detection and tracking application, multiple sensors should be active to gather enough information about the intruder (or malicious node) and track its motion accurately. *Redundant coverage* (or $k$-coverage) ensures higher data redundancy and accuracy. Indeed, sensor nodes may deplete their energy and die. Hence, a network should tolerate the presence of faulty sensors and remains functional. Thus, $k$-coverage provides a high degree of fault tolerance of the network, where the value of $k$ depends on the requirements of the application in terms of coverage.
- *Heterogeneity:* Unlike most related work that considered $k$-coverage with homogeneous sensors, our framework focuses on *heterogeneous* sensors, which do not necessarily have the same capabilities in terms of computation, storage, sensing and communication ranges, and initial energy. In real-world applications, WSNs are composed of heterogeneous sensors that have a potential to increase the network lifetime and reliability without causing significant increase in its cost. Indeed, as we mentioned earlier, the use of heterogeneous sensors helps reduce the probability of simultaneous failure of all the neighbours of a sensor [29].
- *Dimensionality:* In the literature, most of the works on WSNs dealt with two-dimensional settings, where sensors are deployed in a planar field. However, there exist several applications that cannot be effectively modeled in a two-dimensional space. For instance, sensors deployed on the trees of different heights in a forest, or in a building with multiple floors, or underwater applications [4, 5] require the design in a three-dimensional space. Moreover, oceanographic data collection, pollution monitoring, offshore exploration, disaster prevention, and assisted navigation are typical applications of underwater sensor networks [5], which have to be designed using three-dimensional settings, which represent more accurately the network design for real-world applications.
- *Stochastic Features:* Although the majority of studies on wireless sensor networks considered the disk sensing model, where all sensor readings are assumed to be precise and have no uncertainty, the signal attenuation and the presence of noise associated with sensor readings require the use of a more realistic sensing model that reflects the real properties of the sensors. Precisely, the sensing capability of a sensor must be modelled as the probability of successful detection of an event, and hence should depend on the distance between it and the event as well as the type of propagation model being used (free-space vs. multi-path). Indeed, it has been shows that the probability that an event in a distributed detection application can be detected by an acoustic sensor depends on the distance between the event and the sensor [77]. Thus, the protocols designed for WSNs should account for the probabilistic nature of the sensor capabilities, and particularly, their sensing and communication ranges.

- *Data Delivery:* The main goal of the design of WSNs is to monitor a phenomenon in a field of interest and collect data related to that specific phenomenon. It is very important that the sensed data reach the sink for further processing. The accuracy of the decisions and actions taken by the sink depends on the availability of the data. Thus, the sensors should be able to use robust protocols that will enable them to deliver the data they have collected to the sink with high data delivery ratio. This issue depends on the data forwarding and dissemination capabilities of the sensors, which are in turn strongly dependent on the efficiency of the design of the corresponding routing protocols.
- *Delay:* For some time-critical sensing applications, such as forest fire detection and tracking, the sink should receive the sensed data collected by the sensors in a timely manner to avoid any undesirable consequence. It is necessary that the design of protocols for these types of sensing applications be conducted under the delay constraint so that the sensed data reach the sink within a certain time bound. In case the sensors are always *on* (or active), the data forwarding and dissemination protocol are responsible for meeting this delay constraint. However, when the sensors are duty-cycled, the duty-cycling protocols are also responsible to meet this delay bound.

## 1.6 Contributions of This Book

This book aims at investigating the following research problems, which are not totally disjoint. Each of these problems is introduced by a brief statement that is accompanied by a brief solution statement.

- *Almost Sure Connected Coverage*: What is the critical sensor spatial density above which a deployment field (respectively, network) is *almost surely* covered (respectively, connected)?

  We propose continuum percolation-based approaches to study phase transitions in coverage and connectivity in static WSNs in an integrated fashion. Precisely, we propose probabilistic approaches to compute the critical sensor spatial density above which a field is *almost surely* covered and the network is *almost surely* connected. Our proposed solutions consider both two-dimensional and three-dimensional deployment of the sensors. These solutions help the network designers achieve full coverage of a field with a minimum number of connected, active sensors, thus maximizing the network lifetime.
- *Connected k-Coverage*: What is a sufficient condition of the sensor spatial density for full $k$-coverage of a two-dimensional (respectively, three-dimensional) deployment field, where each point in the field is guaranteed to be covered by at least $k$ sensors while maintaining network connectivity using static homogeneous sensors only under the assumption of a deterministic sensing model?

To solve this problem, thus supporting different applications and environments with diverse requirements in terms of coverage and connectivity using static sensors only, we extend our above analysis to $k$-coverage using a deterministic approach so the network self-configures to meet these requirements while considering a deterministic sensing model. To this end, we compute the active sensor spatial density that is necessary to achieve full $k$-coverage of a field while guaranteeing connectivity between all active sensors. Our analysis is based on *Helly's Theorem* [44] and the geometric properties of the *Reuleaux triangle* (respectively, *Reuleaux tetrahedron*) in two-dimensional (respectively, three-dimensional) deployment fields. Using this analysis, we design randomized centralized protocols, pseudo-distributed protocols as well as fully-distributed protocols for connected $k$-coverage configurations in WSNs.

- *Heterogeneous Connected k-Coverage*: Given a field to be monitored, a positive integer $k \geq 3$, and a set $S$ of heterogeneous sensors, select a minimum subset of sensors $S' \subseteq S$ to stay *on* (or *active*) such that each point in the field is *k-covered* while the network induced by all the sensors in $S'$ is guaranteed to be *connected.*

  We exploit the results obtained with the homogeneous model to solve the connected $k$-coverage problem for heterogeneous WSNs, where the sensors do not necessarily have the same sensing range, communication range, and initial energy. We show that while it is possible to design distributed protocols to guarantee connected $k$-coverage of a field using heterogeneous sensors while achieving good performance overall, it is impossible that a centralized protocol could be designed efficiently due to sensor deployment randomness and sensor heterogeneity. Thus, we propose a pseudo-random deployment approach, where the sensors are deployed in different layers in a circular deployment field with respect to the sink according to the strengths of their sensing and communication ranges as well as their initial energy. Based on this deployment strategy, we propose centralized and distributed protocols for generating energy-efficient connected $k$-coverage configurations using heterogeneous sensors.
- *Mobile Connected k-Coverage*: How to guarantee connected $k$-coverage in mission-oriented mobile WSNs under the following requirements? (i) *On-demand k-coverage*: A region of interest in a field should be $k$-covered whenever needed, where $k \geq 3$. Consequently, a region of interest to be $k$-covered does not have to be the same all the time and hence may change. (ii) *Network connectivity*: The sensors should be maintained connected for the correct network operation. (iii) *Sensor mobility*: The sensors should be able to move to designated locations in a region of interest to ensure its $k$-coverage whenever necessary.

  We divide the problem of $k$-coverage in mission-oriented mobile WSNs into two sub-problems, namely *sensor placement* and *sensor selection*. The sensor placement problem is to compute the minimum number of sensors and their locations in a region of interest so that this region is $k$-covered. The sensor selection problem is to determine which sensors should move to the above-computed locations in the region while minimizing the total energy consumption due to the mobility of the sensors and their communication. Specifically, we propose centralized and distributed approaches to solve the $k$-coverage problem in mission-oriented mobile WSNs. In the centralized approach, the sink designates a set of sensors to move towards specific locations in the region of interest to be $k$-covered. These locations are computed by the sink. In the distributed approach, the sensors compute the target locations, where the sensors should move to, and coordinate between themselves to $k$-cover a region of interest with *as small number of sensors as possible* (or simply *small number of sensors*). Our approach enables the sensors to move towards a region of interest and $k$-cover it while minimizing their mobility energy based on their closeness to the target locations in the region and the availability of other sensors.

- *Stochastic Connected k-Coverage*: Find a tight sufficient condition so that every point in a field is probabilistically covered by at least $k$ sensors with a probability no less than $p_{th}$, called *threshold probability*, under a stochastic sensing model and compute the required number of sensors. Then, select and schedule the sensors while providing stochastic $k$-coverage of a two-dimensional deployment field as well as connectivity between all the selected sensors.

  We adapt the results of the sensor scheduling problem for $k$-coverage of a two-dimensional deployment field under the deterministic sensing model, which is discussed earlier, to solve the sensor scheduling problem for stochastic $k$-coverage under *probabilistic* (or *stochastic*) sensing and communication models. It has been found that a stochastic sensing mode is more realistic than a deterministic (or binary) sensing model. Indeed, the former takes into account not only the distance between the sensors and the target locations but also the type of propagation model being used, i.e. free-space model or multi-path model. Under a stochastic sensing model, a point $p$ in a field is said to be *probabilistically k-covered* if the detection probability of an event occurring at $p$ by at least $k$ sensors is at least equal to some *threshold probability*, $0 < p_{th} < 1$. We propose a distributed stochastic $k$-coverage protocol, where each sensor runs a *k-coverage candidacy algorithm* to check whether it is eligible to turn itself *on* (or active). This helps us design a global framework for $k$-coverage in WSNs that considers both of the deterministic and stochastic sensing models.

- *Geographic Forwarding on Always-on Sensors*: How can data be forwarded in always-on WSNs while minimizing the total energy consumption of the sensors, thus maximizing the network lifetime?

  There is an ongoing debate on short-range versus long-range data forwarding in multi-hop wireless networks. This book supports the short-range data forwarding strategy for WSNs, where energy should be given the highest priority. More precisely, we propose an energy-efficient data forwarding protocol for WSNs so they remain operational as long as possible. Our protocol, called *Weighted Localized Delaunay Triangulation-based data forwarding* (WLDT), uses 1-lookahead scheme to guarantee data delivery to the sink. WLDT aims to minimize the average energy consumption of the sensors during data forwarding towards the sink. It exploits the geometric properties of the Delaunay triangulation [36] to build an energy-efficient path between a source and the sink as a sequence of sub-paths whose endpoints are called *checkpoints*. These checkpoints are selected based on their locations in the field and their remaining energy. A sub-path between a pair of checkpoints consists of a *series of forwarders*, which are the endpoints of short Delaunay edges and are selected based on their locations and remaining energy to forward the data between their corresponding checkpoints.

- *Energy-Delay Trade-off in Geographic Forwarding*: How to achieve minimum energy consumption of the sensors while ensuring uniform battery power (or energy) depletion of the sensors and meeting the required delay constraints in geographic forwarding in static WSNs?

  We propose a communication range slicing-based approach to trade-off between conflicting objectives of sensing applications, namely minimum energy consumption, minimum delay, and uniform energy depletion. Our approach aims to slice the communication range of the sensors into *concentric circular bands* and classify them with a goal to satisfy specific requirements of sensing applications in terms of energy consumption, delay, and energy depletion. First, we formulate the trade-off between these three conflicting goals as a multi-objective optimization problem which is solved using a *weighted scale-uniform-unit sum* approach. Then, we propose a data forwarding protocol for WSNs, which exploits our solution to the multi-objective optimization problem to find an optimum trade-off between three conflicting goals. For tractability, we consider a unit-disk communication model, where the communication ranges of the sensors are supposed to be circular and have the same radius. Then, we will discuss ways of relaxing these assumptions.

- *Energy Sink-Hole*: How and to what extent can a uniform energy depletion of all the sensors be guaranteed so as to avoid the energy sink-hole problem in always-on, static WSNs, where the sensors nearer the sink are heavily used in forwarding data to the sink on behalf of all other sensors, thus depleting their energy very quickly compared to all other sensors in the network? How can this problem be addressed in homogeneous, always-on WSNs?

  We show that static WSNs suffer from the energy sink-hole problem regardless of how efficient a geographic forwarding protocol is. We propose a solution to this problem by enabling sensors to adjust their transmission range when sending/forwarding sensed data to the sink. However, we prove that this solution imposes a severe restriction on the size of the deployment field. To alleviate this shortcoming, we propose another solution that suggests a sensor deployment strategy exploiting energy heterogeneity with a goal that the sensors in the network deplete their energy uniformly. When all the sensors have the same initial energy, we propose greedy, localized protocol, called energy aware Voronoi diagram-based data forwarding (EVEN), which exploits sink mobility and uses our proposed new concept of Voronoi diagram, called *energy aware Voronoi diagram*, where the locations of the sensors are time varying and are locally and virtually computed based on their remaining energy.

- *Geographic Forwarding on Duty-Cycled Sensors*: How to design energy-efficient geographic forwarding protocols with and without data aggregation in duty-cycled, $k$-covered WSNs in two-dimensional and three-dimensional deployment fields?

  We design a unified framework for geographic forwarding on duty-cycled sensors. More specifically, we propose different approaches for both cases of two-dimensional and three-dimensional sensor deployment. In the former case, using a potential fields-based approach, we propose energy-efficient clustering-based geographic forwarding protocols for duty-cycled, $k$-covered WSNs with different levels of data aggregation. In the latter case, we focus on finding a trade-off between *uncertainty* due to duty-cycling with deterministic forwarding and *contention* due to opportunistic forwarding. Then, we propose a hybrid forwarding approach based on this trade-off. Indeed, in deterministic forwarding, a next best forwarder is determined *a priori*. Hence, duty-cycling introduces uncertainty at the sender side which is not totally certain that its selected next best forwarder would remain awake after data is being forwarded. In opportunistic forwarding, however, a next best forwarder is decided on-the-fly and after the data is transmitted. Thus, several active sensors may hear the transmitted data, thus creating high contention at the receiver side to select a next best forwarder.
- *Network Connectivity and Fault-Tolerance Measures*: What are the unconditional and conditional network connectivity and fault tolerance of $k$-covered WSNs that are deployed in two-dimensional and three-dimensional fields?

  We benefit from our characterization of $k$-coverage in WSNs stated earlier and compute the traditional (or *unconditional*) connectivity of $k$-covered WSNs. Furthermore, we compute the *conditional* connectivity of $k$-covered WSNs based on the concept of *forbidden faulty set*. The latter shows that the classical connectivity used to capture network fault tolerance underestimates the resilience of large-scale networks, such as $k$-covered WSNs. Our measures consider both two-dimensional and three-dimensional deployment fields. We show that our measures of connectivity for the latter case are not a straightforward generalization of those for the former case.

## 1.7 Summary

In this chapter, we reviewed the main challenges in the design of WSNs as well as their potential applications. Then, we described the design metrics that drive the design process of these types of networks. Moreover, we presented the motivations behind writing this book. Furthermore, we gave an overview of the main contributions of this book by stating the research problems being investigated along with a brief description of our proposed solutions.

# Chapter 2
# Background and Fundamentals

This chapter introduces the terminology and background that are necessary for the description of all the protocols discussed in this book. Precisely, it gives some key definitions and describes a percolation model to study coverage and connectivity in two-dimensional and three-dimensional wireless sensor networks. In addition, it presents an energy consumption model and defines the default network model.

## 2.1 Introduction

In this book, we use some terminology and different models, such as Voronoi diagram model, energy model, sensing model, and continuum percolation model, to describe our proposed approaches and protocols for connected $k$-coverage, duty-cycling, and geographic forwarding in wireless sensor networks. Furthermore, our work is based on a specific network model. The goal of this chapter is to present the different terms and models we use in this book.

The remainder of this chapter is organized as follows: Section 2.2 presents the key definitions and fundamental concepts that are used in this book. Section 2.3 presents deterministic and stochastic sensing models. Section 2.4 discusses different types of network connectivity. Section 2.5 describes the energy model while Sect. 2.6 presents the percolation model. Section 2.7 presents the network model that we use in the design of our energy-efficient framework for joint $k$-coverage, duty-cycling, and geographic forwarding in wireless sensor networks. Section 2.8 summarizes the chapter.

## 2.2 Terminology

In this section, we give the key definitions and describe some fundamental concepts used throughout this book.

*Definition 2.1:* The *sensing range* (or *detection range*) of a sensor $s_i$ is a region where every event that takes place in this region can be detected by $s_i$. The *sensing neighbour set* of a sensor $s_i$, denoted by $SN(s_i)$, is the set of all the sensors that are located in its sensing range. ∎

*Definition 2.2:* The *communication range* of a sensor $s_i$ is a region such that $s_i$ can communicate with any sensor located in this region. The *communication*

H.M. Ammari: Challenges and Opportunities of Connected $k$-Covered WSNs, SCI 215, pp. 15–27.
springerlink.com 

*neighbour set* of a sensor $s_i$, denoted by $CN(s_i)$, is a set of all the sensors that are located in its communication range. ■

Throughout this book, we use *communication range* and *transmission range* interchangeably. To illustrate the following definitions, we assume that the sensing ranges and communication ranges of the sensors in a two-dimensional (respectively, three-dimensional) wireless sensor network are represented by disks (respectively, spheres). Furthermore, the sensors are assumed to have the same sensing range and the same communication range. Figure 2.1 shows the sensing and communication ranges of the sensors of radii $r$ and $R$, respectively.

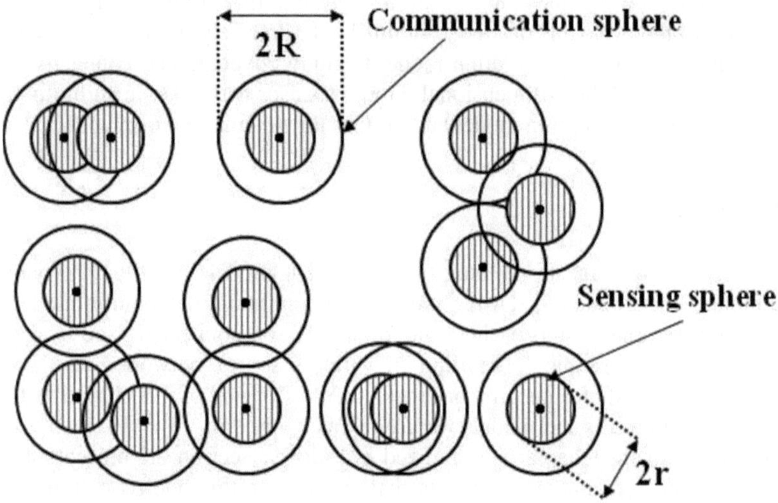


**Fig. 2.1** Schematic of overlapping concentric disks (respectively, spheres)

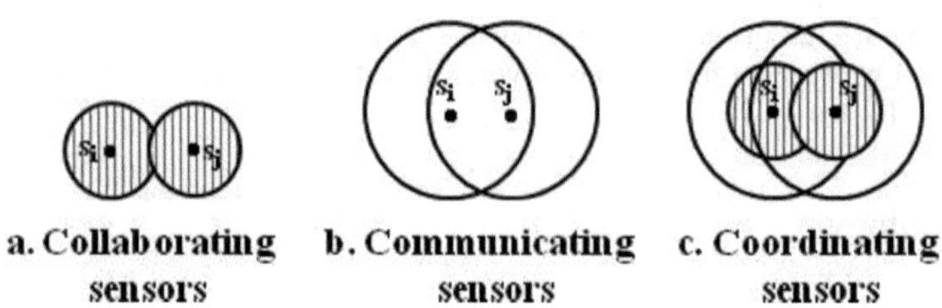


**Fig. 2.2** **(a)** Collaborating, **(b)** communicating, and **(c)** coordinating sensors

*Definition 2.3:* Two sensors $s_i$ and $s_j$ in a two-dimensional (respectively, three-dimensional) wireless sensor network are said to be *collaborating* if the Euclidean distance between the centres of their sensing disks (respectively, spheres) satisfies $|\xi_i - \xi_j| \le 2r$, where $r$ is the radius of the sensing disks (respectively, spheres) of

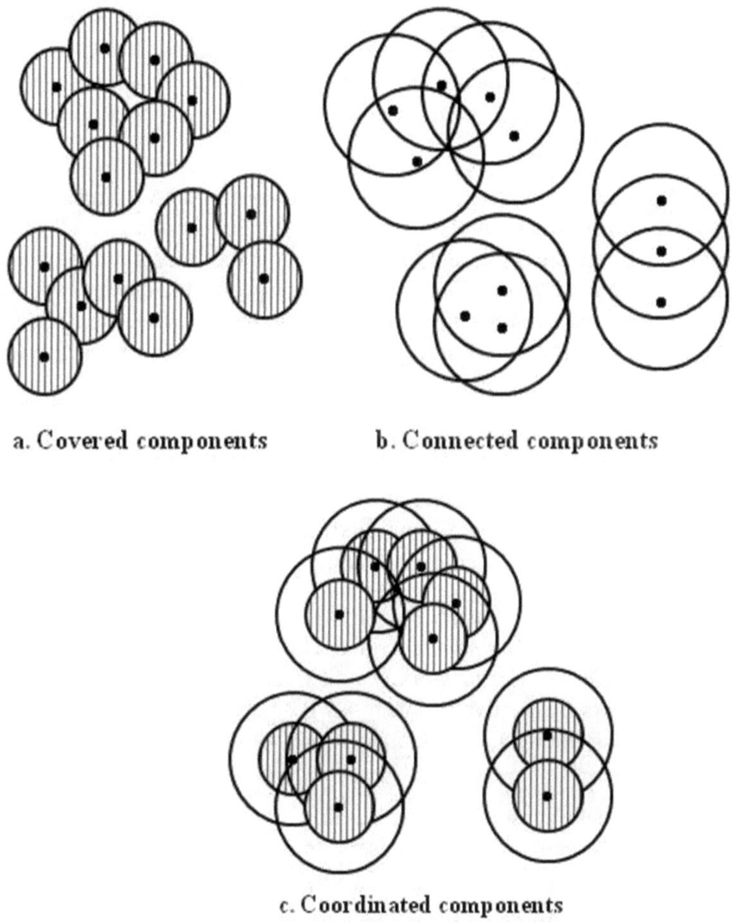


**Fig. 2.3** **(a)** Covered, **(b)** connected, and **(c)** coordinated components

the sensors. Intuitively, the two sensing disks (respectively, spheres) centred at $\xi_i$ and $\xi_j$ are either *tangential* or *overlapping* (see Fig. 2.2a). The *collaborating set* of a sensor $s_i$, denoted by $Col(s_i)$, includes all the sensors it can collaborate with, i.e., $Col(s_i) = \{ s_j : |\xi_i - \xi_j| \leq 2r \}$. ■

*Definition 2.4:* Two sensors $s_i$ and $s_j$ in a two-dimensional (respectively, three-dimensional) wireless sensor network are said to be *communicating* if the Euclidean distance between the centres of their communication disks (respectively, spheres) satisfies $|\xi_i - \xi_j| \leq R$, where $R$ is the radius of the communication disks (respectively, spheres) of the sensors (see Fig. 2.2b). The *communicating set* of a sensor $s_i$ is the set of sensors it can communicate with, i.e., $Com(s_i) = \{ s_j : |\xi_i - \xi_j| \leq R \}$. ■

*Definition 2.5:* Two sensors $s_i$ and $s_j$ in a two-dimensional (respectively, three-dimensional) wireless sensor network are said to be *coordinating* if and only if they both collaborate and communicate (Fig. 2.2c). The *coordinating set* of a sensor $s_i$ is the set of sensors it can collaborate and communicate with at the same time. ■

*Definition 2.6:* A *covered component* (or *covered region*) in a two-dimensional (respectively, three-dimensional) wireless sensor network is a *maximal* set of sensing disks (respectively, spheres), i.e., not included in any other subset except when it is equal to the original entire set of sensing disks (respectively, spheres) whose corresponding sensors are *collaborating* directly or indirectly (Fig. 2.3a shows three covered components). A *covered k-component*, denoted by $CC_k$, is a covered component having *k* sensing disks (respectively, spheres). ■

*Definition 2.7:* A *connected component* in a two-dimensional (respectively, three-dimensional) wireless sensor network is a *maximal* set of communication disks (respectively, spheres) whose corresponding sensors are *communicating* directly or indirectly (Fig. 2.3b shows three connected components). A *connected k-component*, denoted by $CC_k$, is a connected component with *k* communication disks (respectively, spheres). ■

*Definition 2.8:* A *coordinated component* in a two-dimensional (respectively, three-dimensional) wireless sensor network is a maximal set of concentric sensing and communication disks (respectively, spheres) whose corresponding sensors are *coordinating* directly or indirectly (Fig. 2.3c shows three coordinated components). ■

*Definition 2.9:* A wireless sensor network is said to be *connected* if there is a communication path between any pair of sensors. In other words, any pair of sensors can communicate with each other either directly or indirectly. ■

*Definition 2.10:* A two-dimensional (respectively, three-dimensional) wireless sensor network is said to be *homogeneous* if all of it sensors have the same storage, processing, battery power, sensing, and communication capabilities. In particular, all deployed sensors have the same radius $r$ of their sensing disks (respectively, spheres) and the same radius $R$ of their communication disks (respectively, spheres). Otherwise, the network is said to be *heterogeneous.* ■

*Definition 2.11:* The *width* of a closed convex area is the maximum distance between parallel lines that bound it. ■

*Definition 2.12:* The *breadth* of closed convex volume is the *maximum distance* between tangential planes on opposing faces or edges of the volume. ■

*Definition 2.13:* The *largest enclosed disk* (respectively, sphere) of closed convex area (respectively, volume) *A* is a disk (respectively, sphere) that lies inside *A* and whose diameter is equal to the minimum distance between any pair of points on the boundary of *A*. ■

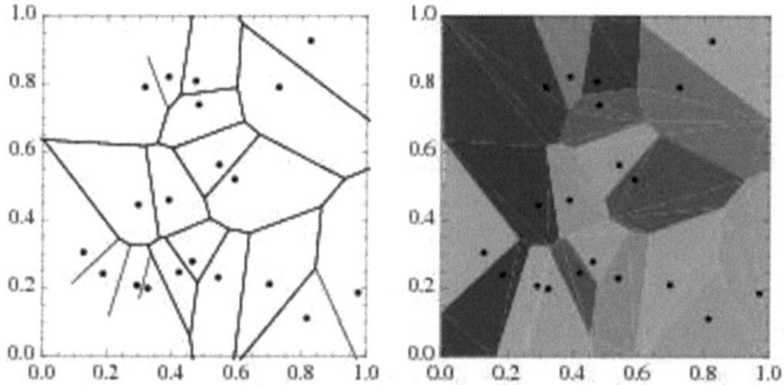


**Fig. 2.4** Voronoi diagram

**Fig. 2.5** The Delaunay triangulation (*bold lines*) on top of the Voronoi diagram (*dotted lines*) of a wireless sensor network

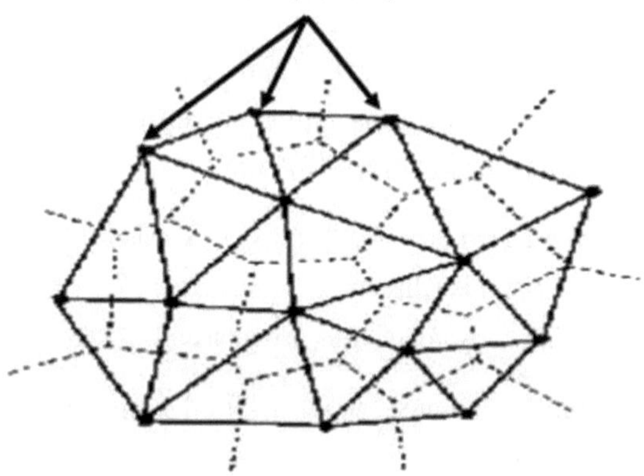


The *Voronoi diagram* [36], also known as *Dirichlet tessellation*, represents one of the most fundamental data structures in computational geometry. It has interesting mathematical and algorithmic properties and potential applications.

*Definition 2.14:* Let $S = \{s_1, \ldots, s_n\}$ be a finite set of $n$ sites (or points) in the plane. The Voronoi diagram of $S$, denoted by $Vor(S)$, is a subdivision of the plane containing $S$ into $n$ cells $VC(s_i)$, $1 \le i \le n$, such that each cell $VC(s_i)$ includes only one site $s_i$ with the property that any point $p$ located in $VC(s_i)$ is closer to $s_i$ than any other site in $S$. The cell $VC(s_i)$ corresponding to site $s_i$ is called the *Voronoi cell* of $s_i$, which is a (possibly unbounded) open convex polygonal region. The edges of a Voronoi cell are called *Voronoi edges* and its endpoints are called *Voronoi vertices*. The Voronoi diagram of $S$ is the union of the Voronoi cells of all sites in $S$. The Delaunay triangulation, denoted by $DT(S)$, is the dual of the Voronoi diagram $Vor(S)$ [36]. A $DT(S)$ graph has an edge between two sites $s_i$ and $s_j$ if and only if their Voronoi cells $VC(s_i)$ and $VC(s_j)$, respectively, share a common edge. ■

Notice that $DT(S)$ is a planar graph whose Delaunay edges are orthogonal to their corresponding Voronoi edges. Figure 2.4 shows a Voronoi diagram while Fig. 2.5 shows a Voronoi diagram and its dual, i.e., Delaunay triangulation.

Let $CN(s_i)$ be the *communication neighbour set* of a sensor $s_i$, which are located in its communication disk whose radius is equal to $R_i$. From $CN(s_i)$, the sensor $s_i$ considers only a subset of sensors, denoted by $SCN(s_i, s_m)$, located between $s_i$ and the sink $s_m$ to act as data *forwarders* to the sink.

*Definition 2.15:* Let $s_j \in SCN(s_i, s_m)$. The Voronoi diagram computed by $s_i$, denoted by $Vor(\{s_i, s_m\} \cup SCN(s_i, s_m))$, is said to be localized. A Voronoi cell $VC(s_m)$ of the sink $s_m$ is said to be *adjacent* to the sensor $s_j$ if $VC(s_m)$ and $VC(s_j)$ of $s_m$ and $s_j$, respectively, have *at least one common* Voronoi edge. ■

*Definition 2.16:* A *localized Delaunay triangulation* of a sensor $s_i$, denoted by $LDT(s_i)$, is the Delaunay triangulation computed by $s_i$ with respect to $SCN(s_i) \cup \{s_i, s_m\}$. ■

*Definition 2.17:* The *candidate checkpoints* of a sensor $s_i$, denoted by $CCP(s_i, s_m)$, are the sensors that are adjacent to the sink $s_m$ in the localized Delaunay triangulation of $s_i$, $LDT(s_i)$. ■

*Definition 2.18:* Let $s_j \in SCN(s_i, s_m)$. A sensor $s_j$ is said to be a *candidate forwarder* of $s_i$ if $VC(s_m)$ is adjacent to $s_j$, where $VC(s_m) \in Vor(\{s_i, s_m\} \cup SCN(s_i, s_m))$. The set of candidate forwarders of $s_i$ is denoted by $CF(s_i, s_m)$. ■

Assume that a source $s_0$ wishes to disseminate its data to the sink $s_m$. Fig. 2.6 shows the localized Voronoi diagram of $s_0$, where $SCN(s_0, s_m) = \{s_i \mid 1 \le i \le 18\}$. Notice that $CF(s_0, s_m) = \{s_9, s_{10}, s_{15}, s_{17}, s_{18}\}$, where the Voronoi cell of each of

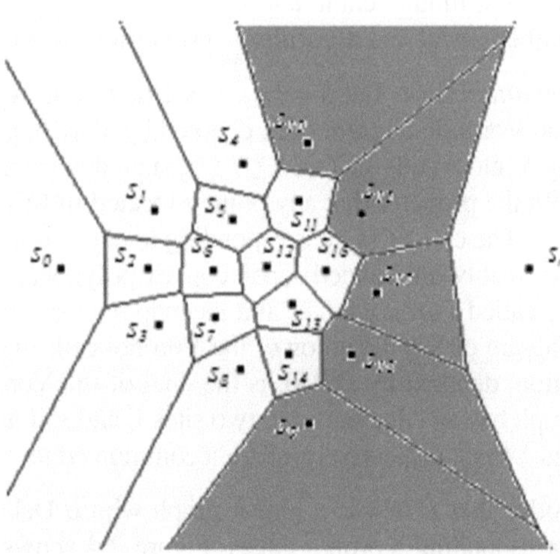

**Fig. 2.6** Localized Voronoi diagram

those sensors shares one Voronoi edge with that of $s_m$. These Voronoi cells are shaded in green.

*Definition 2.19:* Let $s_{ref} \in SCN(s_i, s_m)$. A sensor $s_{ref}$ is said to be a *reference sensor* of $s_i$ if $s_{ref}$ has the highest remaining energy amongst all sensors in $SCN(s_i, s_m)$. ■

*Definition 2.20:* A *data forwarding path* is a path that is traversed by a data packet originated from a source sensor (or simply *source*) and destined to the sink. It includes all the sensors (including the source) that forwarded the data packet to the sink on behalf of the source. ■

*Definition 2.21:* A forwarding scheme is said to be *long range* if each sensor in any data forwarding path can use *at most one* of its one-hop neighbours to forward a data packet towards its ultimate destination. A forwarding scheme is said to be *short range* if each sensor in any data forwarding path can use *multiple* one-hop neighbours to forward a data towards its destination. ■

## 2.3 Deterministic and Stochastic Sensing Models

In the *deterministic sensing model* (also known as *binary*), a point (or event) $p$ in a field is sensed/covered (or detected) by a sensor $s_i$ based on the Euclidean distance $\delta(p,s_i)$ between $p$ and $s_i$. Throughout this chapter, we use "coverage of a point" and "detection of an event" interchangeably. Formally, the coverage $Cov(p,s_i)$ of a point $p$ by a sensor $s_i$ under the deterministic sensing model is defined as follows:

$$Cov(p,s_i) = \begin{cases} 1 & \text{if } \delta(p,s_i) \le r \\ 0 & \text{otherwise} \end{cases} \tag{2.1}$$

As can be seen, the deterministic sensing model for two-dimensional (respectively, three-dimensional) wireless sensor networks considers the sensing range of a sensor as a disk (respectively, sphere), and hence all sensor readings are precise and have no uncertainty. However, it has been found that the communication range of the radios is highly probabilistic and irregular [224]. Thus, the deterministic sensing model does not reflect the real behaviour of the sensing units of the sensors, which are irregular in nature. Hence, given the signal attenuation and the presence of noise associated with sensor readings, it is necessary to consider a more realistic sensing model by defining the coverage $Cov(p,s_i)$ using some probability function. In other words, the sensing capability of a sensor needs to be modelled as the probability of successful detection of an event. Specifically, the sensor's sensing capability should depend on the distance between it and the event as well as the type of propagation model being used (free-space vs. multi-path). Indeed, it has been demonstrated that the probability that an event in a distributed detection application can be detected by an acoustic sensor depends on the distance between the event and sensor [77]. A realistic sensing model for passive infrared (PIR) sensors that reflects their non-isotropic range is presented in [54].

This sensing irregularity of PIR sensors has been verified by simulations [54]. Thus, in a *stochastic sensing model*, the coverage $Cov(p,s_i)$ is defined as the *probability of detection* $P(p,s_i)$ of an event occurring at point $p$ by sensor $s_i$ as follows:

$$P(p,s_i) = \begin{cases} e^{-\beta\,\delta(\xi,s_i)^{\alpha}} & \text{if } \delta(p,s_i) \le r \\ 0 & \text{otherwise} \end{cases} \tag{2.2}$$

where $\beta$ represents the physical characteristic of the sensing units of the sensors and $2 \le \alpha \le 4$ is the path-loss exponent. Precisely, $\beta$ measures the uncertainty introduced by the sensing unit of the sensors. Also, for the free-space model, we have $\alpha = 2$ and for the multi-path model, $2 < \alpha \le 4$. Our stochastic sensing model is motivated by the one introduced by Elfes [80], where the sensing capability of a sonar sensor is modeled by a Gaussian probability density function. Moreover, a probabilistic sensing model for coverage and target localization in wireless sensor networks is proposed in [229]. This sensing model is similar to ours except that it considers $\delta(p,s_i) - (r - r_e)$ instead of $\delta(p,s_i)$, where $r$ is the detection range of the sensors and $r_e < r$ is a measure of uncertainty in the sensor detection capability. Our stochastic sensing model is also similar to the one in [228], except that ours uses $\alpha$, and reduces to the deterministic sensing model if we set $\beta = 0$.

*Definition 2.22:* Under the deterministic sensing model, a point $p$ in a field is said to be *k-covered* if it belongs to the intersection of the sensing ranges of at least $k$ sensors. ■

*Definition 2.23:* Under the stochastic sensing model, a point $p$ in a field is said to be probabilistically *k-covered* if the detection probability of an event occurring at $p$ by at least $k$ sensors is at least equal to some *threshold probability* $0 < p_{\text{th}} < 1$. ■

*Definition 2.24:* For both sensing models, a region $A$ is said to be *k-covered* if every point $p \in A$ is *k-covered.* A *k-covered* wireless sensor network is a network that fully $k$-covers a field. ■

*Definition 2.25:* We call *degree of coverage* provided by a wireless sensor network the *maximum value* of $k$ such that a field is fully $k$-covered. ■

*Definition 2.26:* A given $k$-coverage is said to be *connected* if the *network induced* by all the sensors that are selected to achieve $k$-coverage is connected. ■

*Definition 2.27:* A region of interest $A$ is said to be *energy-efficiently k-covered* if a small number of sensors is selected to $k$-cover $A$. ■

## 2.4 Network Connectivity and Fault Tolerance

*Definition 2.28:* A *communication graph* of a homogeneous (respectively, heterogeneous) wireless sensor network is an *undirected* (respectively, *directed*) graph, $G = (S,E)$, where $S$ is a set of sensors and $E$ is a set of edges (respectively, *arcs*) between them such that for all $s_i$, $s_j \in S$, $(s_i,s_j) \in E$ if $\delta(p_i,p_j) \le R_i$, where $p_i$ and $R_i$ stand for the location and radius of the communication disk of sensor $s_i$, respectively. The *vertex connectivity* (or simply, *connectivity*) of $G$ is equal to $K$ if $G$ can be disconnected by the removal (or failure) of at least $K$ nodes. The *fault tolerance* of the underlying network is equal to $K - 1$. ∎

*Definition 2.29:* A *forbidden faulty set* of a graph $G = (S,E)$ of a wireless sensor network is a set of sensors $F \subset S$ that cannot fail at the same time. ∎

Let $P$ be "The faulty set cannot include the neighbour set of any sensor", $F_P \subset S$ a faulty set satisfying property $P$, and $G = (S,E)$ a communication graph representing a wireless sensor network.

According to our conditional fault-tolerance model, a faulty sensor set $F_P$ is given by $F_P = \{U \subset S \mid \forall\, s_i \in S\colon CN(s_i) \not\subset U\}$, where $CN(s_i)$ is the communication neighbour set of sensor $s_i$. Thus, the communication neighbour set of a sensor cannot fail simultaneously, and hence it is a forbidden faulty set. More precisely, any faulty sensor set that includes the communication neighbour set of at least one sensor $s_i$ is considered as a forbidden faulty set.

*Definition 2.30:* The *conditional connectivity* of $G = (S,E)$ with respect to the property $P$, denoted by $\kappa(G:P)$, is the minimum size of $F_P$ such that the resulting graph $G_{dis} = (S - F_P, E_{dis})$ is disconnected into components each satisfying $P$. ∎

Another generalization of connectivity, called *restricted connectivity*, was proposed by Esfahanian [81] in which the restriction is on the faulty set (i.e., set of nodes that can fail). Restricted connectivity uses the concept of *forbidden faulty set* in which the entire neighbour set of a node cannot be faulty at the same time.

*Definition 2.31:* The *conditional fault tolerance* of $G = (S,E)$ with respect to the property $P$ is given by $\eta(G:\mathrm{P}) = \kappa(G:\mathrm{P}) - 1$. ∎

## 2.5 Energy Consumption Model

The energy consumption of the sensors is *dominated* by data transmission and reception. Let $s_i$ and $s_j$ be two neighbouring sensors. According to the energy consumption model specified by Heinzelman et al. [102], when data is sent from a transmitter to a receiver, there is energy consumption incurred at both ends (i.e., transmitter and receiver). While at the receiver end, the energy consumption is due to only one component, called the *transceiver*, the energy consumption at the transmitter end depends on two components, namely the *transceiver* and the *transmitter amplifier*. The energy consumed by the latter component depends on the size of the data packet that has been sent, the distance between the transmitter and the receiver, and another constant, called *transmitter amplifier* and denoted by $\varepsilon$. The value of this constant depends on whether the free-space or the multi-path model is being considered. Formally, according to [102], the energy spent in data transmission and reception are given by $E_{tx}(s_i, s_j) = a\,(E_{elec} + \varepsilon\, \delta^{\alpha}(s_i, s_j))$ and $E_{rx}(s_i) = a\, E_{elec}$, respectively, where $\delta$ is the Euclidean distance function, $a$ is data size in bits, $E_{elec}$ is the electronics energy and is equal to 50 nJ/bit,

$\varepsilon \in \{\varepsilon_{fs}, \varepsilon_{mp}\}$ is the transmitter amplifier in the free-space ($\varepsilon_{fs} = 10 pJ/bit/m^2$) or the multi-path ($\varepsilon_{mp} = 0.013 pJ/bit/m^2$) model, and $2 \le \alpha \le 4$ is the path-loss exponent. When the identities of sender and receiver are not important, we simply write $E_{tx} = a(E_{elec} + \varepsilon d^{\alpha})$ and $E_{rx} = a E_{elec}$, where $d$ stands for the *transmission distance* used by a sender.

*Definition 32:* The *delay* is defined as the time elapsed between the departure of the data from a source $s_0$ and its arrival to the sink $s_m$. This delay is given by

$$D(s_0,s_m) = (qd + td + pd)\, N_f(s_0,s_m)$$

where *qd* is the average queuing delay per intermediate forwarder, *td* is the average transmission delay, *pd* is the average propagation delay, and $N_f(s_0, s_m)$ is the number of intermediate forwarders between $s_0$ and $s_m$. ■

Given that the size of the field is in the order of a few miles, the average propagation delay is negligible. Thus, the delay $D(s_0,s_m)$ is proportional to $N_f(s_0,s_m)$, i.e.,

$$D(s_0, s_m) = c\, N_f(s_0, s_m) \propto N_f(s_0, s_m)$$

where $c = qd + td$. A similar result can be found in [120].

## 2.6 Percolation Model

Assume that the sensing and communication ranges of the sensors in a two-dimensional (respectively, three-dimensional) wireless sensor network are modelled by disks (respectively, spheres).

Let $X_{\lambda} = \{\xi_i : i \ge 1\}$ be a two-dimensional (respectively, three-dimensional) homogeneous Poisson point process of density $\lambda$, where $\xi_i$ represents the location of a sensor $s_i$. Let $X_{\lambda}(\mathrm{A})$ be a random variable representing the number of points in an area (respectively, volume) A. The probability that there are *k* points inside A is computed as

$$P(X_{\lambda}(\mathrm{A}) = k) = \frac{\lambda^k |\mathrm{A}|^k}{k!} e^{-\lambda|\mathrm{A}|} \tag{2.3}$$

for all $k \ge 0$, where $|\mathrm{A}|$ is the size of the area (respectively, volume) of A.

*Definition 2.33:* In a two-dimensional (respectively, three-dimensional) wireless sensor network, the *covered area* (respectively, *volume*) *fraction* of a Poisson Boolean model $(X_{\lambda}, \{B_i(r) : i \ge 1\})$ given by $A(r) = 1 - e^{-b\lambda}$ [100] is the mean fraction of area (respectively, volume) covered by the sensing disks (respectively, spheres) $B_i(r)$, for $i \ge 1$, in a region of unit area (respectively, volume), where

$b = \pi r^2$ (respectively, $b = 4\pi r^3/3$) is the area (respectively, volume) of the sensing disks (respectively, spheres) of the sensors and $\lambda$ is the density of the Poisson point process $X_\lambda$. ■

Assume that $\lambda$ is not a constant as the sensors could appear and disappear independently of one another. We want to compute the density $\lambda_c$, called *critical percolation density* (or *critical density*) such that there exists an infinite covered component when $\lambda > \lambda_c$, and hence the Boolean model $(X_\lambda, \{D_i(r) : i \geq 1\})$ is said to be *percolating*. Otherwise, there is no infinite covered component and hence the Boolean model $(X_\lambda, \{D_i(r) : i \geq 1\})$ does not percolate.

*Definition 2.34:* The *critical covered area* (respectively, *volume*) *fraction* of $(X_\lambda, \{D_i(r) : i \geq 1\})$, computed as $A_c(r) = 1 - e^{-a\lambda_c}$, is the fraction of area (respectively, volume) covered at critical percolation, where $\lambda_c$ is the associated density of $X_\lambda$. ■

Percolation processes were introduced by Broadbent and Hammersley [51] to model the random flow of a fluid through a medium. Because of their simplicity of description and display of critical behaviour, where the behaviour of a model changes abruptly (phenomenon known as *phase transition*) as the value of a parameter crosses a threshold, percolation models are attractive in several areas of mathematics, physical science, and engineering. A *percolation model* can be viewed as an ensemble of points distributed in space, where some pairs are adjacent (or connected) [83]. We consider a Boolean model [155] which is defined as follows:

*Definition 2.35:* A *Boolean model* consists of two components, namely *point process* $X_\lambda$ and *connection function* $h$. The set $X_\lambda = \{\xi_i : i \geq 1\}$ is a homogeneous Poisson point process of density $\lambda$ in a two-dimensional Euclidean plane $\mathrm{IR}^2$ (respectively, three-dimensional Euclidean space $\mathrm{IR}^3$), where the elements of $X_\lambda$ are the locations of the sensors used to cover a field. The connection function, $h$, is defined such that two points $\xi_i$ and $\xi_j$ are *adjacent* with probability $h(|\xi_i - \xi_j|) = 1$ independent of all other points if $|\xi_i - \xi_j| \leq d$ and $h(|\xi_i - \xi_j|) = 0$ if $|\xi_i - \xi_j| > d$, where $d \geq 0$ and $|\xi_i - \xi_j|$ is the Euclidean distance between $\xi_i$ and $\xi_j$. In other words, $h(|\xi_i - \xi_j|)$ given by

$$h(|\xi_i - \xi_j|) = \begin{cases} 1 & \text{if } |\xi_i - \xi_j| \leq d \\ 0 & \text{otherwise} \end{cases}$$ ■

### *2.6.1 Why a Continuum Percolation Model?*

We consider a *continuum* percolation model rather than a discrete percolation model for the following reason. In discrete percolation [95], also known as *lattice model*, the sites, which are randomly occupied in a discrete lattice, may have different configurations, namely square, triangle, honeycomb, etc. In continuum percolation [155], the positions of the sites are randomly distributed and thus there is no need to have different analysis for each of these regular lattices. Precisely, we consider a continuum percolation model, which consists of homogeneous disks (respectively, spheres) whose centres representing the locations of the sensors are randomly distributed in two-dimensional (respectively, three-dimensional), according to a spatial Poisson point process of density $\lambda$. In percolation theory, we are interested in the *critical density* $\lambda_c$ above which an *infinite cluster of overlapping disks* (respectively, *spheres*) first appears. The density $\lambda_c$ is the critical value for the density $\lambda$ such that there exists no infinite cluster of overlapping disks (respectively, spheres) *almost surely* when $\lambda < \lambda_c$, but there is an infinite cluster of overlapping disks (respectively, spheres) *almost surely* when $\lambda > \lambda_c$ and we say that *percolation* occurs.

## 2.7 Network Model

In this section, we specify the default network model used in this book unless stated otherwise.

We assume that all the sensors in a two-dimensional (respectively, three-dimensional) wireless sensor network are static and isotropic. In other words, all the sensors do not move and have the same sensing and communication ranges. Furthermore, the latter follow the *unit disk model* (respectively, *sphere model*). That is, the sensing range of a sensor $s_i$ is modelled by a disk (respectively, sphere) whose radius is $r$, and its communication range is represented by a disk (respectively, sphere) with radius equal to $R$. Moreover, both of the sensing and communication ranges of the sensor $s_i$ are centred at $p_i$, i.e., the location of the sensor $s_i$. Therefore, each sensor $s_i$ is characterized by two concentric disks (respectively, spheres) associated with its sensing and communication ranges, respectively, as shown in Fig. 2.1.

Moreover, all the sensors are always-on, meaning that they constantly report their sensed data to a single static sink. Hence, the sensors cannot be turned off while monitoring a physical phenomenon. Also, each sensor has a unique *id* (an integer, for instance) and is aware of its own location information through Global Positioning System (GPS) or some localization technique [114]. The sensors advertise their location information only once when they start their sensing task. In addition, each sensor advertises its remaining energy by piggybacking it on the data sent to the sink. The sensors are randomly, uniformly deployed in a field whose size is much larger than that of the sensing and communication ranges of the sensors. Moreover, the sensors are supposed to be *densely* deployed. As indicated in Chap. 1, the limited battery power of the sensors and the difficulty of replacing and/or recharging batteries on the sensors in hostile environments require that the sensors be deployed with high density to extend the network lifetime. Also, we assume that each sensor has transmit-power control and hence can adjust its transmission distance so it can transmit its data over a distance that is less than or equal to the radius of its communication range. The communication links between the sensors are perfectly reliable while the sensors can fail or die independently due to low battery power.

We should mention that some of these assumptions will be further relaxed so as to promote the use of our proposed protocols in real-world sensing applications. These relaxations will be discussed later.

## 2.8 Summary

In this chapter, we defined useful terms that are used throughout this book. Also, we discussed the notion of unconditional (or traditional) connectivity as well as conditional connectivity, and derived the network connectivity and fault tolerance. Moreover, we described the energy consumption and percolation models. Furthermore, we specified our default network model, which is used in the design and description of all the protocols proposed in this book.

## 2.8 Summary

# Chapter 3
# Phase Transitions in Coverage and Connectivity in Two-Dimensional Deployment Fields

This chapter addresses the problems of *almost sure* integrated coverage and connectivity in two-dimensional wireless sensor networks from the perspective of percolation theory. Specifically, it focuses on finding the critical sensor density above which the network is *almost surely* connected and the deployment field is *almost surely* covered. It proposes our solution to this problem using a probabilistic approach. Precisely, each of the above problems is discussed separately. Then, both are investigated in an integrated manner using a suitable integration model.

## 3.1 Introduction

In wireless sensor networks, *sensing coverage* reflects the surveillance quality provided by active sensors in a field, whereas *network connectivity* enables active sensors to communicate with each other in data forwarding to a central gathering node, called the *sink*. For the correct operation of the network, it is necessary that both sensing coverage and network connectivity be maintained. Assuming perfectly reliable wireless links, both sensing coverage and network connectivity are affected by the sensor spatial density. In this chapter, we compute the value of this sensor density to *provide* sensing coverage and network connectivity.

Given a field that is initially uncovered, as more and more sensors are continuously added to the network, the size of the partially covered areas increases. At some point, the situation abruptly changes from small fragmented covered areas to a single large covered area in the field. We call this abrupt change as the *sensing-coverage phase transition* (SCPT) [16]. The SCPT problem can be stated as follows:

> Given a field that is initially uncovered, what is the sensor spatial density corresponding to the first appearance of a single large covered component that spans the entire network?

Likewise, given a network that is originally disconnected, the number of connected components changes with the addition of sensors such that the network suddenly becomes connected at some point. We call this sudden change in the network topology as the *network-connectivity phase transition* (NCPT) [16]. The NCPT problem can be expressed as follows:

> Given a network that is initially disconnected, what is the sensor spatial density corresponding to the first appearance of a single large connected component that spans the entire network?

H.M. Ammari: Challenges and Opportunities of Connected *k*-Covered WSNs, SCI 215, pp. 29–49.
springerlink.com 

The nature of such phase transitions is a central topic in percolation theory of Boolean models. The process of the *ground getting wet* during a period of rain [155] gives us a better analogy with the SCPT and NCPT problems. A circular wet patch forms whenever a point of the ground is hit by a raindrop. At the start of the rain, one can see a small wet area within a large dry area. After some time and as many raindrops continue to hit the ground, the situation suddenly changes and one can see a small dry area within a large wet area. This phase transition phenomenon occurs at a given density of the raindrops. This example helps us approach the SCPT and NCPT problems from a perspective of continuum percolation. In this chapter, we propose a probabilistic approach to compute the *covered area fraction* at critical percolation for both the SCPT and NCPT problems. Then, we derive the corresponding critical sensor density for each of these problems taken separately. In addition, we compute the critical sensor density when both these problems are considered together in an integrated manner. In [12], we propose a different percolation theory-based approach to study coverage and connectivity for three-dimensional wireless sensor networks. This study will be discussed in Chap. 4.

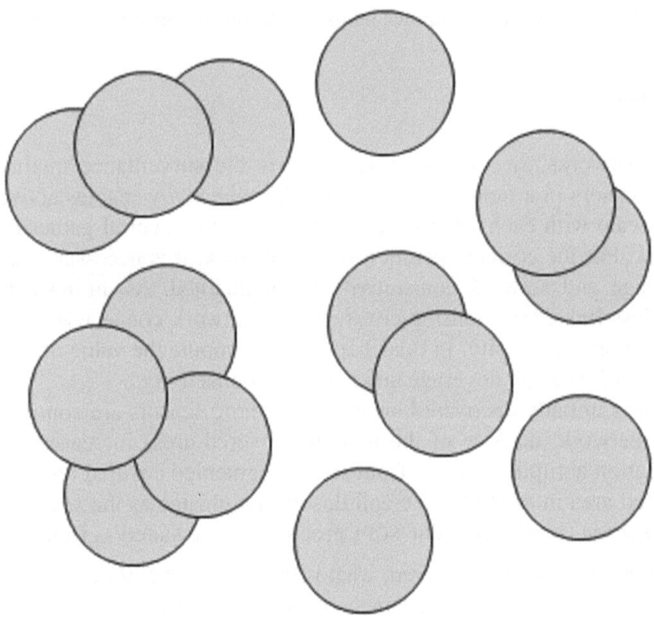

**Fig. 3.1** Schematic of overlapping disks (three covered components of size 1, two of size 2, one of size 3, and one of size 4)

As will be discussed later, the specific connection function used in the NCPT problem has not been studied before and hence no bound on the critical covered area fraction is known. Furthermore, given that sensing coverage and network connectivity are not totally orthogonal [30, 205], we propose a new model for percolation in wireless sensor networks, called *correlated disk model*, which

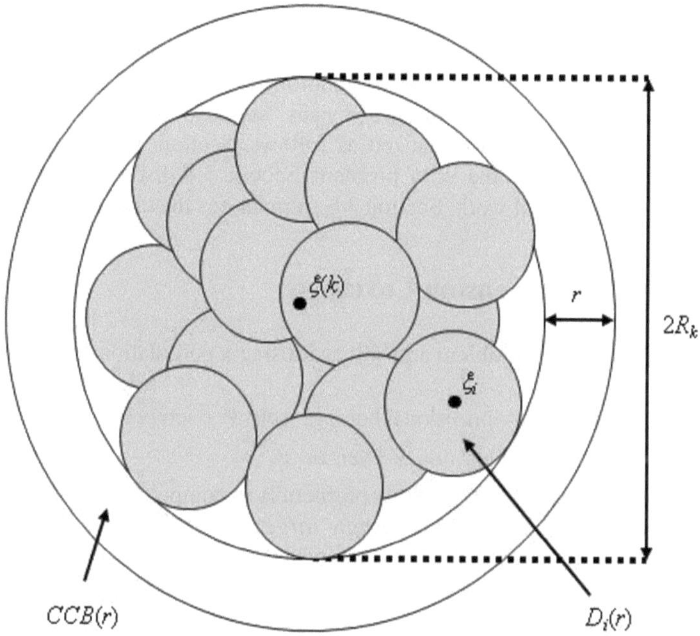


**Fig. 3.2** Shape of a covered component

allows network connectivity and sensing coverage to be studied together in an integrated fashion. We show that the SCPT and NCPT problems have the same solution (i.e., same critical covered area fraction). Precisely, we solve the SCPT and NCPT problems together, where the radii of the sensing disks $(r)$ of the sensors and the radii of their transmission disks $(R)$ are related by $R = \alpha r$ with $\alpha \geq 1$. We show that network–connectivity phase transition occurs provided that sensing–coverage phase transition arises and the ratio $R/r$ has certain value.

This study is of practical use for WSN designers to build up more reliable sensing applications in terms of their sensing coverage and network connectivity. For several real-word sensing applications, and in particular, intruder detection and tracking, it is required that each location in a target field be covered (or *sensed*) by at least one sensor. This would definitely imply better information gathering about the intruder, thus leading to accurate analysis and processing of the situation. On one hand, sensing coverage is associated with all locations in a target field, and hence would guarantee that any event about the intruder is sensed by sensors. On the other hand, network connectivity would enable gathered data about an intruder to reach a central control unit for further analysis and processing. Thus, both sensing coverage and network connectivity should be maintained for high intruder detection and tracking accuracy. Network connectivity, however, depends on sensing coverage. Thus, it is necessary to compute the critical sensor spatial density above which the target field is *almost surely* guaranteed to be covered and the network is *almost surely* guaranteed to be connected.

It is worth noting that the exact value of the critical density at which an infinite (or *single large*) cluster of overlapping disks first appears is still an open problem, and its approximation is either predicted by simulations [166, 173, 188] or computed analytically [86]. From now on, "infinite" means "single large".

The remainder of this chapter is organized as follows. Section 3.2 solves the SCPT problem. Section 3.3 solves the NCPT problem. Section 3.4 discusses our results. Section 3.5 reviews related work. Section 3.6 summarizes the chapter.

## 3.2 Phase Transition in Sensing Coverage

This section discusses the SCPT problem and solves it using a percolation-theoretic approach.

Let $X_\lambda = \{\xi_i : i \geq 1\}$ be a two-dimensional homogeneous Poisson point process of density $\lambda$, where $\xi_i$ represents the location of a sensor $s_i$.

Given an initially uncovered field, the SCPT problem is to compute the probability of the first appearance of an *infinite (or single large) covered component* that spans the entire network. In particular, we are interested in the limiting case of an infinite field, where there exists no single large covered component for sufficiently small density $\lambda$ and it suddenly appears at a *critical percolation density* $\lambda_c$.

### *3.2.1 Estimation of the Shape of Covered Components*

Each covered $k$-component $CC_k$ (Chap. 2, Definition 2.6) is characterized by a reference point, called *centre* and denoted by $\xi(k)$. Figure 3.1 shows various covered components of different sizes. Using the Poissonness argument stated in [100, pp. 200–202], as the centres $\{\xi_i : i \geq 1\}$ form a Poisson process with density $\lambda$, the centres of all covered $k$-components also form a Poisson process with density $\lambda(k)$. In other words, the covered components are randomly and independently distributed according to a Poisson process with a density of $\lambda(k)$ centres per unit area. We want to determine the smallest shape enclosing a covered $k$-component. In fact, the shape of the covered components varies depending on the number of its overlapping sensing disks. For tractability of the problem, we assume that the geometric form that encloses a covered $k$-component is a circle (Fig. 3.2), which tends to minimize the area of uncovered region around the covered component. Indeed, the circle is the most compressed shape. Let $R_k$ be the radius of a circle, denoted by $C(R_k, k)$, which encloses a covered $k$-component. Thus, there is no other sensing disk that could overlap with the boundary of the circle. In other words, the concentric circular band of width $r$, denoted by $CCB(r)$ and which surrounds the circle, should not include any other sensing disk. Hence, the annulus between radii $R_k$ and $R_k + r$ around the centre $\xi(k)$ must be empty.

Let $\mathrm{P}(k)$ be the conditional probability that the circle encloses only one covered $k$-component. This probability is given by

$$\mathrm{P}(k) = \mathrm{Prob}[\, C\,(R_k, k) \mid CCB(r) \text{ is empty } ]$$

By definition, this conditional probability is computed as

$$\mathrm{P}(k) = \frac{\mathrm{Prob}[\, C\,(R_k,k) \wedge CCB\,(r) \text{ empty } ]}{\mathrm{Prob}[\, CCB\,(r) \text{ empty } ]} \tag{3.1}$$

where $\mathrm{Prob}[\, C\,(R_k, k) \wedge CCB\,(r) \text{ empty } ]$ can be interpreted as the probability that the circle of radius $R_k + r$ encloses only one covered $k$-component. Thus,

$$\mathrm{Prob}[\, C\,(R_k, k) \wedge CCB\,(r) \text{ empty } ] = \mathrm{Prob}[\, C(R_k + r, k)\ ]$$

Using Eq. 2.3 (see Chap. 2), we obtain the following results:

$$\mathrm{Prob}[\, C\,(R_k + r, k)\,] = \frac{(\lambda\,\pi\,(R_k + r)^2)^k}{k!} e^{-\lambda\,\pi\,(R_k + r)^2}$$

$$\mathrm{Prob}[\, CCB\,(r) \text{ empty } ] = e^{-\lambda\,\pi\,((R_k + r)^2 - r^2)}$$

Therefore, Eq. 3.1 becomes

$$\mathrm{P}(k) = \frac{(\lambda\,\pi\,(R_k + r)^2)^k}{k!} e^{-\lambda\,\pi\,R_k^{\,2}} \tag{3.2}$$

It is worth mentioning that the analysis of SCPT and NCPT problems will be based on the form of conditional probability given in (3.2).

### 3.2.2 *Critical Density of Covered Components*

Although there exists a few definitions of the average distance between clusters (i.e., covered components), one of them is more appropriate. It is defined as the average of the minimum distance between all pairs of sensing disks, each from one covered component. Indeed, two covered components could be merged together into a single one if there is at least a pair of sensing disks, one from each covered component, such that the distance between their centres is at most equal to $2r$. Lemma 3.1 computes the mean distance between neighbouring covered $k$-components at critical percolation.

*Lemma 3.1:* Let $\{CC_k\}$ be a set of covered $k$-components with density $\lambda(k)$ and $Y$ a random variable representing distances between them. The mean distance $d^1_{avg}$ between two neighbouring covered components at critical percolation is computed as follows:

$$d^1_{avg} = \frac{1}{2\sqrt{\lambda_c(k)}} \tag{3.3}$$

where $\lambda_c(k)$ is the density of $\{CC_k\}$ at critical percolation.

*Proof:* Let $\omega_k$ be the mean number of covered $k$-components in a circular field of radius $\Re$. Denote by $p(\sigma)$ the probability that there is a covered component whose centre is located at a distance upper bounded by $\sigma$ from the centre, say $\xi(k)$, of a given covered component. We denote by $P(\sigma)d\sigma$ the probability that a nearest centre of a covered component to a given centre $\xi(k)$ is located at a distance between $\sigma$ and $\sigma + d\sigma$. Hence, $P(\sigma)d\sigma$ can be viewed as the probability that there exists one of the $\omega_k - 1$ covered components at a distance between $\sigma$ and $\sigma + d\sigma$ from the centre $\xi(k)$ and the other $\omega_k - 2$ covered components are at a distance larger than $\sigma$ from $\xi(k)$. Thus,

$$\begin{aligned} P(\sigma)\, d\sigma &= \begin{pmatrix} \omega_k - 1 \\ 1 \end{pmatrix} \frac{\partial\, p(\sigma)}{\partial\, \sigma} d\sigma \left(1 - p(\sigma)\right)^{(\omega_k - 2)} \\ &= (\omega_k - 1) \frac{\partial p(\sigma)}{\partial \sigma} d\sigma \left(1 - p(\sigma)\right)^{(\omega_k - 2)} \end{aligned} \tag{3.4}$$

where $\frac{\partial\, p(\sigma)}{\partial\, \sigma} d\sigma$ stands for the probability that there is a covered component whose centre lies within a circular band located at a distance $\sigma$ from the centre $\xi(k)$ and whose width is $d\sigma$. Notice that $p(\sigma)$ can be computed as the ratio of the number of covered components within the circle of radius $\sigma$ to the total number of covered components within the field. Thus, we obtain:

$$p(\sigma) = \frac{\lambda_k\, \pi\, \sigma^2}{\lambda_k\, \pi\, \Re^2} = \frac{\sigma^2}{\Re^2}$$

and

$$\frac{\partial\, p(\sigma)}{\partial\, \sigma} = \frac{2\,\sigma}{\Re^2} \tag{3.5}$$

Substituting Eq. 3.3 in Eq. 3.4 gives

$$P(\sigma)\, d\sigma = (\omega_k - 1) \frac{2\,\sigma}{\Re^2} d\sigma \left(1 - \frac{\sigma^2}{\Re^2}\right)^{(\omega_k - 2)}$$

where $\omega_k = \lambda_k\, \pi\, \Re^2$. We assume that the circular field contains all covered components. Now, the mean distance between two covered $k$-components can be computed as

$$E[Y] = \int_0^{\Re} \sigma P(\sigma)\, d\sigma$$
$$= \frac{2(\omega_k - 1)}{\Re^2} \int_0^{\Re} \sigma^2 \left(1 - \frac{\sigma^2}{\Re^2}\right)^{(\omega_k - 2)} d\sigma$$

Using the variable change $T = \dfrac{\sigma^2}{\Re^2}$, we obtain

$$E[Y] = (\omega_k - 1)\,\Re \int_0^1 \sqrt{T}\,(1 - T)^{(\omega_k - 2)}\, dT$$

Recall that the beta function [231] is defined by

$$B(\mathrm{m}, \mathrm{n}) = \int_0^1 u^{m-1}\,(1-u)^{n-1} du$$

Thus,

$$E[Y] = (\omega_k - 1)\,\Re\, B\,(3/2, \omega_k - 1) \tag{3.6}$$

where

$$B(\mathrm{m}, \mathrm{n}) = \frac{\Gamma(\mathrm{m})\,\Gamma(\mathrm{n})}{\Gamma(\mathrm{m}+\mathrm{n})}$$

$$\Gamma(m) = (m-1)\Gamma(m-1) = (\mathrm{m}-1)!$$

Hence, Eq. 3.6 becomes

$$E[Y] = (\omega_k - 1)\,\Re \frac{\Gamma(3/2)\,\Gamma(\omega_k - 1)}{\Gamma(\omega_k + 1/2)} = \Re \frac{\Gamma(3/2)\,\Gamma(\omega_k)}{\Gamma(\omega_k + 1/2)}$$

However, Graham et al. [94] proved that

$$\frac{\Gamma(\mathrm{x}+1/2)}{\Gamma(\mathrm{x})} = \sqrt{x}\left(1 - \frac{1}{8x} + \frac{1}{128x^2} + \frac{5}{1024x^3} - \frac{21}{32768x^4} + \ldots\right)$$

Thus,

$$\lim_{x \to \infty} \frac{\Gamma(\mathrm{x}+1/2)}{\Gamma(\mathrm{x})} = \sqrt{x}$$

Notice that at critical percolation, the value of $\omega_k$ should be *large enough* $(\omega_k \to \infty)$ so an infinite covered component spanning the network could form. Since $\Gamma(3/2) = \dfrac{\sqrt{\pi}}{2}$ and $\omega_k = \lambda_k \pi \Re^2$, the mean distance $d^1_{avg}$ between two neighbouring covered $k$-components at critical percolation is given by

$$d^1_{avg} = \lim_{\omega_k \to \infty} \mathrm{E}[Y] = \Re\,\Gamma(3/2)\frac{1}{\sqrt{\omega_k}} = \frac{1}{2\sqrt{\lambda_c(k)}}$$

where $\lambda_c(k)$ is the critical density of covered $k$-components. ■

Lemma 3.2 computes the average distance between neighbouring covered $k$-components at critical percolation using another approach. As can be seen later, Lemma 3.2 will help us compute the density of covered $k$-components at critical percolation.

*Lemma 3.2:* Let $\{CC_k\}$ be a set of covered $k$-components with density $\lambda(k)$, and $Y$ a random variable associated with the distances between them. The mean distance $d^2_{avg}$ between two neighbouring covered components at critical percolation is computed as

$$d^2_{avg} = \frac{erf(2\sqrt{\lambda_c\,\pi}\,r) - 4\sqrt{\lambda_c}\,r\,e^{-4\lambda_c\pi r^2}}{2\sqrt{\lambda_c}} \tag{3.7}$$

where $\lambda_c$ is the density of a set of sensing disks $\{D_i(r) : i \geq 1\}$ at critical percolation.

*Proof:* For a homogeneous Poisson point process, the probability that there is no neighbour within distance $\sigma$ of an arbitrary point is given by $e^{-\lambda\pi\sigma^2}$ [68]. Therefore, the probability that the distance between a point and its neighbour is less than or equal to $\sigma$ is given by

$$\mathrm{P}[Y \leq \sigma] = 1 - e^{-\lambda\pi\sigma^2}$$

Hence, the corresponding probability density function is given by

$$f(Y \mid Y \leq \sigma) = 2\lambda\,\pi\,\sigma\,e^{-\lambda\pi\sigma^2}$$

The mean distance $d^2_{avg}$ between two neighbouring covered $k$-components of $\{CC_k\}$ at critical percolation is obtained when the distance $\sigma$ between two sensing disks, say $D_i(r)$ and $D_j(r)$, each from one covered component, belongs to the interval $[0,2\,r]$. Therefore,

$$d^2_{avg} = \mathrm{E}[Y \mid Y \leq 2r] = \int_0^{2r} \sigma \times f(Y \mid Y \leq \sigma)d\sigma = \frac{erf(2\sqrt{\lambda_c\pi}\,r) - 4\sqrt{\lambda_c}\,re^{-4\lambda_c\pi r^2}}{2\sqrt{\lambda_c}}$$

where $erf(x)$ is the error function [232]. ■

Lemma 3.3, which follows from Lemmas 4.1 and 4.2, computes the density of covered $k$-components at critical percolation.

*Lemma 3.3:* The critical density of a set of covered $k$-components $\{CC_k\}$ is computed as follows:

$$\lambda_c(k) = \frac{\lambda_c}{\left(erf(2\sqrt{\lambda_c\,\pi}\,r) - 4\sqrt{\lambda_c}\,r\,e^{-4\lambda_c\,\pi\,r^2}\right)^2} \tag{3.8}$$

where $\lambda_c$ is the density of sensing disks at critical percolation and $erf(x)$ is the error function [232].

*Proof:* From Lemma 3.1 (Eq. 3.3) and Lemma 3.2 (Eq. 3.7), the mean distance between two covered $k$-components at critical percolation should verify the following equality $d^1_{avg} = d^2_{avg}$, which implies that the density of covered $k$-components at critical percolation $\lambda_c(k)$ is given by

$$\lambda_c(k) = \frac{\lambda_c}{\left(erf(2\sqrt{\lambda_c\,\pi}\,r) - 4\sqrt{\lambda_c}\,r\,e^{-4\lambda_c\,\pi\,r^2}\right)^2}$$ ■

### 3.2.3 Critical Radius of Covered Components

There is a particular value of the radius $R_k$ of the circular shape enclosing a covered component that *almost surely* guarantees the formation of special class of covered $k$-components, called *critical covered k-components.* Any non-empty circle of radius $2r$ should enclose a covered $k$-component. In other words, regardless of the number of sensing disks of radius $r$ located in a circle of radius $2r$, these sensing disks should definitely form a covered $k$-component. Moreover, this covered $k$-component is a *complete graph* in that each pair of sensors, say $s_i$ and $s_j$, whose sensing disks are included in this circle of radius $2r$ are collaborating given that $|\xi_i - \xi_j|_{\max} \le 2r$. Lemma 3.4 computes the density of critical covered $k$-components at critical percolation.

*Lemma 3.4:* At critical percolation, the density of covered $k$-components, which are enclosed in circles whose radii is equal to $2r$, is given by

$$\lambda_c(k) = \lambda_c \frac{(9\lambda_c\,\pi\,r^2)^k}{k!} e^{-4\lambda_c\,\pi\,r^2} \tag{3.9}$$

where $\lambda_c$ and $\lambda_c(k)$ are the densities of sensing disks and covered $k$-components at critical percolation, respectively.

*Proof:* Let N be the total number of sensing disks that are randomly deployed in a circular field of radius $\Re$ according to a spatial Poisson process with density equal to

$$\lambda = \frac{\mathrm{N}}{\pi \Re^2} \tag{3.10}$$

Using $\omega_k = \lambda(k)\,\pi\,\Re^2$, which represents the mean number of covered $k$-components in the circular field, and Eq. 3.10, we obtain

$$\lambda(k) = \lambda \frac{\omega_k}{\mathrm{N}} \tag{3.11}$$

We can approximate $\frac{\omega_k}{\mathrm{N}}$ by the probability $\mathrm{P}[rad(CC_k) = 2r]$ of finding a covered $k$-component whose radius is equal to $2r$. Hence, we have

$$\mathrm{P}[rad(CC_k) = 2r] = \frac{\omega_k}{\mathrm{N}} \tag{3.12}$$

Substituting Eq. 3.12 in Eq. 3.11 gives

$$\lambda(k) = \lambda\,\mathrm{P}[rad(CC_k) = 2r] \tag{3.13}$$

Following the same reasoning as in Sect. 3.2.1, $\mathrm{P}[rad(CC_k) = 2r]$ is the conditional probability of finding $k$ sensing disks enclosed in a circle with radius $2r$ and centred at $\xi(k)$ such that the annulus between circles of radii $2r$ and $2r + r$ around the centre $\xi(k)$ is empty. Substituting $R_k = 2r$ into Eq. 3.2 gives

$$\mathrm{P}(k) = \frac{(9\lambda\,\pi\,r^2)^k}{k!} e^{-4\lambda\,\pi\,r^2}$$

and hence Eq. 3.13 becomes

$$\lambda_c(k) = \lambda_c \frac{(9\lambda_c\,\pi\,r^2)^k}{k!} e^{-4\lambda_c\,\pi\,r^2}$$

where $\lambda_c$ and $\lambda_c(k)$ are the critical densities of sensing disks and covered $k$-components, respectively. ■

### 3.2.4 *Characterization of Critical Percolation*

Now, we generate an equation that characterizes a set of covered $k$-components at critical percolation. By equating Eqs. 3.8 and 3.9, we obtain a new equation $g_1(\lambda_c, r, k) = 0$, where

$$\begin{aligned} g_1(\lambda_c, r, k) = {} & (erf(2\sqrt{\lambda_c\,\pi}\,r) - 4\sqrt{\lambda_c}\,r\,e^{-4\lambda_c\,\pi\,r^2})^2 \\ & \times \frac{(9\lambda_c\,\pi\,r^2)^k}{k!} e^{-4\lambda_c\,\pi\,r^2} - 1 \end{aligned} \tag{3.14}$$

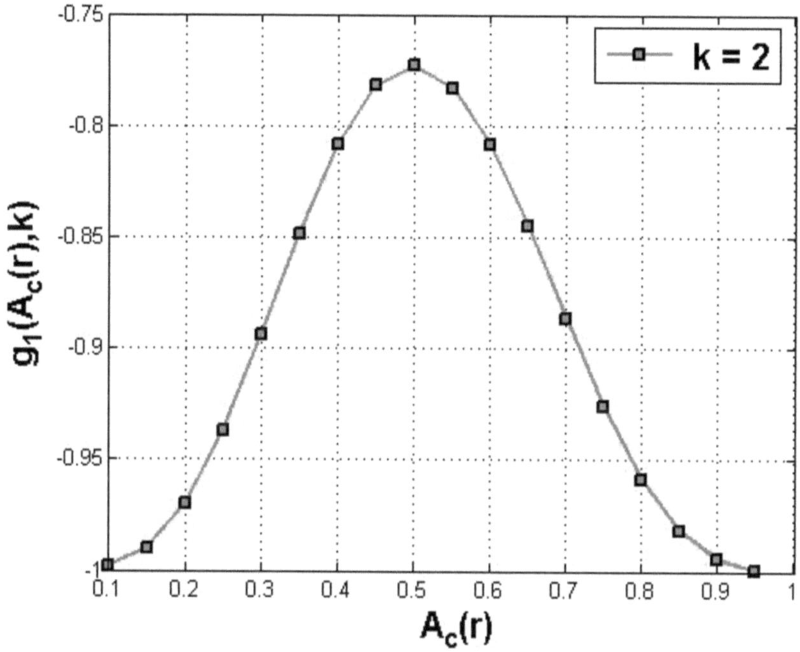


**Fig. 3.3** No critical percolation at $k = 2$

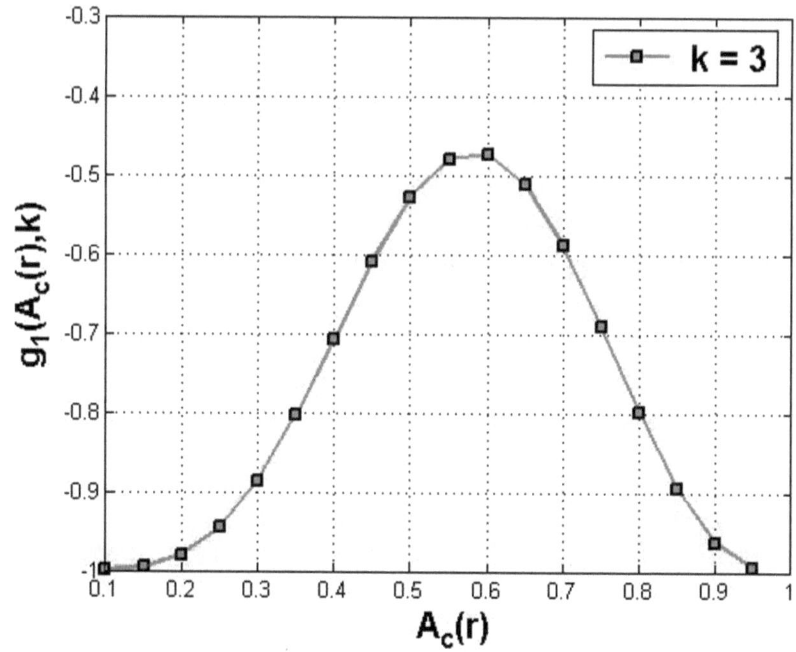


**Fig. 3.4** No critical percolation at $k = 3$

Instead of focusing on finding the critical value of the density $\lambda_c$ of sensing disks at which an infinite covered component first appears, we consider a dimensionless metric, i.e., the *covered area fraction* at critical percolation given by

$$A_c(r) = 1 - e^{-\lambda_c \pi r^2}$$

The benefits of using $A_c(r)$ instead of $\lambda_c$ are two fold: first, the number of unknown parameters is reduced to two, namely $A_c(r)$ and $k$, thus removing any direct dependency of $g_1(\lambda_c, r, k)$ on $r$. Hence, the parameter $r$ will not have any direct impact on the critical percolation density. Second, we know the exact domain of $A_c(r)$ is [0,1], which helps us study exactly the entire behaviour of the function $g_1(\lambda_c, r, k) = 0$ for all values of $A_c(r)$. Substituting $A_c(r)$ into Eq. 3.14 and let $\mu = -\log(1 - A_c(r))$ gives a new function $g_1(A_c(r), k)$ given by

$$g_1(A_c(r), k) = \left( erf(2\sqrt{\mu}) - \frac{4\sqrt{\mu} e^{-4\mu}}{\sqrt{\pi}} \right)^2 \times \frac{(9\mu)^k}{k!} e^{-4\mu} - 1 \tag{3.15}$$

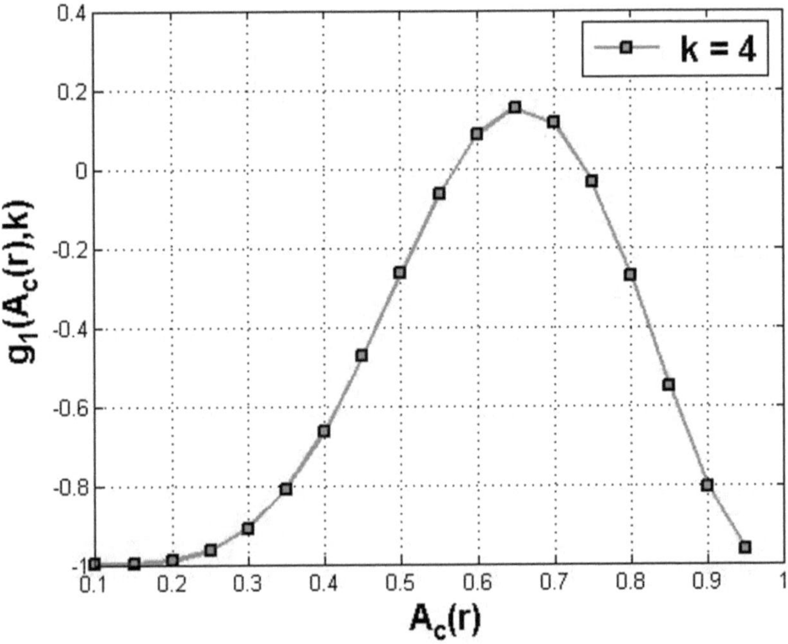


**Fig. 3.5** Critical percolation at $k = 4$ and $A_c(r) = 0.575$

### 3.2.5 Numerical Results

Figures 3.3–3.5 plot the function $g_1(A_c(r), k)$ given in Eq. 3.15 with respect to different values of $k$ and $A_c(r)$. Notice that for $k < 4$, the function $g_1(A_c(r), k)$ cannot be equal to zero (Figs. 3.3 and 3.4). Thus, percolation first occurs at $k = 4$ and $A_c(r) = 0.575$ (Fig. 3.5), which is a bit smaller than the values 0.688 of Vicsek and Kertesz [188], 0.68 of Pike and Seager [166], and 0.62 of Roberts [173] (all predicted by Monte Carlo experiments), and the value 0.67 as calculated by Fremlin [86] for studying the percolation of overlapping homogeneous disks. Thus, when the number of collaborating sensors of a sensor is larger than four $(k \geq 5)$, it is *almost surely* that an infinite covered component that spans the entire network will appear for the first time.

## 3.3 Phase Transition in Network Connectivity

Let $X_\lambda = \{\xi_i : i \geq 1\}$ be a two-dimensional homogeneous Poisson point process of density $\lambda$, where $\xi_i$ represents the location of sensor $s_i$.

Given a network that is originally disconnected, the *network-connectivity phase transition* (NCPT) problem is to compute the sensor spatial density corresponding to the first appearance of an *infinite* (*or single large*) *connected component* that spans the network.

Notice that both of the SCPT and NCPT problems have similar structure although the difference of the concepts of collaboration (SCPT) and communication (NCPT) between the sensors in the SCPT and SCPT problems, respectively, as stated earlier in Sect. 2.2 (see Chap. 2). In the SCPT problem, two sensing disks belong to the same covered component if the distance between them is at most equal to one diameter (2*r*). However, the NCPT problem requires that two communication disks be at a distance of at most half their diameter (i.e., *R*) from each other so they belong to the same connected component, where $r$ and $R$ stand for the radii of the sensing and communication disks of the sensors, respectively. To our knowledge, the connection function of the NCPT problem has not been studied previously in the literature.

Some sensing applications require that every location in the field be covered by at least one sensor and that the active sensors be also able to communicate with each other so the sensed data could reach the sink. Indeed, sensed data would be meaningless if connectivity between the sensors is not maintained. Thus, we are mainly interested in the formation of an *infinite* (or single large) *connected covered component* that spans the entire network. Next, we study the SCPT and NCPT problems *together* using percolation theory.

### 3.3.1 Integrated Sensing Coverage and Network Connectivity

We propose a new model for percolation in wireless sensor networks, called *correlated disk model*. Each sensor is associated with two concentric disks of radii $r$

and $R$ representing the radii of its sensing and communication disks, respectively. This kind of structure reveals a double behaviour of the sensors that can be described by their *collaboration* and *communication*. The collaboration between sensors depends on the relationship between the radii of their sensing disks, whereas communication is related to the relationship between the radii of their communication disks. Previous studies by Wang et al. [197] and Ammari and Das [30] showed the existence of certain dependency between the concepts of sensing coverage and network connectivity. Our proposed correlated disk model allows us to study these two concepts together from a percolation-theoretic viewpoint to account for their correlation. This problem can be viewed as a *correlated continuum percolation* problem. Next, we study the simultaneous percolation of the sensing and communication disks of the sensors based on the ratio $R/r$.

#### 3.3.1.1 Simultaneous Phase Transitions When $R \geq 2r$

As it will be discussed later in this chapter, Wang et al. [197] proved that if a wireless sensor network is configured to be covered and the radius $R$ of the communication disk of the sensors is at least double the radius $r$ of their sensing disk, then the network is guaranteed to be connected. Ammari and Das [30] provided a tighter relationship between $R$ and $r$ while achieving network connectivity provided that sensing coverage is guaranteed. In fact, the "worst-case" behaviour is when the sensing disks of the sensors are tangential, i.e., the distance between their corresponding centres is equal to $2r$. Hence, when $R \geq 2r$, there is a dependency between sensing coverage and network connectivity in that the former implies the latter. In other words, collaboration between the sensors will lead to their communication. In this case, the SCPT and NCPT problems are equivalent, and thus have the same critical covered area fraction. Thus, a set of communication disks percolates at $k = 4$ with a covered area fraction $A_c(R) = 0.575$ at critical percolation. Therefore, when the number of communicating sensors of a given sensor is larger than four ($k = 4$), an infinite connected component spanning the network will *almost surely* form.

#### 3.3.1.2 Simultaneous Phase Transitions When $r \leq R < 2r$

The interesting case is when the radii of the sensing and communication disks of the sensors are related by $R = \alpha r$, where $1 \leq \alpha < 2$. Precisely, we focus on the study of the percolation of the sensing disks of the sensors, where two sensors collaborate if and only if the distance between the centres of their sensing disks is equal to $\alpha r$, where $1 \leq \alpha < 2$. The communication disks of the sensors will also percolate given that $R = \alpha r$. Thus, our goal is to compute the critical covered area fraction above which both the sensing and communication disks of the sensors percolate when $r \leq R < 2r$. It is a valid assumption that the radius of the communication disks of the sensors cannot be less than the radius of their sensing disks as shown in Tables 2 and 3 for a wide spectrum of sensor devices [218].

We consider the previous analysis in Sect. 3.2, where we replace $2r$ by $\alpha r$, with $1 \le \alpha < 2$. Without repeating those details, we obtain a new equation that characterizes a set of covered $k$-components at critical percolation, which is given by $g_2(\lambda_c, r, \alpha, k) = 0$, where

$$g_2(\lambda_c, r, \alpha, k) = (erf(\sqrt{\lambda_c \pi}\, \alpha\, r) - \sqrt{\lambda_c}\, \alpha^2\, r\, e^{-\lambda_c \pi \alpha^2 r^2})^2 \times \frac{(9\lambda_c \pi \alpha^2 r^2/4)^k}{k!} e^{-\lambda_c \pi \alpha^2 r^2} - 1$$

Let $\mu = -\log(1 - A_c(r))$. We substitute $A_c(r)$ in $g_2(\lambda_c, r, \alpha, k)$ to obtain a new function $g_2(A_c(r), \alpha, k)$ given by

$$g_2(A_c(r), \alpha, k) = \left( erf(\alpha\sqrt{\mu}) - \frac{\alpha^2 \sqrt{\mu} e^{-\alpha^2 \mu}}{\sqrt{\pi}} \right)^2 \times \frac{(9\alpha^2 \mu/4)^k}{k!} e^{-\alpha^2 \mu} - 1 \tag{3.16}$$

Figures 3.6–3.9 show the plots of the function $g_2(A_c(r), \alpha, k)$ given in Eq. 3.16 for different values of $k$ and $\alpha$, where $2 \le k \le 5$ and $1 \le \alpha < 2$. As can be seen from Figures 3.6 and 3.7, the function $g_2(A_c(r), \alpha, k)$ cannot be

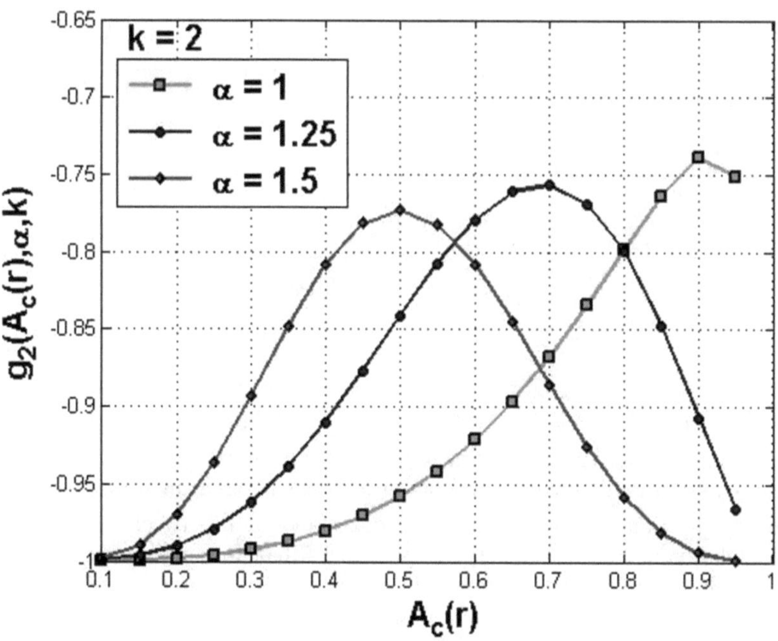


**Fig. 3.6** Plot of the function $g_2(A_c(r), \alpha, k)$ for different values of $\alpha$ $(1 \le \alpha < 2)$. No critical percolation occurs at $k = 2$

equal to zero for $k < 4$, regardless of the value of $\alpha$. Furthermore, a set of sensing disks percolates (which occurs when $g_2(A_c(r), \alpha, k) = 0$) faster for large values of $\alpha$. For instance, when $\alpha = 1$ (which corresponds to $R = r$), critical percolation occurs at $k = 5$ and $A_c(r) = 0.925$ (Fig. 3.9). Thus, when $\alpha = 1$, it is *almost surely* that an infinite covered component spanning the entire network will appear when the number of collaborating sensors of a sensor is larger than five $(k \geq 6)$. However, when $\alpha = 1.5$ (i.e., $R = 1.5\,r$), critical percolation occurs at $k = 4$ and $A_c(r) = 0.580$ (Fig. 3.8).

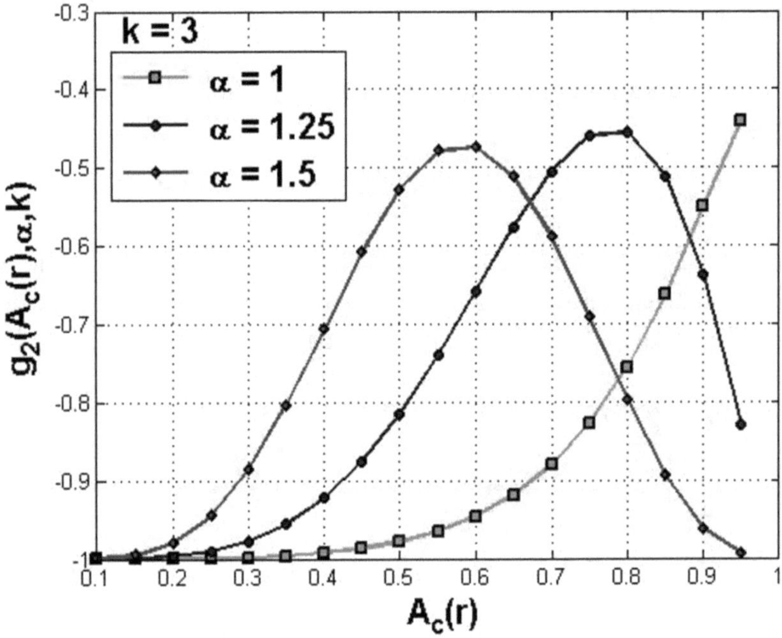


**Fig. 3.7** Plot of the function $g_2(A_c(r), \alpha, k)$ for different values of $\alpha$ $(1 \leq \alpha < 2)$. No critical percolation occurs at $k = 3$

Finally, for $\alpha = 1.25$ (i.e., $R = 1.25\,r$), critical percolation occurs at $k = 4$ and $A_c(r) = 0.760$ (Fig. 3.8). For the last two cases $(\alpha = 1.25$ and $\alpha = 1.5)$, it is *almost surely* that an infinite covered component that spans the entire network will appear when the number of collaborating sensors of a sensor is larger than four $(k \geq 5)$. Given the connection function defined for the collaboration between the sensing disks, percolation should be quicker for large disks than for smaller ones. In all cases, the value of the corresponding *critical covered area fraction* will *almost surely* guarantee the appearance of an infinite connected component that spans the underlying network provided that an infinite covered component arises and spans the entire network. Moreover, the value of *critical covered area fraction* depends on the ratio $R/r$.

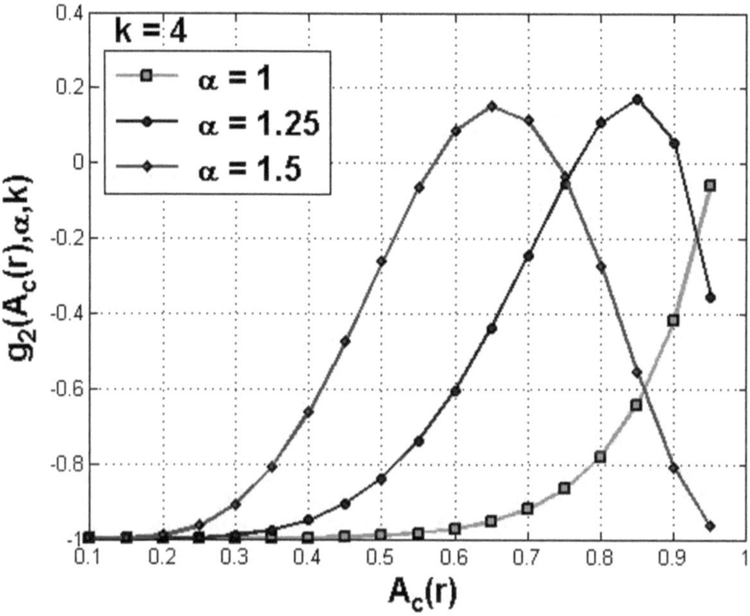


**Fig. 3.8** Plot of the function $g_2(A_c(r),\alpha,k)$ for different values of $\alpha$ $(1 \le \alpha < 2)$. For $k = 4$, critical percolation depends on the value of $\alpha$

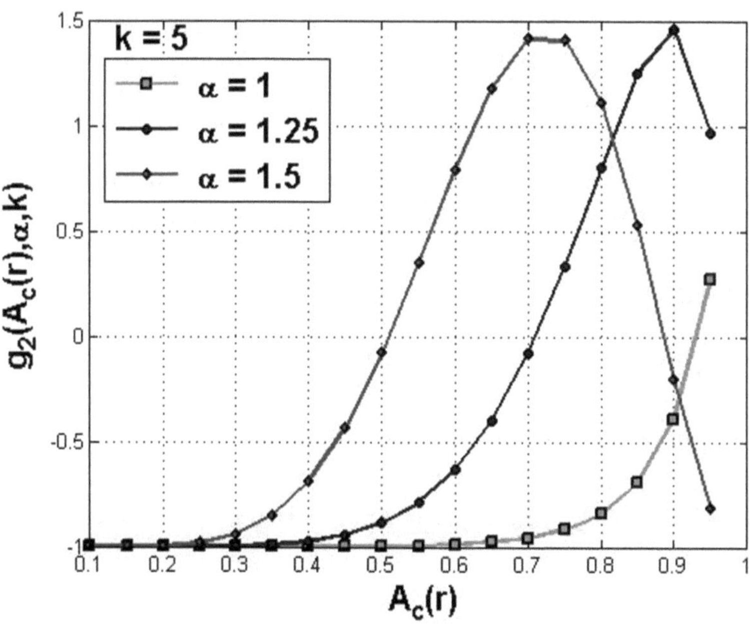


**Fig. 3.9** Plot of the function $g_2(A_c(r),\alpha,k)$ for different values of $\alpha$ $(1 \le \alpha < 2)$. For $k = 4$, critical percolation depends on the value of $\alpha$

## 3.4 Discussion

It is worth noting that both values of critical covered area fractions for sensing coverage and network connectivity represent only lower bounds. In other words, if the actual covered area fraction is higher than $A_c(r)$, it is *almost surely* that there exists an infinite (or single large) covered component that spans the entire network. That is, a large portion of the deployment field is guaranteed to be covered. Otherwise, there are only a few small fragmented regions of the field that are covered. However, there is no guarantee on the size of the region of the field being covered. Similarly, if the actual covered area fraction is higher than $A_c(R)$, it is *almost surely* that there exists an infinite connected component that spans the entire network. Otherwise, it is *almost surely* that the network is disconnected. However, there is no guarantee neither on the number of nodes being connected in this infinite component nor whether the sink belongs to the infinite connected component.

There appears to be little disagreement between our theoretical calculation of the critical covered area fraction ($A_c(r) = 0.575$) compared to the values previously obtained by approximate calculation and Monte Carlo simulation (between 0.62 and 0.688). Our analysis of phase transitions in both sensing coverage (respectively network connectivity) is mainly based on an estimation of the smallest shape enclosing a covered (respectively connected) $k$-component. We assumed that this shape is a circle. Although it may not be always true that a circle is the smallest shape enclosing covered (and connected) $k$-component, we used it to simplify the analysis enough and make it mathematically tractable. Also, we considered this shape as an ellipse with minor axis $a_k$ and major axis $b_k$. Maximizing $\mathrm{P}(k) = \frac{(\lambda \pi (a_k + r)(b_k + r))^k}{k!} e^{-\lambda \pi a_k b_k}$, the probability that an ellipse encloses a covered $k$-component, leads however to a unique solution $a_k = b_k$ representing a degenerate ellipse or *circle*.

## 3.5 Related Work

Adlakha and Srivastava [2] showed that the number of sensors required to cover an area of size $A$ is in the order of $\mathrm{O}(A/\hat{r}_2{}^2)$, where $\hat{r}_2$ is a good estimate of the radius $r$ of the sensing disk of the sensors. Specifically, $r$ lies between $\hat{r}_1$ and $\hat{r}_2$, where $\hat{r}_1$ overestimates the total number of sensor required to cover an area of size $A$ while $\hat{r}_2$ underestimates it. Franceschetti et al. [85] investigated the number of disks of given radius $r$, centered at the vertices of an infinite square grid, which are required to entirely cover an arbitrary disk of radius $r$ placed on the plane. Their result depends on the ratio of $r$ to the grid spacing.

Kumar et al. [128] proved that for random deployment with uniform distribution, if there exists a slowly growing function $\phi(np)$ such that

$n\,p\pi\,r^2 \geq \log(np) + k\log\log(np) + \phi(np)$, then a square unit area is $k$-covered with high probability when $n$ sensors are deployed in it, where $p$ is the probability that a sensor is active. It is worth noting that $n$ also represents the sensor spatial density given that the area of the square region is equal to 1. Hence, the above inequality can be written as $n\,p \geq \frac{\log(np) + k\log\log(np) + \phi(np)}{\pi\,r^2}$, which means that the minimum sensor density required for $k$-coverage of a unit square region is equal to $\frac{\log(np) + k\log\log(np) + \phi(np)}{\pi\,r^2}$. If we set $p = 1$ (i.e., every sensor is active), we obtain $\frac{\log(n) + k\log\log(n) + \phi(n)}{\pi\,r^2}$. Recently, Balister et al. [38] computed the sensor density necessary to achieve both sensing coverage and network connectivity in finite region, such as thin strips (or annuli) whose lengths are finite. Balister et al. [38] applied this result to achieve barrier coverage [127] and connectivity in thin strips. In this type of deployment, the sensors act as a barrier to ensure that any moving object or phenomenon that crosses the barrier of sensors will be detected.

Zhang and Hou [217, 219] proved that the required density for $k$-coverage of a square field, where sensors are distributed according to a Poisson point process and always active, depends on both the side length of the field and $k$. Precisely, Zhang and Hou [217, 219] found that a necessary and sufficient condition of complete $k$-coverage of a square field with side length $l$ is that the sensor density is equal to $\lambda = \log l^2 + (k+1)\log\log l^2 + c(l)$, where $c(l) \to +\infty$ as $l \to \infty$. Also, given a wireless sensor network deployed as a Poisson point process with density $\lambda$ and every sensor is active, Zhang and Hou [216] provided a sufficient condition for $k$-coverage of a square region with area $A$. Precisely, they proved that assuming $\lambda = \log A + 2\,k\log\log A + c(A)$, if $c(A) \to \infty$ as $A \to \infty$, then the probability of $k$-coverage of the square region approaches 1. Furthermore, Zhang and Hou [216] provided the same result in the case where sensors are deployed according to a uniformly random distribution. Both results are based on the following statement: the square region is divided into square grids with side length $s = \frac{\sqrt{2}\,r}{\log A}$, where $r$ stands for the radius of the sensing range of the sensors. For a grid $i$ to be completely $k$-covered, it is sufficient that there are at least $k$ sensors within a disk centred at the centre of the grid and with radius $(1-u)\,r$, denoted by $B_i\,((1-u)\,r)$, where $u = 1/\log A$.

Wan and Yi [189] showed that with boundary effect, the asymptotic $(k+1)$-coverage of a square with area $s$ by Poisson point process with unit-area coverage range requires that the sensor density be equal to $\log s + 2\,(k+1)\log\log s + \xi(s)$ with $\lim_{s\to\infty} \xi(s) = \infty$. Without the boundary effect, however, the asymptotic $(k+1)$-coverage requires that the sensor density be computed as $\log s + (k+2)\log\log s + \xi(s)$ with $\lim_{s\to\infty} \xi(s) = \infty$.

The concept of *continuum percolation* originally due to Gilbert [91], is to find the critical density of a Poisson point process at which an unbounded connected component *almost surely* appears so the network can provide long-distance multi-hop communication. For random plane networks, Gilbert claimed that the filling factor, representing the critical value of the expected number of points in a circle of radius $R$ should be around 3.2. Since then, Gilbert's model has become the basis for studying continuum percolation in wireless networks. Booth et al. [45] discussed different classes of covering algorithms and determined the critical density of a Poisson point process (centres of spheres of radius $r$) above which an unbounded connected component arises. They also discussed the almost sure existence of an unbounded connected component based on the ratio of the connectivity range of the base stations to the connectivity range of the clients. Bertin et al. [42] proved the existence of site percolation and bond percolation in the Gabriel graph [87] for both Poisson and hard-core stationary point processes. The critical bounds corresponding to the existence of a path of opens sites and a path of open bonds were found by simulation. Glauche et al. [92] proposed a distributed protocol, which guarantees strong connectivity *almost surely* of ad hoc nodes. They translated the problem of finding the critical communication range of mobile devices to that of determining the critical node neighbourhood degree above which an ad hoc network graph is *almost surely* connected. To achieve a little above this degree, each node needs to adjust its transmission power locally. Jiang and Bruck [115] proposed the concept of monotone percolation based on the local adjustment of the communication radii of the nodes for efficient topology control of the network. Their proposed algorithms guarantee the existence of relatively short paths between any pair of source and destination nodes, which makes monotonic progress. Liu and Towsley [146] considered both Boolean and general sensing models, each with a variety of network scenarios, to characterize fundamental coverage properties of large-scale sensor networks, namely area coverage, node coverage, and detectability. According to their simulation setting, the critical density is about $3.53\times10^{-3}$.

## 3.6 Summary

In this chapter, we investigated two phase transition problems for *sensing-coverage* and *network-connectivity* in wireless sensor networks using a probabilistic approach [16]. Our goal is to determine when an infinite covered component (SCPT problem) and an infinite connected component (NCPT problem) could form for the first time. To achieve this objective, we computed the covered area fraction for SCPT and NCPT problems at critical percolation. The problem of overlapping disks has been studied extensively in percolation theory and is similar to the SCPT problem. We found that the value of the covered area fraction is close to the one found by Monte Carlo simulations. The specific connection function of the Boolean model associated with the NCPT problem, however, has not been studied before and hence no bound exists in the literature. We proposed a *correlated disk model* to study SCPT and NCPT problems in an integrated way from a continuum percolation perspective. Precisely, we considered the physical correlation between them, which is based on the ratio of the radius of the communication disks of the sensors to the radius of their sensing disks. Thus, when an infinite covered component arises for the first time, an infinite connected component will *almost surely* appear based on the ratio $\alpha = R / r$.

# Chapter 4
# Phase Transitions in Coverage and Connectivity in Three-Dimensional Deployment Fields

This chapter focuses on the problem of *almost sure* integrated coverage and connectivity in three-dimensional wireless sensor networks. Precisely, it investigates the critical sensor density above which coverage percolation and connectivity percolation in three-dimensional deployment fields will *almost surely* occur. It discusses our solution to this problem using an approach that is totally different from the one we proposed for two-dimensional deployment fields. Also, it addresses the problem of integrated coverage and connectivity in a two-dimensional space.

## 4.1 Introduction

Although most of the studies on *coverage* and *connectivity* in wireless sensor networks considered two-dimensional settings, such networks can in reality be accurately modelled in a three-dimensional space. While extensive work has been done on coverage and connectivity in two-dimensional wireless sensor networks, the analysis for three-dimensional wireless sensor networks has hardly received any sustained attention. In particular, only a few references on three-dimensional wireless sensor networks exist in the literature [8, 132, 167, 169, 172], although there are several scenarios of real-world set up where the sensors are deployed in the space rather than in a planar field. For instance, sensor networks deployed on the trees of different heights in a forest, in a building with multiple floors, and underwater, require design in three-dimensional rather than two-dimensional space.

While coverage creates *collaboration* between the sensors in covering a target field for monitoring specific phenomena, connectivity enables *communication* between them. The concepts of continuum percolation theory best fit the problem of connectivity in wireless sensor networks that aims to find out whether the network can provide long-distance multi-hop communication [155]. We say the network exhibits *coverage percolation* when a *giant covered region almost surely* spans the entire deployment field for the first time. Also, we say the network presents *connectivity percolation* when a *giant connected component almost surely* spans the entire network for the first time. While coverage depends on the sensing capability of the sensors, connectivity depends on their communication capability. However, these two concepts are not totally independent of one another and thus can be studied in combination from a perspective of percolation theory. Indeed, it has been proved that when a wireless sensor network is configured to provide coverage and the radius of the communication ranges of the sensors is at least

H.M. Ammari: Challenges and Opportunities of Connected $k$-Covered WSNs, SCI 215, pp. 51–71.
springerlink.com 

double the radius of their sensing ranges, the network is guaranteed to be connected [205].

This chapter investigates the problems of coverage percolation, connectivity percolation, and integrated coverage and connectivity percolation in three-dimensional wireless sensor networks using a continuum percolation theory-based approach [12]. In particular, we want to address the following questions:

- What is the critical density above which a giant covered region of a field will *almost surely* appear for the first time? What is the corresponding critical network degree?
- What is the critical density above which a giant connected component will *almost surely* appear for the first time? What is the corresponding critical network degree?
- What is the critical density above which a giant covered region of a field and a giant connected component will *almost surely* appear simultaneously for the first time? What is the corresponding critical network degree?

As described in Chap. 3, a few approaches for studying coverage and connectivity in wireless networks, including two-dimensional wireless sensor networks, and using percolation theory have been proposed [42, 45, 91, 92, 115, 146]. In this chapter, we investigate the *critical density* for percolation in coverage and connectivity in three-dimensional wireless sensor networks as well as the corresponding *critical network degree*. Our proposed approach is based on Baxter's factorization [40] of Ornstein–Zernike equation [165] and the pair-connectedness theory [65], which were initially proposed to understand the structure and dynamics of simple liquids.

The remainder of this chapter is organized as follows. Section 4.2 introduces our *integrated-concentric-sphere model*, discusses coverage percolation, connectivity percolation, and integrated coverage and connectivity percolation problems in three-dimensional wireless sensor networks, and computes their corresponding critical density. It also computes their corresponding network degree. Section 4.3 discusses the results presented in Sect. 4.2 and generalizes them by relaxing some widely used assumptions in coverage and connectivity in wireless sensor networks. Section 4.4 reviews related work and Sect. 4.5 summarizes the chapter.

## 4.2 Three Percolation Problems

In this section, we discuss coverage percolation, connectivity percolation, and integrated coverage and connectivity percolation problems in three-dimensional wireless sensor networks. Precisely, we compute their corresponding critical densities and network degrees.

Let $X_\lambda = \{\xi_i : i \geq 1\}$ be a homogeneous Poisson point process of density $\lambda$ in $\mathbb{R}^3$, where $\xi_i$ represents the center of two concentric spheres corresponding to the sensing and communication spheres of a sensor $s_i$. Assume that $\lambda$ is not a constant as the sensors could appear and disappear independently of one another.

### 4.2.1 Sensing Coverage Percolation

*Problem formulation:* The *coverage percolation* (COVP) problem can be stated as follows: Given a field that is initially uncovered (or consists of small fragmented covered areas), compute the density $\lambda_c^{\text{cov}}$, called *critical percolation density* (or *critical density*) such that there *surely* exists a *giant covered region* that spans the entire deployment field when $\lambda > \lambda_c^{\text{cov}}$, and hence the Boolean model $(X_\lambda, \{B_i(r) : i \geq 1\})$ *percolates*. Otherwise, there is no giant covered region and hence $(X_\lambda, \{B_i(r) : i \geq 1\})$ does not percolate.

Precisely, we are interested in the limiting case of an *infinite* (or *very large*) field, where at a density lower than some density, called *critical percolation density* $\lambda_c$, there are several small covered regions in a deployment field and that at a density higher than $\lambda_c^{\text{cov}}$, there appears a giant covered region. In this problem, we only consider the sensing spheres of the sensors of radius $r$.

The concept of physical clustering of particles was first investigated by Hill [106]. In this section, the sensors will be identified with particles. According to Hill [106], the probability of connectedness and disconnectedness depends on the Boltzmann factor $e(\sigma)$ that is separated into connected and disconnected parts $e^+(\sigma)$ and $e^*(\sigma)$, respectively, such that

$$e(\sigma) = e^+(\sigma) + e^*(\sigma) = e^{-\beta u(\sigma)} \tag{4.1}$$

where

$$\begin{cases} e^+(\sigma) = e^{-\beta u^+(\sigma)} \\ e^*(\sigma) = e^{-\beta u^*(\sigma)} \end{cases} \tag{4.2}$$

$$\begin{cases} u^+(\sigma) : \text{interaction potential for connected pair of} \\ \qquad\qquad \text{particles separated by a distance, } \sigma \text{, apart} \\ u^*(\sigma) : \text{interaction potential for disconnected pair of} \\ \qquad\qquad \text{particles separated by a distance, } \sigma \text{, apart} \end{cases} \tag{4.3}$$

and $\beta = 1/(k_B T)$, where $T$ is the temperature and $k_B$ the Boltzmann constant. Following Hill's definition of physical clusters based on inter-particle distance [106], we set up the interaction potentials $\boldsymbol{u}^+(\sigma)$ and $\boldsymbol{u}^*(\sigma)$ as follows:

$$\begin{cases} u^+(\sigma) = u(\sigma) & \text{and} \quad u^*(\sigma) = +\infty & \text{for } 0 < \sigma < 2r \\ u^+(\sigma) = +\infty & \text{and} \quad u^*(\sigma) = u(\sigma) & \text{for } \sigma > 2r \end{cases} \tag{4.4}$$

Hill's work was the basis for other theories, such as theory of percolation in liquids [72, 73] and theory of pair connectedness [65]. Coniglio et al. [64] extended Hill's work with a general theory of the equilibrium distribution of physical clusters, which was then extended with a theory of the pair connectedness. The pair-connectedness function [65] $H(\sigma)$ is defined so that $\lambda^2 H(\sigma)d^2\sigma$ is the probability of finding a pair of particles (or sensors) in the same cluster (or covered component) and separated by a distance $\sigma$ apart, where $\lambda$ is the particle density. Our correlated continuum percolation-based approach is driven by the concept of physical clustering [106] and can be formulated in terms of an integral equation of Ornstein–Zernike [165]. Hence, we need to compute the pair-connectedness function $H(\sigma)$. It is worth mentioning that the Ornstein–Zernike relation was originally proposed for a homogeneous fluid to deal with inter-particle correlation in order to find out how the position of a particle affects that of its neighbours by virtue of inter-particle interaction [105, 106]. Coniglio et al. [65] considered another metric, namely *connectedness*, and proposed an Ornstein–Zernike equation for the pair-connectedness function in analogy with that for the pair-correlation function [165]. According to Coniglio et al. [65], the Ornstein–Zernike relation can be established from the pair-connectedness function $H(\sigma)$ and the direct pair-connectedness function $C^+(\sigma)$ as follows:

$$H(\sigma) = C^+(\sigma) + \lambda \int C^+(\sigma')H(|\sigma - \sigma'|)d\sigma' \tag{4.5}$$

Following Baxter's transformation [40], Eq. 4.5 can be formulated as

$$\tilde{H}(k) = \tilde{C}^+(k) + \lambda \tilde{C}^+(k)\tilde{H}(k) \tag{4.6}$$

where

$$\tilde{\boldsymbol{H}}(\boldsymbol{k}) = 4\pi \boldsymbol{k}^{-1} \int_0^\infty \boldsymbol{d\sigma} \sin(\boldsymbol{k\sigma})\boldsymbol{\sigma} \boldsymbol{H}(\boldsymbol{\sigma})$$

and

$$\tilde{\boldsymbol{C}}^+(\boldsymbol{k}) = 4\pi \boldsymbol{k}^{-1} \int_0^\infty \boldsymbol{d\sigma} \sin(\boldsymbol{k\sigma})\boldsymbol{\sigma} \boldsymbol{C}^+(\boldsymbol{\sigma})$$

are the Fourier transforms of $\boldsymbol{H}(\boldsymbol{\sigma})$ and $C^+(\sigma)$, respectively. Equation 4.6 implies that

$$\tilde{\boldsymbol{H}}(\boldsymbol{k})(1 - \lambda \tilde{\boldsymbol{C}}^+(\boldsymbol{k})) = \tilde{\boldsymbol{C}}^+(\boldsymbol{k})$$

and

$$\tilde{\boldsymbol{C}}^+(\boldsymbol{k})(1 + \lambda \tilde{\boldsymbol{H}}(\boldsymbol{k})) = \tilde{\boldsymbol{H}}(\boldsymbol{k})$$

Thus, it follows that

$$1 + \lambda \tilde{\boldsymbol{H}}(\boldsymbol{k}) = \frac{1}{1 - \lambda \tilde{\boldsymbol{C}}^+(\boldsymbol{k})} \tag{4.7}$$

Using Baxter's factorization [40], we introduce a new function $Q(\sigma)$ such that

$$\tilde{\boldsymbol{Q}}(\boldsymbol{k})\tilde{\boldsymbol{Q}}(-\boldsymbol{k}) = 1 - \lambda \tilde{\boldsymbol{C}}^+(\boldsymbol{k}) \tag{4.8}$$

where

$$\tilde{Q}(k) = 1 - 2\pi\lambda \int_0^{2r} d\sigma\, e^{ik\sigma} Q(\sigma) \tag{4.9}$$

and

$$Q(\sigma) = 0 \text{ for } \sigma > 2r$$

Hence, the Ornstein–Zernike relation for the pair-connectedness function (Eq. 4.5) can be written as follows [40]:

$$\begin{cases} \sigma C^{+}(\sigma) = -\dfrac{\partial Q(\sigma)}{\partial \sigma} + 2\pi\lambda \displaystyle\int_{\sigma}^{2r} dt \frac{\partial Q(t)}{\partial t} Q(t-\sigma) & (4.10) \\ \text{for } 0 < \sigma < 2r & \\ \sigma H(\sigma) = -\dfrac{\partial Q(\sigma)}{\partial \sigma} + 2\pi\lambda \displaystyle\int_{0}^{2r} dt\, (\sigma - t)\, H(|\sigma - t|)\, Q(t) & (4.11) \\ \text{for } \sigma > 0 & \end{cases}$$

where

$$Q(\sigma) = 0 \text{ for } \sigma < 0 \text{ and } \sigma > 2r$$

Notice that the direct pair-connectedness function $C^{+}(\sigma)$ vanishes outside the associated range parameter $(C^{+}(\sigma) = 0)$ for $\sigma > 2r$ ($2r$ is the diameter of sensing spheres). Furthermore, Coniglio et al. [65] showed that

$$H(\sigma) = e^{-\beta u^{+}(\sigma)} g(\sigma) e^{\beta u(d)} + e^{-\beta u^{*}(\sigma)} (H(\sigma) - C^{+}(\sigma)) \tag{4.12}$$

Substituting Eq. 4.4 for $\sigma \le 2r$ in Eq. 4.12 implies

$$H(\sigma) = g(\sigma) \tag{4.13}$$

where $g(\sigma) = g_s(\sigma) = 1 - \omega_s$ [65] stands for the Percus–Yevick approximation [41] of the pair-connectedness function, and $\omega_s$ is the minimum overlap volume fraction between two sensing spheres whose range is given by $0 < \omega_s < 1$. Indeed, two collaborating sensing spheres need only intersect. Thus, $H(\sigma) = 1 - \omega_s$. Substituting the value of $H(\sigma)$ in Eq. 4.7 for $0 < \sigma < 2r$, it follows that

$$\sigma (1 - \omega_s) = -\frac{\partial Q(\sigma)}{\partial \sigma} + 2\pi\lambda \int_{\sigma}^{2r} dt\, (\sigma - t)(1 - \omega_s) Q(t) \tag{4.14}$$

$$\text{for } 0 < \sigma < 2r$$

That is,

$$\begin{aligned} \frac{\partial Q(\sigma)}{\partial \sigma} &= \left( (\omega_s - 1) + 2\pi\lambda (1 - \omega_s) \int_0^{2r} Q(t)\, dt] \right) \sigma \\ &\quad + 2\pi\lambda (\omega_s - 1) \int_0^{2r} t\, Q(t)\, dt = a\sigma + b \end{aligned} \tag{4.15}$$

where $a$ and $b$ are constant determined by $Q(\sigma)$ and given by

$$\begin{cases} a = (\omega_s - 1) + 2\pi\,\lambda\,(1-\omega_s)\displaystyle\int_0^{2r} Q(t)\,dt \\ b = 2\pi\,\lambda(\omega_s - 1)\displaystyle\int_0^{2r} tQ(t)\,dt \end{cases} \tag{4.16}$$

The integration of Eq. 4.15 under the assumption that $Q(\sigma) = 0$ for $\sigma < 0$ and $\sigma > 2\boldsymbol{r}$ gives

$$\boldsymbol{Q}(\sigma) = \frac{\boldsymbol{a}}{2}\,(\sigma^2 - 4\boldsymbol{r}^2) + \boldsymbol{b}\,(\sigma - 2\boldsymbol{r}) \tag{4.17}$$

Substituting Eq. 4.17 in Eq. 4.16 for $a$ and $b$, and solving them, produces a unique solution given by

$$\begin{cases} a = \dfrac{-12\,(1-\omega_s)\,(3 - \lambda\,\pi\,(1\text{-}\omega_s)\,(2\,r)^3)}{\lambda^2\,\pi^2\,(1\text{-}\omega_s)^2\,(2\,r)^6 + 12\,\lambda\,\pi\,(1-\omega_s)\,(2\,r)^3 + 36} \\ b = \dfrac{-9\,\lambda\,\pi\,(1\text{-}\omega_s)^2\,(2\,r)^4}{\lambda^2\,\pi^2\,(1\text{-}\omega_s)^2\,(2\,r)^6 + 12\,\lambda\,\pi\,(1-\omega_s)\,(2\,r)^3 + 36} \end{cases} \tag{4.18}$$

According to [65], the *mean cluster (or covered component) size* is computed as

$$MCS = 1 + \lambda \int H(\sigma)\,d\sigma \tag{4.19}$$

Using the Fourier transform $\tilde{H}(k)$ of the pair connectedness, we obtain

$$MCS = 1 + \lambda\tilde{H}(0) \tag{4.20}$$

According to Eq. 4.7,

$$MCS = \frac{1}{1 - \lambda\,\tilde{C}^{+}(0)} \tag{4.21}$$

At critical percolation, the cluster mean size diverges (i.e., $MCS \to \infty$) and this occurs only when $\lambda_c\,\tilde{c}^{+}(0) = 1$, where $\lambda_c$ is the percolation critical density. Using Eq. 4.8, we find that

$$\boldsymbol{MCS} = \frac{1}{\tilde{\boldsymbol{Q}}^2(0)} \tag{4.22}$$

Substituting Eq. 4.9 in Eq. 4.22 and using Eqs. 4.17 and 4.18, we find that

$$\begin{aligned} MCS &= \left(1 - 2\pi\,\lambda \int_0^{2r} Q(\sigma)\,d\sigma\right)^{-2} \\ &= \left(\frac{\lambda^2\pi^2(1\text{-}\omega_s)^2(2r)^6 + 12\lambda\pi(1-\omega_s)(2r)^3 + 36}{-12\lambda\pi(1-\omega_s)(2r)^3 + 36}\right)^2 \end{aligned} \tag{4.23}$$

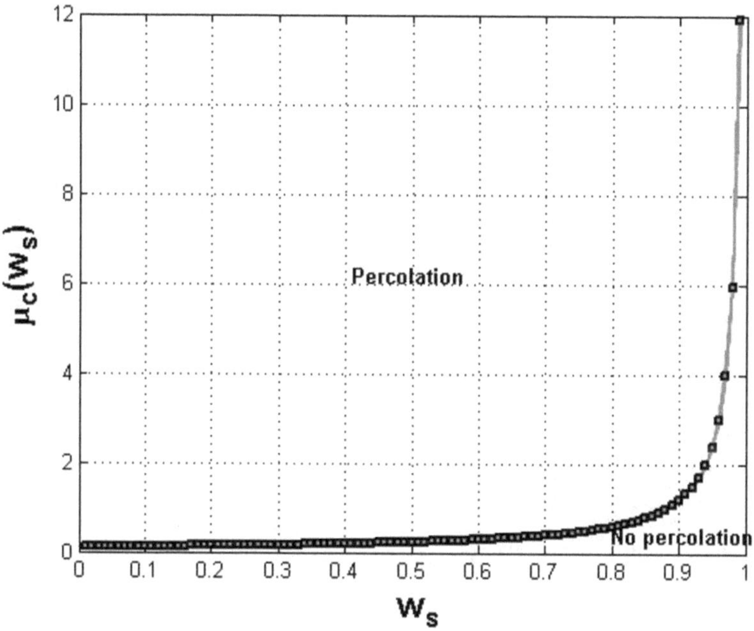


**Fig. 4.1** Plot of the function $\mu_c(\omega_s)$ for $0 < \omega_s < 1$

Hence, the critical percolation density $\lambda_c^{\text{cov}}(r,\omega_s)$ of the sensing spheres of radius $r$ that corresponds to the divergence of $MCS$ (i.e., $MCS \to \infty$ or $-12\,\lambda\,\pi\,(1-\omega_s)\,(2\,r)^3 + 36 = 0$) is given by

$$\lambda_c^{\text{cov}}(r,\omega_s) = \frac{0.119}{(1-\omega_s)\,r^3} \tag{4.24}$$

where $0 < \omega_s < 1$ is the minimum overlap volume fraction between sensing spheres. Notice that $\lambda_c$ is a three-dimensional variable.

*Theorem 4.1:* The *critical sensor density* $\lambda_c^{\text{cov}}(r,\omega_s)$ above which a giant covered region that span a three-dimensional deployment field will *almost surely* form is given by

$$\lambda_c^{\text{cov}}(r,\omega_s) = \frac{0.119}{(1-\omega_s)\,r^3}$$

where $0 < \omega_s < 1$ is the minimum overlap volume fraction between the sensing spheres of the sensors of radius $r$. ■

To remove the dependability on $r$, we consider the function $\mu_c(\omega_s) = \lambda_c^{\text{cov}}(r,\omega_s)\,r^3 = \dfrac{0.119}{1-\omega_s}$. Figure 4.1 shows the plot of the function, $\mu_c(\omega_s)$

for $0 < \omega_s < 1$. As can be seen, the percolation density depends monotonically on the minimum overlap volume fraction $\omega_s$ between the sensing spheres of the sensors. Higher overlap volume fraction requires high percolation density and vice versa. Notice that the percolation density on Fig. 4.1 corresponds to the minimum density above which a giant covered component *almost surely* appears for the first time. However, Kong and Yeh [122] found that for three-dimensional Poisson random geometric graphs (RGGs), the analytical lower bound for the critical density is 0.4494. Although their analysis is elegant, we believe that the critical density should depend on the coverage capability of the sensors (i.e., sensing range). Let us now compute the *critical network degree* $\eta_c$, which in this case coincides with the average size of the collaborating sets of the sensors (Chap. 2, Definition 2.3) at the percolation density.

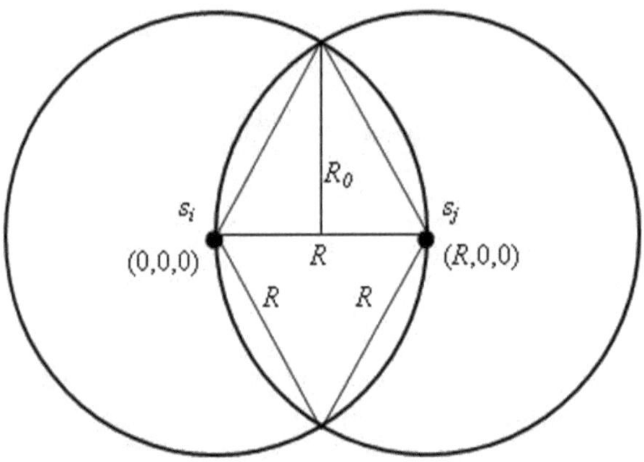


**Fig. 4.2** Minimum overlap volume for communication

According to [105], the average network degree at percolation (or *critical network degree*) is given by

$$\eta_c = \text{avg}\{|\boldsymbol{Col}(s_i)|_c : i \geq 1\} = 4\pi\lambda_c \int_0^{2r} \sigma^2 g(\sigma) d\sigma \tag{4.25}$$

where $g(\sigma) = g_s(\sigma) = 1 - \omega_s$. Using Eqs. 4.24 and 4.25, we find that $\eta_c = 4$. Thus, when the sensing spheres of the sensors percolate, every sensor's sensing sphere overlaps with exactly other four sensing spheres. However, Dall and Christensen [69] found that in RGGs, the lowest connectivity at which the fraction of vertices in the largest cluster is larger than 0 in the macroscopic limit is equal to 2.74, which is lower than ours.

## 4.2.2 *Network Connectivity Percolation*

*Problem formulation:* The *connectivity percolation* (CONP) problem can be stated as follows: Given a network that is originally disconnected (or consists of multiple connected components), compute the critical percolation density $\lambda_c^{con}$, such that there exists a *giant connected component* that spans the entire network when $\lambda > \lambda_c^{con}$, and hence the Boolean model $(X_\lambda, \{B_i(R) : i \geq 1\})$ *percolates*. Otherwise, there is no giant connected component and hence $(X_\lambda, \{B_i(R) : i \geq 1\})$ does not percolate.

Specifically, we are interested in the limiting case of an infinite field, where at a density lower than $\lambda_c^{con}$, there are several small connected components and that at a density higher than $\lambda_c^{con}$, a giant connected component appears. Now, we consider the communication spheres of radius $R$.

The main difference between the COVP and CONP problems is the connection function (Definitions 2.3 and 2.4), which has an impact on the overlap volume fraction between the sensing spheres $(\omega_s)$ and between the communication spheres $(\omega_t)$ of the sensors. For the COVP problem, it is sufficient that a pair of sensing spheres intersect for collaboration to occur. Connectivity between a pair of sensors, however, requires that at least the halves of their communication spheres overlap. Let $\omega_t$ be the minimum overlap volume fraction between two communication spheres so their corresponding sensors communicate with each other. Lemma 4.1 computes this minimum overlap volume fraction.

*Lemma 4.1:* Two sensors communicate with each other if the *minimum overlap volume fraction* of their communication spheres is equal to

$$V_{\min} = (5/12) \times \pi \times R^3$$

where $R$ is the radius of their communication spheres.

*Proof:* Let us consider the diagram in Fig. 4.2, which corresponds to the minimum overlap volume fraction of the communication spheres of two communicating sensors $s_i$ and $s_j$ when they are located at a distance $R$ from each other. Without loss of generality, assume that one of the two spheres is located at the center $(0,0,0)$ of the Euclidean space $\mathbb{R}^3$, while the other one is located at $(R,0,0)$. Recall that the equations of these two spheres are given by $x^2 + y^2 + z^2 = R^2$ and $(x - R)^2 + y^2 + z^2 = R^2$, respectively. Substituting the result of the first equation into the second one gives $(x - R)^2 + x^2 = 0$ whose unique solution is given by $x_0 = R/2$. It means that the intersection of the two spheres is a curve, which lies in a plane parallel to the $yz$ plane at a single $x$ coordinate. Substituting the value of $x_0$ into the first equation of the sphere gives

$$y^2 + z^2 = R^2 - R^2/4 = 3R^2/4$$

which corresponds to a circle whose radius is $R_0 = \sqrt{3}R/2$. The intersection volume of the two spheres is the sum of the two spherical caps, where the distance between the center of the first sphere and the center of the intersection volume (or base of the left spherical cap) is $x_0$ and the distance between the center of the second sphere and base of the right spherical cap is $R - x_0 = x_0$ too. Hence, the height of these spherical caps is equal to $R - x_0 = x_0$ and thus the volume of these two spherical caps is given by

$$V = 2\times(1/3)\times\pi\times(R/2)^2\times(3R - R/2) = (5/12)\times\pi\times R^3$$

Hence, the minimum intersection volume of the communication spheres of two communicating sensors is equal to $V_{\min} = (5/12)\times\pi\times R^3$. ■

From Lemma 4.1, it follows that the minimum overlap volume fraction of the communication spheres of two communicating sensors is equal to

$$\omega_t = V_{\min}/V_0 = (5/12)\times(3/4) = 0.3125$$

where $V_0 = (4/3)\times\pi\times \boldsymbol{R}^3$ is the volume of a communication sphere of radius $R$. Hence, the range of $\omega_t$ is given by $0.3125 \le \omega_t < 1$. Following the same analysis as in the COVP problem, which is discussed in Sect. 4.2.1, we find that the critical density $\lambda_c^{con}(R,\omega_t)$ for the CONP problem is given by

$$\lambda_c^{con}(R,\omega_t) = \frac{0.955}{(1-\omega_t)\,R^3} \tag{4.26}$$

where $0.3125 \le \omega_t < 1$.

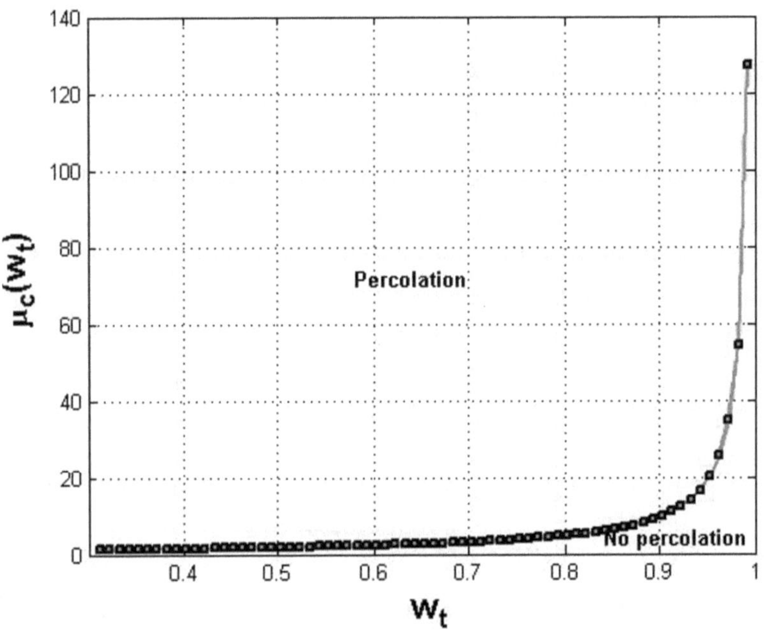


**Fig. 4.3** Plot of the function $\mu_c(\omega_t)$ for $0.3125 \le \omega_t < 1$

*Theorem 4.2:* The *critical sensor density* $\lambda_c^{con}(R,\omega_t)$ above which a giant connected component that span a three-dimensional wireless sensor network will *almost surely* form is computed as

$$\lambda_c^{con}(R,\omega_t) = \frac{0.955}{(1-\omega_t)\,R^3}$$

where $0.3125 \leq \omega_t < 1$ is the overlap volume fraction between the communication spheres of the sensors of radius $R$. ■

Figure 4.3 shows the plot of the function $\mu_c(\omega_t) = \lambda_c^{con}(R,\omega_t)\,R^3 = \frac{0.955}{1-\omega_t}$, for $0.3125 \leq \omega_t < 1$. Notice that the percolation density in the case of the CONP problem is higher than its counterpart for the COVP problem. This is mainly due to the difference in the concepts of collaboration and communication between the sensors. Recall that two collaborating sensors require that the distance between their sensing spheres be at most equal to $2r$, whereas the distance between the communication spheres of two communicating sensors should be at most equal to $R$. Thus, it is expected that the network that results from connectivity percolation be denser than the network which results from coverage percolation. Also, the *critical network degree*, which in this case coincides with the average size of the communicating sets of the sensors (Definition 2.4) at the percolation density, which is given by

$$\eta_c = avg\{\,|\,Com(s_i)\,|_c : i \geq 1\} = 4\pi\,\lambda_c \int_0^R \sigma^2\, g(\sigma)\, d\sigma$$

is equal to 4 where $g(\sigma) = g_t(\sigma) = 1-\omega_t$. In other words, the communication sphere of a sensor overlaps with four other communication spheres at the percolation density so that communication between it and the corresponding four sensors is guaranteed.

### 4.2.3 Coverage and Connectivity Percolation

*Problem formulation:* The *integrated coverage and connectivity percolation* (ICCP) problem can be expressed as follows: Given a field that is initially uncovered and a network that is originally disconnected, compute the critical percolation density $\lambda_c^{cov-con}$ such that there *almost surely* exists a *giant coordinated component* that span the entire deployment field and network when $\lambda > \lambda_c^{cov-con}$, and hence the Boolean model $(X_\lambda, \{B_i(r) : i \geq 1\} \cup \{B_i(R) : i \geq 1\})$ *percolates*. Otherwise, there is no giant coordinated component and hence $(X_\lambda, \{B_i(r) : i \geq 1\} \cup \{B_i(R) : i \geq 1\})$ does not percolate.

In this problem, we are interested in the limiting case of an infinite field, where at a density lower than $\lambda_c^{cov-con}$, there are several small coordinated components and that at a density higher than $\lambda_c^{\text{cov}-con}$, there appears a giant coordinated component.

#### 4.2.3.1 Two-Concentric-Sphere Model

The model that we propose to address coverage and connectivity in three-dimensional wireless sensor networks in an integrated fashion is called *two-concentric-sphere* (TCS). Recall that each sensor is represented by two concentric spheres of radii $r$ and $R$, respectively, namely sensing sphere and communication sphere. In this section, we study a two-phase transition problem, where both coverage percolation and connectivity percolation are considered together. A dependency between both problems stems from the fact that coverage percolation is capable of leading to connectivity percolation when specific conditions, which will be discussed later, hold. Our proposed TCS model accounts for this inherent relationship. Precisely, we investigate integrated coverage and connectivity percolation of the sensing and communication spheres of the sensors based on their ratio $R/r$.

Our TCS model is different from existing models for studying two-phase systems, such as the *permeable sphere* model [40], the *adhesive sphere* model [41], and the *penetrable-concentric-shell* (PCS) model [187]. DeSimone et al. [72, 73] proposed a model of *extended spheres*, which is very similar to the PCS model. The TCS and PCS models have some similarity except that the notion of hard-core, which exists in the PCS model, does not exist in the TCS model. Indeed, according to our TCS model, any pair of sensing spheres of the sensors can be penetrable, i.e., the volume of their overlap region can be greater than zero.

#### 4.2.3.2 Integrated Continuum Percolation

Our TCS model exhibits in general two types of interactions: one between the sensing spheres of the sensors and the other one between their communication spheres. Also, these interactions depend on the distance between the centers of the concentric spheres of the sensors. Lemma 4.2 states the condition under which a pair of sensor coordinates. As will be discussed, this lemma will be the basis of our study of percolation of integrated sensing coverage and network connectivity in three-dimensional wireless sensor networks.

*Lemma 4.2:* A pair of sensors, say $s_i$ and $s_j$, coordinates if $|\xi_i - \xi_j| \leq \tau r$, where $\tau = \min\{2, R/r\}$. Furthermore, the coordination process depends on the ratio $R/r$ of the radius of communication spheres $R$ to the radius of sensing spheres $r$ of the sensors.

*Proof:* By definition, two sensors $s_i$ and $s_j$ collaborate if $|\xi_i - \xi_j| \leq 2\boldsymbol{r}$ and they communicate if $|\xi_i - \xi_j| \leq R$. Thus, $s_i$ and $s_j$ coordinate if $|\xi_i - \xi_j| \leq \min\{2r, R\}$, i.e., $|\xi_i - \xi_j| \leq \min\{2, R/r\} \times r$, where the values of $R$ correspond to varying degrees of sensing sphere overlap. For instance, if $R = 2r$, then the coordination between two sensors requires that the latter be at most $2r$ apart from each other. Hence, it would be sufficient if the sensing spheres of the sensors are tangential. Now, if $R = r$, the overlap area between the sensors increases since the distance between them decrease as the coordination requires $|\xi_i - \xi_j| \leq r$. As can be seen, the overlap volume fraction between the sensing spheres of the sensors depends on the ratio $R/r$. When this ratio is high, the size of overlap volume of the sensing spheres is small and increases when this ratio decreases. ∎

By Lemma 4.2, a solution to the problem of percolation of integrated coverage and connectivity in three-dimensional wireless sensor networks would depend on the ratio $R/r$ of the radius of communication spheres to the radius of sensing spheres of the sensors.

Without loss of generality and because of the physical properties of the sensors, it is reasonable to assume that the radius of the communication spheres of the sensors is at least equal to the radius of their sensing spheres, i.e., $R \geq r$. As a matter of fact, it is always the case that the radius of the communication range of the sensors is higher than that of their sensing range, according to the data sheet regarding the specification of the architectural features of the sensors, such as those manufactured by CrossBow [230]. By Lemma 4.2, the integrated coverage and connectivity percolation of the network will depend on the distance between the centers of the concentric spheres as well as the ratio $R/r$. For the general case, we can set up $R = \alpha r$, where $\alpha \geq 1$. Xing et al. [205] proved that when the network is configured to provide sensing coverage and $R \geq 2r$, the network is guaranteed to be connected. Since we are interested in the co-appearance of a giant covered component and a giant connected component, we will consider two ranges of values of $\alpha$, namely $1 \leq \alpha < 2$ and $\alpha \geq 2$. Notice that the relationship $R = 2r$ corresponds to the worst-case behaviour, where the sensing spheres of a pair of sensors are tangential. As studied earlier, the distance between a pair of overlapping sensing spheres cannot exceed $2r$. Hence, if we apply the same reasoning to the case where $\alpha \geq 2$, we obtain a critical percolation density $\lambda_c^{\text{cov}-con}(r, \omega_s)$ equal to

$$\lambda_c^{\text{cov}-con}(r, \omega_s) = \frac{0.119}{(1-\omega_s)\, r^3}$$

where $0 < \omega_s < 1$.

In the sequel, we only focus on the case $R = \alpha r$, where $1 \leq \alpha < 2$. Thus, the maximum distance between the centers of overlapping sensing spheres is equal to $\alpha r$ with $1 \leq \alpha < 2$. The following analysis will consider only the sensing spheres to compute the percolation density. With this in mind, the Ornstein–Zernike relation for the pair-connectedness function can be expressed as follows:

$$\begin{cases} \sigma\, C^{+}(\sigma) = -\dfrac{\partial Q(\sigma)}{\partial \sigma} + 2\pi\,\lambda \displaystyle\int_{\sigma}^{\alpha\, r} dt \dfrac{\partial Q(t)}{\partial t}\, Q(t-\sigma) & (4.27) \\ \quad \text{for } 0 < \sigma < \alpha\, r & \\ \sigma\, H(\sigma) = -\dfrac{\partial Q(\sigma)}{\partial \sigma} + 2\pi\,\lambda \displaystyle\int_{0}^{\alpha\, r} dt\,(\sigma - t)\, H(|\sigma - t|)\, Q(t) & (4.8) \\ \quad \text{for } \sigma > 0 & \end{cases}$$

where $Q(\sigma) = 0$ for $\sigma < 0$ and $\sigma > \alpha\, r$. We also have $H(\sigma) = g(\sigma)$ with $g(\sigma) = g_s(\sigma) = 1 - \omega_s$ for $\sigma < \alpha$ r. Following the same analysis as in Sect. 4.2.1, we find that the minimum volume fraction of the intersection of two sensing spheres whose centers are located at a distance $\alpha\, r$ from each other is equal to $V_{\min} = (1/12)\,\pi\,(4+\alpha)\,(2-\alpha)\,r^3$. Therefore, the minimum overlap volume fraction of the sensing spheres of the sensors is equal to $\vartheta(\alpha) = V_{\min}/V_0 = (4+\alpha)\,(2-\alpha)/16$, where $V_0 = (4/3)\,\pi\, r^3$ is the volume of a sensing sphere of radius equal to $r$, and hence $\vartheta(\alpha) \le \omega_s < 1$. Substituting the value of $H(\sigma)$ in Eq. 4.24 for $0 < \sigma < \alpha\, r$, it follows that

$$\sigma\,(1-\omega_s) = -\frac{\partial Q(\sigma)}{\partial \sigma} + 2\pi\,\lambda \int_0^{\alpha\, r} dt\,(\sigma - t)\,(1-\omega_s)\,Q(t) \qquad (4.29)$$

$$\text{for } 0 < \sigma < \alpha\, r$$

Similarly, we find $\dfrac{\partial Q(\sigma)}{\partial \sigma} = a\,\sigma + b$ whose integration gives

$$Q(\sigma) = \frac{a}{2}\,(\sigma^2 - (\alpha\, r)^2) + b\,(\sigma - \alpha\, r) \qquad (4.30)$$

with $Q(\sigma) = 0$ for $\sigma < 0$ and $\sigma > \alpha\, r$, where

$$\begin{cases} a = (\omega_s - 1) + 2\pi\,\lambda \displaystyle\int_0^{\alpha\, r} (1-\omega_s)\,Q(t)\,dt \\ b = 2\pi\,\lambda \displaystyle\int_0^{\alpha\, r} (\omega_s - 1)\,t\,Q(t)\,dt \end{cases} \qquad (4.31)$$

Using the same procedure as earlier, solving the two integral Eq. 4.31 with two unknowns, namely $a$ and $b$, defines the function $Q(\sigma)$ given in Eq. 4.30. Using Eqs. 4.9, 4.22, 4.30, and 4.31, we find that the critical percolation density $\lambda_c^{\text{cov}-con}(r, \omega_s, \alpha)$ for integrated coverage and connectivity is given by

$$\lambda_c^{\text{cov}-con}(r, \omega_s, \alpha) = \frac{0.955}{(1-\omega_s)\,\alpha^3\, r^3} \qquad (4.32)$$

where $\vartheta(\alpha) \le \omega_s < 1$, with $\vartheta(\alpha) = (4+\alpha)\,(2-\alpha)/16$ being the minimum overlap volume fraction between sensing spheres, and $1 \le \alpha < 2$.

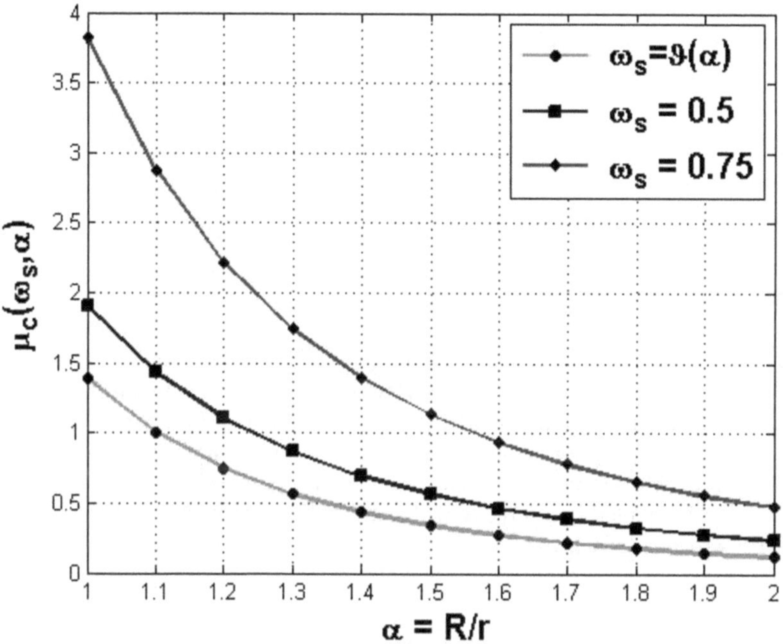


**Fig. 4.4** Plot of the function $\mu_c(\omega_s,\alpha)$ for $\vartheta(\alpha) \le \omega_s < 1$ and $1 \le \alpha < 2$

*Theorem 4.3:* Under the assumption that $R = \alpha r$, a giant coordinated component spans the entire three-dimensional wireless sensor network at a critical percolation sensor density $\lambda_c^{\text{cov}-con}$ equal to

$$\lambda_c^{\text{cov}-con}(r,\omega_s,\alpha) = \frac{0.955}{(1-\omega_s)\,\alpha^3\, r^3}$$

where $\omega_s$ is the overlap volume fraction between sensing spheres of the sensors with $\vartheta(\alpha) \le \omega_s < 1$ and $\vartheta(\alpha) = (4+\alpha)(2-\alpha)/16$, $1 \le \alpha < 2$, and $\boldsymbol{r}$ is the radius of the sensing spheres of the sensors. ∎

Figure 4.4 plots $\mu_c(\omega_s,\alpha) = \lambda_c^{\text{cov}-con}(r,\omega_s,\alpha)\, r^3 = \dfrac{0.955}{(1-\omega_s)\,\alpha^3}$ depending on the value of $\alpha$ ($1 \le \alpha < 2$), which represents the ratio of the radius $R$ of the communication spheres to the radius $r$ of the sensing spheres of the sensors, and the value of the overlap volume fraction between the sensing spheres $\omega_s$, where $\vartheta(\alpha) \le \omega_s < 1$. We find that the percolation density increase with the overlap volume fraction between the sensing spheres. More importantly, we notice that the percolation density decreases when the ratio $\alpha = \boldsymbol{R}/\boldsymbol{r}$ increases. Indeed, when $\alpha$ increases, the radius of the communication spheres increases, and hence the percolation of the communication spheres of the sensors will almost surely occur for lower percolation density. Furthermore, we find that the critical network degree, which in this case corresponds to the average size of the collaborating sets of the sensors at percolation and given by

$$\eta_c = \text{avg}\{|\,\boldsymbol{Col}(s_i)\,|_c : \boldsymbol{i} \geq 1\} = 4\pi\lambda_c \int_0^{\alpha r} \sigma^2 g(\sigma) d\sigma$$

where $g(\sigma) = 1 - \omega_s$, is equal to 4.

## 4.3 Further Discussion

In this chapter, we considered the problem of wireless sensor networks that need to provide sensing coverage and network connectivity in a three-dimensional field of interest. Further, rather than the placement (or deployment) problem of finding constraints on where sensors could be placed to provide the necessary coverage and connectivity, we addressed a more theoretical question: Given that the sensors are randomly distributed in a three-dimensional field, what is the necessary minimum sensor spatial density that *almost surely* guarantees that the sensors will provide the desired coverage properties?

### *4.3.1 Practicality and Generalizability Issues*

Concerning the practicality and generalizability of our results, we make the following points. First, as we mentioned earlier, there are some situations where the two-dimensional assumption is not valid anymore. The design of wireless sensor networks in three dimensions is indeed more realistic than that in two dimensions. For instance, wireless sensor networks deployed underwater require design in three-dimensional space. Oceanographic data collection, disaster prevention, assisted navigation, offshore exploration, and pollution monitoring are typical applications of underwater wireless sensor networks. The success of these applications depends on how well the three-dimensional volume under consideration is covered. In particular, every point in the three-dimensional volume should be covered. Hence, it is important to investigate the situation when full coverage of a three-dimensional volume would be achieved. The critical density that we computed corresponds to the density above which coverage percolation in three-dimensional wireless sensor networks will *almost surely* occur. In this case, the underlying three-dimensional wireless sensor network would be able to detect any object that would penetrate the three-dimensional volume.

Second, the measures of critical percolation density that we provide in this chapter for sensing coverage and network connectivity are more general than the ones corresponding to the definitions of collaboration and communication between the sensors. Indeed, for the sensing coverage, it is required that sensors collaborate. By Definition 2.3 in Chap. 2, this implies that the sensing spheres of these two sensors must be at least tangential. So, the degree of overlap between the sensing spheres tends towards 0 given that these spheres must be at least tangential. Therefore, if we set $\omega_s = 0$, we obtain a critical density for sensing coverage that depends only on the radius of the sensing spheres of the sensors. Also, by Definition 2.4 in Chap. 2, if we set $\omega_t = 0.3125$, we also obtain critical density for network connectivity that depends only on the radius of the communication spheres of the sensors.

### 4.3.2 Sensor Deployment in Three-Dimensional Fields

In the case of wireless sensor networks deployed on the trees of different heights in a forest, sensors could be seen almost everywhere in the space. Also, Pompili et al. [169] proposed different deployment strategies for two-and three-dimensional communication architectures for underwater acoustic sensor networks, where the sensors are anchored at the bottom of the ocean for the two-dimensional design and float at different depths of the ocean to cover the entire three-dimensional region. Indeed, oceanographic data collection, pollution monitoring, offshore exploration, disaster prevention, and assisted navigation are all typical applications of underwater sensor networks [4, 5]. For wireless sensor networks deployed in buildings with multiple floors, sensors are placed on the ground and/or the wall, but the networks seldom contain sensors floating in the middle of the room. The first examples show that our proposed three-dimensional model is valid and can be applied to choose the sensor density in practical problems. The last example, however, shows the limited validity of our model due to the restriction imposed on the placement of sensors inside buildings or rooms.

### 4.3.3 Relaxations of Assumptions

For practical applicability of our results, we provide the following relaxations of the assumptions that we made, namely the *unit sphere model* for sensing and communication ranges of the sensors (Assumption 1) and *homogeneous sensor model* (Assumption 2). Although these two assumptions are the basis for most of the coverage and connectivity protocols for wireless sensor networks, they may not be valid in practice and hence need to be relaxed.

#### 4.3.3.1 Relaxing the Unit Sphere Model

Zhao and Govindan [221] found that the communication range of MICA motes is asymmetric and environment-dependent. Also, Zhou et al. [224] found that the communication range of radios is highly probabilistic and irregular. In this section, we consider convex sensing and communication models, where the sensors have the same sensing and communication ranges.

Corollaries 4.1, 4.2, and 4.3 correspond to Theorems 4.1, 4.2, and 4.3, respectively. Their proof is the same as that in Sects. 4.2.1, 4.2.2, and 4.2.3, respectively, using the notion of largest enclosed sphere. Let $R_{led}$ and $r_{led}$ be the radii of the largest enclosed spheres of the sensing and communication ranges of the sensors, respectively.

*Corollary 4.1:* The *critical sensor density* $\lambda_c^{\text{cov}}(r_{led}, \omega_s)$ above which a giant covered component that span a three-dimensional deployment field will *almost surely* form is computed as

$$\lambda_c^{\text{cov}}(r_{led}, \omega_s) = \frac{0.119}{(1-\omega_s)\, r_{led}^{\;3}}$$

where $0 < \omega_s < 1$ is the overlap volume fraction of the largest enclosed spheres of the sensors' convex sensing ranges. ■

*Corollary 4.2:* The *critical sensor density* $\lambda_c^{con}(R_{led}, \omega_t)$ above which a giant connected component that span a three-dimensional wireless sensor network will *almost surely* form is given by

$$\lambda_c^{con}(R_{led}, \omega_t) = \frac{0.955}{(1-\omega_t)\, R_{led}^{\;3}}$$

where $0.3125 \le \omega_t < 1$ is the overlap volume fraction of the largest enclosed spheres of the convex communication ranges of the sensors. ■

*Corollary 4.3:* Under the assumption that $R_{led} = \alpha\, r_{led}$, a giant coordinated component spans the entire three-dimensional wireless sensor network at a critical percolation sensor density equal to

$$\lambda_c^{\text{cov}-con}(r_{led}, \omega_s, \alpha) = \frac{0.955}{(1-\omega_s)\,\alpha^3\, r_{led}^{\;3}}$$

where $\omega_s$ is the overlap volume fraction of the largest enclosed spheres of the homogeneous sensors' convex sensing ranges of with $\vartheta(\alpha) \le \omega_s < 1$, $\vartheta(\alpha) = (4+\alpha)(2-\alpha)/16$, and $1 \le \alpha < 2$. ■

#### 4.3.3.2 Relaxing the Homogeneous Sensor Model

Real-world applications may require heterogeneous sensors in terms of their sensing and communication capabilities in order to enhance the reliability of the network and extend its lifetime [209]. Even sensors equipped with identical hardware may not always have the same sensing model. Now, we relax this assumption by considering heterogeneous sensors with different yet convex sensing and communication ranges.

Corollaries 4.4, 4.5, and 4.6 correspond to Theorems 4.1, 4.2, and 4.3, respectively. Their proof is literally the same as that in Sects. 4.2.1, 4.2.2, and 4.2.3 using the notion of largest enclosed sphere. Our reasoning is based on the least powerful sensors in terms of their sensing and communication capabilities. Let $R_{led}^{\min}$ and $r_{led}^{\min}$ be the minimum radii of the largest enclosed spheres of the sensing and communication ranges of the sensors, respectively.

*Corollary 4.4:* The *critical sensor density* $\lambda_c^{\text{cov}}(r_{led}^{\min},\omega_s)$ above which a giant covered component that span a three-dimensional deployment field will *almost surely* form is computed as

$$\lambda_c^{\text{cov}}(r_{led}^{\min},\omega_s)=\frac{0.119}{(1-\omega_s)\,{r_{led}^{\min}}^3}$$

where $0<\omega_s<1$ is the overlap volume fraction between the smallest largest enclosed spheres of the convex sensing ranges of the sensors. ■

*Corollary 4.5:* The *critical sensor density* $\lambda_c^{con}(R_{led}^{\min},\omega_t)$ above which a giant connected component that span a three-dimensional wireless sensor network will *almost surely* form is given by

$$\lambda_c^{con}(R_{led}^{\min},\omega_t)=\frac{0.955}{(1-\omega_t)\,{R_{led}^{\min}}^3}$$

where $0.3125\le\omega_t<1$ is the overlap volume fraction between the smallest largest enclosed spheres of the convex communication ranges of the sensors. ■

*Corollary 4.6:* Under the assumption that $R_{led}^{\min}=\alpha\, r_{led}^{\min}$, a giant coordinated component spans the entire three-dimensional wireless sensor network at a critical percolation density equal to

$$\lambda_c^{\text{cov}-con}(r_{led}^{\min},\omega_s,\alpha)=\frac{0.955}{(1-\omega_s)\,\alpha^3\,{r_{led}^{\min}}^3}$$

where $\omega_s$ is the overlap volume fraction of the smallest largest enclosed spheres of the heterogeneous sensors' convex sensing ranges with $\vartheta(\alpha)\le\omega_s<1$, $\vartheta(\alpha)=(4+\alpha)(2-\alpha)/16$, and $1\le\alpha<2$. ■

Thus, the assumptions of the sphere model for the sensing and communication ranges of the sensors and the homogeneous sensor model can be relaxed with slight updates to the results in Theorems 4.1, 4.2, and 4.3 using the notion of largest enclosed sphere. We should mention that because of these relaxations, the critical density may be a bit higher than necessary in some regions of a deployment field.

## 4.4 Related Work

In this section, we review related work on coverage and connectivity in three-dimensional wireless sensor networks. We also summarize existing works that are based on percolation theory.

The study of coverage and connectivity in three-dimensional wireless sensor networks, such as underwater sensor networks [4], has gained relatively less attention in the literature compared to that of two-dimensional wireless sensor networks. Alam and Haas [8] investigated coverage and connectivity issues in three-dimensional wireless networks. They proposed a placement strategy based on Voronoi tessellation of a three-dimensional space, which creates truncated octahedral cells. This strategy uses a minimum number of nodes to guarantee 100% coverage of a three-dimensional space and the minimum ratio of the communication range to the sensing range of such a placement strategy. Huang et al. [109] addressed the coverage problem in three-dimensional wireless sensor networks by reducing the geometric problem from a three-dimensional space to a two-dimensional space, and proposed a polynomial-time algorithm to solve the $\alpha$-Ball-Coverage ($\alpha$-BC) problem whose goal is to check $\alpha$-coverage of a three-dimensional region, i.e., all locations in the region are $\alpha$-covered or not. Pompili et al. [169] proposed a deployment strategy for three-dimensional communication architecture for underwater acoustic sensor networks, where sensors float at different depths of the ocean to cover the entire three-dimensional region. Poduri et al [167] discussed some difficulties encountered in the design of three-dimensional wireless sensor networks, such as ensuring network connectivity in the case of uniform random deployment and restrictions imposed by the environment structure on sensor deployment. In particular, they discussed a few issues concerning the geometry of three-dimensional wireless sensor networks and possible extensions of existing two-dimensional designs for deployment and configuration to three-dimensional design. In [172], Ravelomanana investigated fundamental properties of randomly deployed three-dimensional wireless sensor networks for connectivity and coverage, such as the required sensing range to guarantee certain degree of coverage of a region, the minimum and maximum network degrees for a given communication range, and the network hop-diameter.

## 4.5 Summary

In this chapter, we considered three-dimensional wireless sensor networks and investigated the critical percolation densities above which a giant covered component *almost surely* span a deployment field (coverage percolation problem) and a giant connected component *almost surely* span the entire network (connectivity percolation problem) [12]. In order to deal with the problem of integrated coverage and connectivity percolation, i.e., COVP and CONP at the same time, we proposed a TCS model, where every sensor is modelled by two concentric spheres, namely sensing sphere and communication sphere. For each of these problems, we computed the corresponding critical sensor spatial density as well as the network degree. We found that the critical density corresponding to the connectivity percolation is higher than its counterpart for coverage percolation. This difference is due to the inherent properties of collaboration and communication between the sensors. Indeed, collaboration between the sensors requires that their sensing spheres be at a distance at most equal to $2r$ from each other, while communication between the sensors necessitates that their communication spheres be at a distance at most equal to $R$. For the ICCP problem, we considered a more general analysis, where $R = \alpha r$ with $1 \le \alpha < 2$ and $\alpha \ge 2$. We found that the percolation density depends monotonically on $\alpha$, which represents the ratio $\boldsymbol{R/r}$ and depends on the overlap volume fraction of the sensing spheres. We believe that these findings can be useful for the design of energy-efficient topology control protocols for three-dimensional wireless sensor networks in terms of coverage and connectivity.

# Chapter 5
# Connected *k*-Coverage in Two-Dimensional Deployment Fields

This chapter focuses on the problem of connected *k*-coverage in densely deployed wireless sensor networks. Precisely, it investigates how to achieve *k*-coverage of a two-dimensional deployment field with a minimum number of sensors while maintaining network connectivity. It considers static and homogeneous sensors under the assumption of a deterministic sensing model. Also, it proposes centralized, pseudo-distributed, and fully distributed sensor scheduling protocols, where the sensors are duty-cycled to save energy while ensuring connected *k*-coverage.

## 5.1 Introduction

*Sensing coverage* is an essential functionality of wireless sensor networks. However, it is also well-known that coverage alone in wireless sensor networks is not sufficient because data originated from *source sensors* are not guaranteed to reach the *sink* for further analysis. Thus, *network connectivity* should also be considered for a network to function correctly. In wireless sensor networks, *coverage* and *connectivity* have been jointly addressed in an integrated framework. While coverage is a metric that measures the quality of surveillance provided by a network, connectivity provides a means to the *source sensors* to report their sensed data to the *sink*. Some real-world applications, such as intrusion detection, may require high degree of coverage (or redundant coverage), and hence large number of sensors to enable accurate tracking of intruders. For such highly dense deployed and energy-constrained sensors, it is necessary to duty-cycle them to save energy. Thus, the design of coverage configuration protocols for wireless sensor networks should minimize the number of *active sensors* to guarantee the degree of coverage of a field required by an application while maintaining connectivity between all active sensors. Hence, the first challenge is the determination of the number of sensors required to remain active to *k*-cover a deployment field. Given that sensors have limited battery power and wireless sensor networks are generally randomly and hence highly dense deployed, the second challenge is the design of an efficient scheduling protocol that decides which sensors to turn *on* (active) or *off* (inactive) for *k*-coverage of a field.

In this chapter, we study *duty-cycling* to achieve both *k*-coverage and connectivity in highly dense deployed wireless sensor networks [15, 22], where each location in a deployment field is *covered* by at least $k \geq 3$ active sensors while maintaining connectivity between *all* active sensors. Indeed, the limited battery power of the sensors and the difficulty of replacing and/or recharging batteries on

H.M. Ammari: Challenges and Opportunities of Connected *k*-Covered WSNs, SCI 215, pp. 73–109.
springerlink.com

the sensors in hostile environments require that the sensors be deployed with high density [179] in order to extend the network lifetime. Also, the sensed data originated from the *source sensors* (or simply *sources*) should be able to reach a central gathering node, called the *sink*, for further analysis and processing. Thus, network connectivity should be guaranteed so sources can be connected to the sink via multiple communication paths. Finally, wireless sensor networks suffer from scarce energy resources. A more practical deployment strategy requires that all the sensors be duty-cycled to save energy. With duty-cycling, sensors can be turned *on* or *off* according to some scheduling protocol, thus reducing the number of active sensors required for $k$-coverage and helping all sensors deplete their energy as slowly and uniformly as possible.

Our study of the connected $k$-coverage problem with $k \geq 3$ is motivated by at least the following three applications: *Space and planet exploration* [184], *multiple-sensor data fusion* [121], and *triangulation-based positioning systems* [162], which are briefly described in Chap. 1. Furthermore, there are several applications, such as intruder detection and tracking, which require that each location in a field be sensed by at least one sensor. In these types of applications, the decision-making process accuracy depends on the availability of the data gathered during monitoring. With connected $k$-coverage, the sink would gather enough data for a successful decision regarding any event that would occur in the monitored field.

An important problem in the design of such network configurations is computing the *minimum active sensor spatial density* required to guarantee $k$-coverage of a field. For tractability of the problem, first we analyze the intersection of $k$ sensing disks so we can characterize $k$-coverage provided by a wireless sensor network regardless of whether the sensors have identical sensing ranges and whether the sensing range of each sensor follows the *unit disk model*. Based on this analysis, we derive a tight sufficient condition of the spatial density of active sensors to achieve complete $k$-coverage of a field. Although the problem of $k$-coverage has been well-studied in the literature, only a few elegant approaches characterized $k$-coverage of a field [205, 218]. However, none of them guarantees the deployment of a minimum number of sensors to achieve $k$-coverage of a field and hence they would result in shorter operational network lifetime. Previous works [205, 218] only characterized $k$-coverage. According to [205], a field is $k$-covered if all *intersection points* between the boundaries of sensing ranges of the sensors and all those between the boundaries of sensing ranges of the sensors and the boundary of a field are $k$-covered. This is a generalization of the result for 1-coverage [100]. Thus, if two sensing ranges intersect, one more is needed to cover their intersection point. A point in a field that coincides with an intersection point would be 3-covered instead of 1-covered. Hence, more than enough sensors are required to $k$-cover a field. However, our approach for characterizing $k$-coverage of a field is different from the ones proposed in [205, 218]. Precisely, our approach is able to quantify the minimum spatial density of active sensors to fully $k$-cover a field, thus computing the corresponding minimum number of sensors.

The remainder of this chapter is organized as follows. Section 5.2 discusses the connected $k$-coverage problem in wireless sensor networks and shows how to solve it. Sections 5.3–5.6 present our connected $k$-coverage configuration protocols. Section 5.7 enhances the applicability of these protocols by relaxing some widely used assumptions in coverage protocols. Section 5.8 presents simulation results of our protocols and compares them with another existing one. Section 5.9 reviews related work while Sect. 5.10 summarizes the chapter.

## 5.2 Achieving Connected $k$-Coverage

In this section, we propose our approach for obtaining connected $k$-coverage configurations in wireless sensor networks. First, we model the connected $k$-coverage problem in wireless sensor networks. Then, we derive a tight sufficient condition of the active sensor spatial density such that a field is fully $k$-covered during the network lifetime while all active sensors are being connected to each other. In Sects. 5.2–5.4, we propose four protocols to solve the problem of connected $k$-coverage in wireless sensor networks based on the results of this section.

### 5.2.1 Connected k-Coverage Problem Modeling

Solving the connected $k$-coverage problem in wireless sensor networks requires finding a sensor deployment strategy such that each location in a field is covered by at least $k$ active sensors while *all* active sensors are connected. Our approach solution to the connected $k$-coverage problem in wireless sensor networks consists of decomposing it into two sub-problems, namely *deployment field slicing* and *sensor selection*, and solving them. The *deployment field slicing* problem is to slice a field into small regions of particular shape (to be defined later), each of which is guaranteed to be $k$-covered provided that at least $k$ sensors are randomly deployed in it. The *sensor selection* problem is to select a minimum subset of sensors to remain active and connected such that each location in a field is guaranteed to be $k$-covered. Besides selecting a *minimum* number of active sensors, all selected sensors should have the *maximum* remaining energy. Hence, our *connected k-coverage* problem can be formulated as follows:

> *Problem*: Minimum connected $k$-coverage
> *Instance*: A field, a set $S$ of sensors, and a positive integer $k \geq 3$.
> *Question*: Select a minimum subset $S_{min} \subseteq S$ of sensors such that each point in the field is $k$-covered while all the sensors in $S_{min}$ are connected to each other.

Because the problem of selecting a minimum subset of sensors to $k$-cover a field is NP-hard [223], we propose efficient centralized, clustered, and distributed approximation algorithms to solve the connected $k$-coverage problem.

### 5.2.2 Sufficient Condition to Ensure k-Coverage

We want to compute the *minimum active sensor spatial density* required to $k$-cover a field. To do so, we should compute the maximum size of a convex area $A$ of a field so that $A$ is $k$-covered with exactly $k$ sensors. Intuitively, the distance between any point in $A$ and each of the $k$ sensors should be at most equal to the radius of their sensing disks. Lemma 5.1 gives an upper bound on the *width* of such a convex area. We omit its proof since it is trivial.

*Lemma 5.1:* Let $r$ be the radius of the sensing disks of the sensors and $k \geq 3$. A convex area $A$ is guaranteed to be $k$-covered when exactly $k$ homogeneous sensors are deployed in it, if the width of $A$ does not exceed $r$. ■

Now, we present *Helly's Theorem* [44] (p. 76), a fundamental result of convexity theory, which characterizes the intersection of convex sets.

*Helly's Theorem* [44]*:* Let $E$ be a family of convex sets in $R^n$ such that for $m \geq n+1$ any $m$ members of $E$ have a non-empty intersection. Then, the intersection of all members of $E$ is non-empty. ■

Theorem 5.1, which is an instance of Helly's Theorem [44], will help us compute the minimum sensor spatial density required to $k$-cover a field. More specifically, Helly's Theorem [44] together with a geometric structure, called *Reuleaux triangle* [233], will be used to characterize $k$-covered wireless sensor networks.

*Theorem 5.1:* Let $k \geq 3$. The intersection of $k$ sensing disks is not empty if and only if the intersection of any three of those $k$ sensing disks is not empty. ■

Following Theorem 5.1, Lemma 5.2 states a *sufficient condition* for complete $k$-coverage of a field.

*Lemma 5.2:* Let $r$ be the radius of the sensing disks of the sensors and $k \geq 3$. A field is $k$-covered if any Reuleaux triangle region of width $r$ in a field contains at least $k$ active sensors.

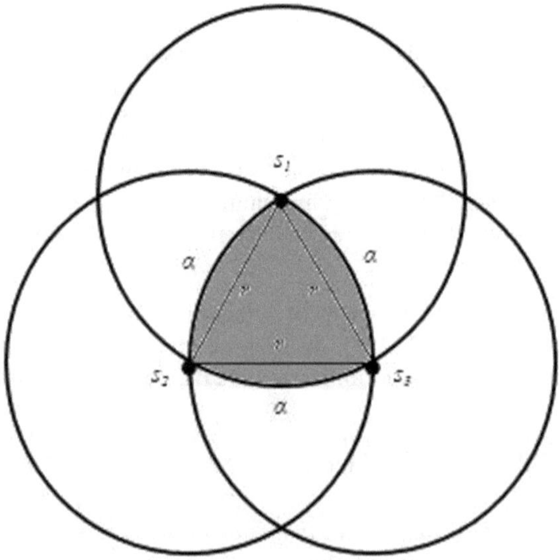

**Fig. 5.1** Symmetric intersection of three sensing disks

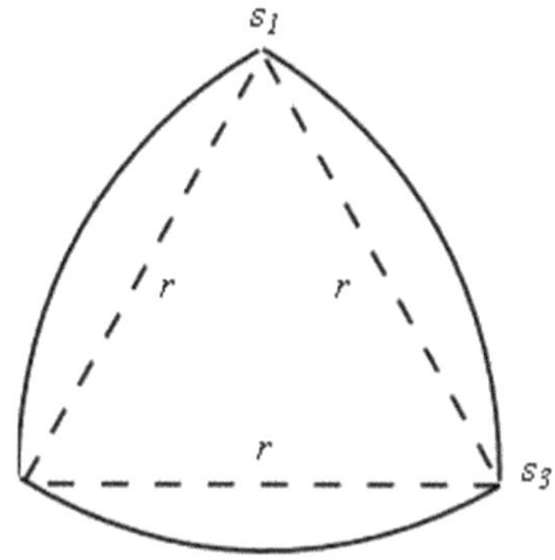


**Fig. 5.2** Reuleaux triangle

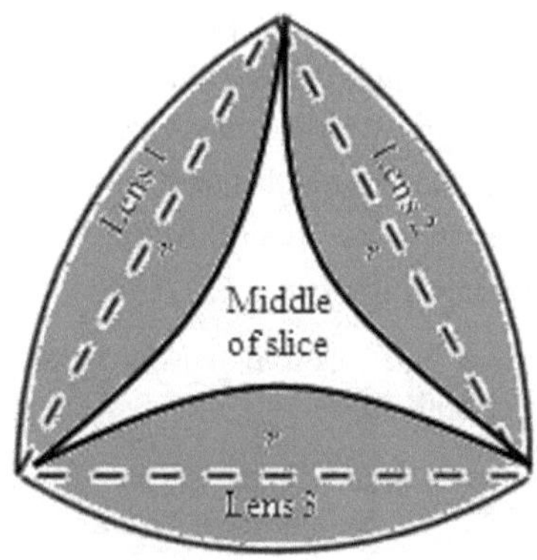


**Fig. 5.3** Three lenses of a slice

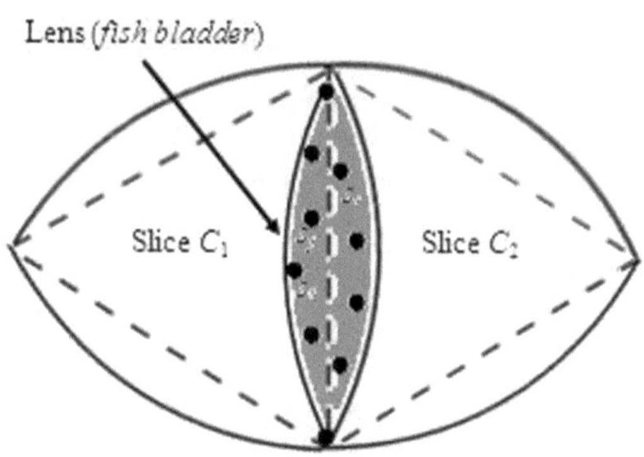


**Fig. 5.4** Adjacent slices

*Proof:* First, we compute the maximum area that is $k$-covered with exactly $k$ sensors. Let $A$ be the intersection area of the sensing disks of $k$ sensors. From Lemma 5.1, the width of $A$ should be upper-bounded by $r$ so that any point in $A$ is $k$-covered by these $k$ sensors. Let us first consider the case of three sensors. Using the Venn diagram given in Fig. 5.1, the maximum size of the intersection area of the sensing disks of the sensors $s_1$, $s_2$, and $s_3$, so that the distance between any pair of sensors is at most equal to $r$, is obtained when $s_1$, $s_2$, and $s_3$ are symmetrically located from each other. This area, called *Reuleaux triangle* [233], is denoted by $RT(r)$ and has a constant width equal to $r$ as shown in Fig. 5.2. Given that the intersection area of $k$ sensing disks is at most equal to that of three sensing disks such that the maximum distance between any pair of sensors is at most equal to $r$, the maximum size of $A$ is equal to the area of $RT(r)$, which we call *slice* (see Fig. 5.3). Thus, any point in $A$ is $k$-covered with exactly $k$ active sensors deployed in $A$. Since this applies to any $RT(r)$ region (or *slice*) in a field, it is guaranteed that the field is $k$-covered. ■

As will be discussed in Sect. 5.2, our sensor selection scheme exploits the overlap between adjacent slices in a field to select a minimum number of active sensors for full $k$-coverage of the field. Precisely, two adjacent slices intersect in a region shaped as a *lens* (also known as the *fish bladder*) so that the sides of their associated regular triangles fully coincide as shown in Fig. 5.4. Note that $k$ sensors located in the lens of two adjacent slices, say $C_1$ and $C_2$, $k$-cover the area associated with their union. Indeed, the distance between any of these $k$ sensors located in the lens and any point in the area of the union of both $C_1$ and $C_2$ is at most equal to $r$. Lemma 5.3 states this result.

*Lemma 5.3:* $k$-active sensors located in the lens of two adjacent slices *surely k*-cover both slices. ■

Theorem 5.2, which exploits the results of Lemma 5.3, refines the result of Lemma 5.2 by stating a *tighter sufficient condition* for complete $k$-coverage of a field.

*Theorem 5.2:* Let $k \geq 3$. A field is guaranteed to be $k$-covered if for any slice in a field, there is at least one adjacent slice such that their lens contains at least $k$ active sensors. ■

Theorem 5.3, which exploits the result of Theorem 2, computes the minimum sensor spatial density required for complete $k$-coverage of a field.

*Theorem 5.3:* Let $k \geq 3$. The *active sensor spatial density* required to guarantee $k$-coverage of a field is given by

$$\lambda(r,k) = \frac{6\,k}{(4\pi - 3\sqrt{3})\,r^2}$$

where $r$ is the radius of the sensing disks of the sensors.

*Proof:* It is easy to check that the area $\| Area(r) \|$ of the union of two adjacent slices is computed as

$$\| Area(r) \| = 2\,A_1 + 4\,A_2 = (4\pi - 3\sqrt{3})\,r^2/6$$

where $A_1 = \sqrt{3}\, r^2/4$ is the area of the central equilateral triangle of side $r$ and $A_2 = (\pi/6 - \sqrt{3}/4)\, r^2$ is the area of each of the three curved regions $\alpha$ (Fig. 5.1). By Theorem 5.2, $k$ sensors should be deployed in the lens of two adjacent slices to $k$-cover both of them. Thus, the *minimum sensor spatial density* that guarantees $k$-coverage of a field is equal to

$$\lambda(r,k) = k/\parallel Area(r) \parallel = 6\,k/(4\,\pi - 3\sqrt{3})\, r^2 \qquad \blacksquare$$

It is worth noting that Adlakha and Srivastava [2] showed that the number of sensors required to cover an area of size $A$ is in the order of $O(A/\hat{r}_2{}^2)$, where $\hat{r}_2$ is a good estimate of the radius $r$ of the sensing disks of the sensors. Precisely, $r$ lies between $\hat{r}_1$ and $\hat{r}_2$; $\hat{r}_1$ overestimates the number of sensors required to cover $A$, while $\hat{r}_2$ underestimates it.

One may suggest that the *maximum area* that is guaranteed to be $k$-covered with exactly $k$ sensors is a circle of radius $r/2$. Fortunately, it is easy to check that our density $\lambda(r,k)$ is smaller than the one corresponding to the configuration where $k$-active sensors are deployed in a circle of radius $r/2$. In other words, $\lambda(r,k) = 6\,k/(4\,\pi - 3\sqrt{3})\, r^2 < 4\,k/\pi\, r^2$.

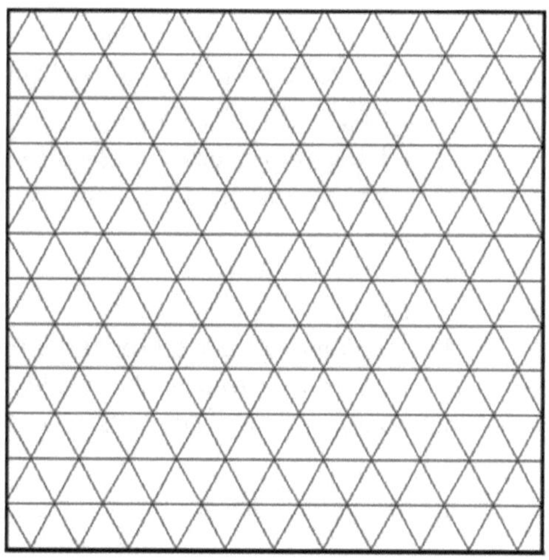

**Fig. 5.5** Slicing grid of a square field

## 5.3 Centralized *k*-Coverage Protocol

In this section, we present our *centralized randomized connected k-coverage* (CERACC$_k$) protocol to fully $k$-cover a field while maintaining connectivity between active sensors.

According to the connected $k$-coverage model mentioned in Sect. 5.2.1, the protocol has two main steps. First, we slice a field into regions whose shape helps characterize the $k$-coverage property of the field, and thus leads to compute the corresponding minimum number of active sensors. Then, we select an appropriate subset of sensors to guarantee $k$-coverage of each slice, and hence $k$-cover the entire field based on the geometric characteristics of those regions.

### 5.3.1 Deployment Field Slicing

This section provides a solution to the deployment field slicing problem, where all sensors have the same sensing and communication disks whose radii are $r$ and $R$, respectively. The goal of this slicing phase is to decompose the sensing range of a sensor into smaller, congruent regions such that each of them is guaranteed to be $k$-covered when exactly $k$ sensors are deployed in it.

---

Algorithm: CERACC$_k$
  Procedure Process_Slice $(i, k, S_{\min})$
  Begin
    1. $j = 1$
    2. $CovD = 0$
    3. While $j \leq 3$ Do
      Begin
        3.1. Randomly select a subset $S'$ of $k/3$
          sensors from $j^{th}$ lens
        3.2. $CovD = CovD + \min\{k/3, |lens(j)|\}$
        3.3. $j = j + 1$
        3.4. $Update\_CovD(slice\_adj(j))$
        3.5. $S_{\min} = S_{\min} \cup S'$
      End
    4. If $CovD < k$ Then
      Begin
    4.1. Randomly select a subset of sensors $S'$
      of $k - CovD$ from middle area of the cell
    4.2. $CovD = CovD + \min\{k - CovD, |mid\_slice(j)|\}$
    4.3. $S_{\min} = S_{\min} \cup S'$
    4.4. If $CovD < k$ Then
      4.3.1. Return ($k$-coverage cannot be provided)
      End
  End
Begin
  1. Randomly slice a field into regular triangles of side $r$
  2. Randomly pick a slice (or Reuleaux triangle) $i$
  3. Call a Breadth-First-Search procedure to $k$-cover slice $i$
End

---

**Fig. 5.6** Sensor selection for $k$-coverage of a field

Let us consider a square field and $k \geq 3$. Based on the result of Theorem 5.2, it is easy to check whether a given network can $k$-cover the field. For this purpose, we propose a *slicing* scheme of a field by dividing it into overlapping Reuleaux triangles of width $r$, called *slices* as shown in Fig. 5.3, such that two adjacent slices intersect in a region shaped as a *lens* (also known as the *fish bladder*) as shown in Fig. 5.4. This implies that the field is mainly sliced into regular triangles of side $r$. The result of this slicing operation is called *slicing grid*. Figure 5.5 shows a slicing grid of a field.

### 5.3.2 Sensor Selection

In this section, we propose a centralized algorithm to select a minimum subset of active sensors that $k$-covers a field. The purpose of the selection phase is to specify *which* sensors turn *on*, *how*, and *when*. We assume that the sink is responsible for this selection process.

Using Lemma 5.3, we start first by selecting sensors located in the three lenses of a given slice as shown in Fig. 5.3 (a slice overlaps with at most three other slices). At every selection, we check whether we already selected $k$ sensors to $k$-cover the underlying slice. At the same time, we update the degree of coverage of the other adjacent slices. We repeat this process until we visit all slices in a field. We assume that each slice has a unique *id*, such as an integer.

Theorem 5.4 is a consequence of Theorem 5.2.

*Theorem 5.4:* Let $k \geq 3$. A field is guaranteed to be $k$-covered with a minimum number of sensors if all the sensors are selected from lenses of adjacent slices. ■

The order in which the slices are treated is critical. It can be easily shown at the end of the sensor selection phase that if the slices are processed randomly, there is no guarantee that each slice is $k$-covered with a minimum number of sensors. Thus, the entire field is not guaranteed to be $k$-covered using a minimum total number of sensors. In order to avoid this problem, it is imperative that slices of a slicing grid be processed in a particular order. Assume that we initially picked slice $C_i$ for processing and let $C_{i1}$, $C_{i2}$, and $C_{i3}$ be its adjacent slices. We use a FIFO (First-In-First-Out) data structure, called *NYP* (Not Yet Processed), to keep track of the slices whose degrees of coverage have been updated but not yet processed. Hence, when we process slice $C_i$, we store the id's $i_1$, $i_2$, and $i_3$ in NYP. When slice $C_i$ has been processed, we consider slice $C_{i1}$ as the next one to be processed. After that, we pick $C_{i2}$ followed by $C_{i3}$ followed by the slices adjacent to $C_{i1}$ and so on. The sensor selection algorithm for $k$-coverage of a field is given in Fig. 5.6.

The sensor selection algorithm described earlier generates only one subset of active sensors to $k$-cover a field. If this algorithm is executed in each round for the same slicing pattern of the field, sensors located in the lenses will be suffering from a severe battery power depletion problem. Hence, those sensors will die very quickly and possibly disconnect the network. Recall that the sensors have several limited resources with energy being the most critical one. Thus, it would be more efficient if in each round a different subset of sensors is selected for $k$-coverage of the field so

**Fig. 5.7** Random slicing grid

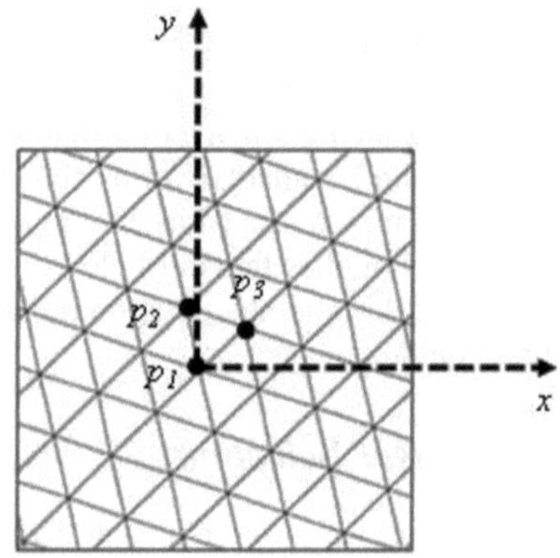


all the sensors are given the same chance to be active. Our objective is to balance the load of $k$-coverage on all sensors so they deplete their energy uniformly. Next, we describe an approach to achieve this goal.

### *5.3.3 Slicing Grid Dynamics*

Our goal is to select different subsets of sensors $S_i$, $i \geq 1$ such that each subset $S_i$ is selected to remain active in the $i^{th}$ round to $k$-cover a field. Notice that to achieve a better load balancing among sensors, we could add a restriction so that selected subsets of sensors are *mutually disjoint*, i.e., $S_i \cap S_j = \emptyset, \forall i \neq j$. However, the disjointness constraint yields a *small number* of mutually disjoint minimum subsets of sensors. Thus, we only require *partially disjoint* subsets of sensors. Given that our selection criterion is based on the remaining energy of sensors, it is guaranteed that sensors with low energy level would be avoided. Furthermore, it is rare that the same sensors participate in several successive rounds to $k$-cover a field. The question that we want to address now is: How would minimum subsets of sensors be selected, each of which k-covers a field?

To address this question, we consider the *dynamics of slicing grid* from one round to another. Recall that the result of slicing a field into slices is called *slicing grid*. The selection of different minimum subsets of active sensors will be determined based on the obtained slicing grids. Since our scheme for selecting active sensors highly prioritizes the ones located in lenses of all slices, it is important that those lenses be able to scan the entire field, and hence include distinct subsets of sensors in different rounds. Thus, it is necessary that the sink be able to generate a slicing grid randomly at each round. Our objective is to obtain as (partially) disjoint minimum subsets of selected sensors as possible. For each obtained slicing grid, the sink applies the selection algorithm in Fig. 5.5. Thus, the slicing grid undergoes some dynamics to achieve balanced load of $k$-coverage among all sensors during the operation of CERACC$_k$. The question that we want to address now is: How would a slicing grid of a field be randomly generated in each round?

First, the sink randomly generates one point $p_1$ in a field as shown in Fig. 5.7. Point $p_1$ is temporarily considered as the centre of the Euclidean plane. To randomly determine a second point $p_2$, the sink generates a random angle $0 \leq \theta \leq 2\pi$ so that the segment $[p_1,p_2]$ forms an angle $\theta$ with the $x$-axis centered at $p_1$ and the length of $[p_1,p_2]$ is $r$. Then, the sink deterministically finds a third point $p_3$ to form the first regular triangle ($p_1$, $p_2$, $p_3$), called *reference triangle*, as shown in Fig. 5.7. As its name indicates, all other regular triangles of the slicing grid are computed based on the reference triangle. Figure 5.7 shows a randomly generated slicing grid of a square field.

Using the RT model described earlier and based on the structure of the clusters of slices, Theorem 5.5 states a condition under which $k$-coverage implies *connectivity*.

*Theorem 5.5: Let* $k \geq 3$. Under the assumption of a centralized or clustered $k$-coverage protocol, a $k$-covered wireless sensor network is guaranteed to be connected if the radius $R$ of the communication disks of the sensors is at least equal to the radius $r$ of their sensing disks, i.e., $R \geq r$.

*Proof:* First, each cluster is connected if $R \geq r$. Also, cluster-heads are connected to each other via active sensors. Thus, A $k$-covered wireless sensor network is guaranteed to be connected if $R \geq r$. ■

Theorem 5.6 states that $CERACC_k$ is a minimum-energy connected $k$-coverage configuration protocol that guarantees maximum network lifetime.

*Theorem 5.6:* $CERACC_k$ fully $k$-covers a field with a minimum number of sensors in each round and maintains connectivity between them. Hence, it consumes minimum energy.

*Proof:* The sink guarantees that each slice of a field is covered by exactly $k$ sensors. Therefore, by Theorem 5.2, each slice of the field is $k$-covered, and hence the entire field is fully covered. Moreover, sensors are selected from lenses so all adjacent slices are $k$-covered with a minimum number of active sensors. Thus, by Theorem 5.4, $CERACC_k$ guarantees that a field is $k$-covered using a minimum total number of active sensors in each round, and hence it consumes a minimum amount of energy in each round. By Theorem 5.5, a $k$-covered wireless sensor network is connected, assuming that $R \geq r$. Given that $CERACC_k$ prioritizes sensors with the highest remaining energy to remain active in each round, and given that it is based on the dynamics of slicing grid, necessarily all sensors deplete their energy slowly and uniformly, thus leading to a maximum network lifetime. ■

In general, the sink is connected to an infinite source of energy, such as a wall outlet, and thus can be viewed as a *line-powered* node [209] that has no energy constraint. Hence, if node failure is due only to low battery power, the problem of single-point failure does not arise at all in this type of centralized wireless sensor network architecture. Also, under the centralized control of the sink, no coordination between sensors is required to select a minimum number of active sensors to $k$-cover a field. Given that sensors have limited energy, this approach would save them energy. Indeed, a schedule that determines which sensors should remain

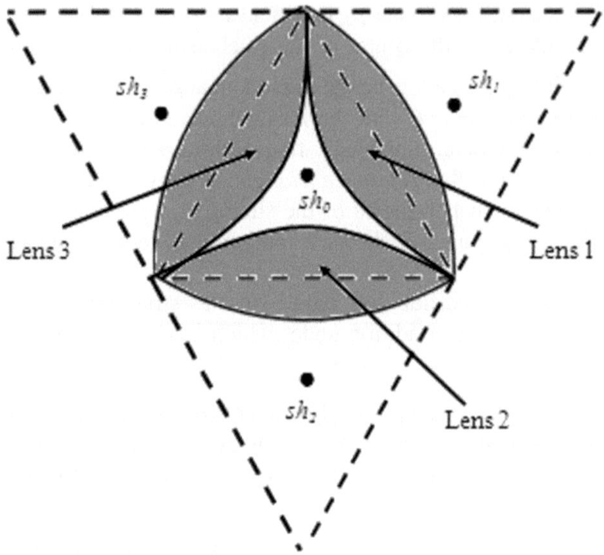


**Fig. 5.8** Adjacent cluster-heads

**Fig. 5.9** Clustering for D-CRACC$_k$

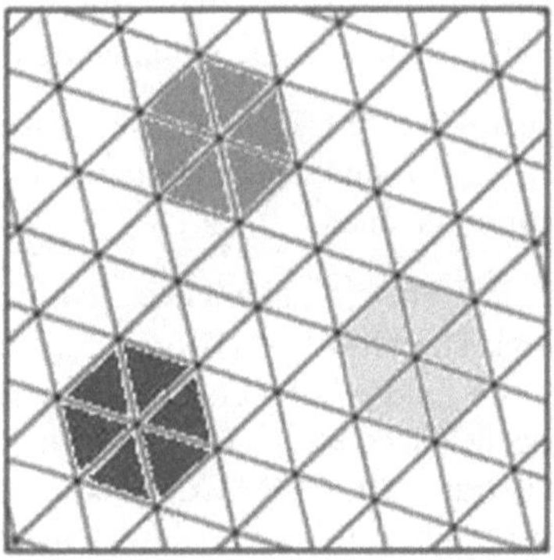

active in each round to $k$-cover a field is computed by the sink and forwarded to the selected ones. Thus, such a centralized approach is intended to gain insight into a lower bound on the number of sensors required for complete $k$-coverage of a field, and hence an upper bound on the network lifetime. In Sect. 5.7, we will show how to relax the centralized approach to implement CERACC$_k$ in a fully distributed manner.

## 5.4 Clustered *k*-Coverage Protocol

In this section, we propose a family of clustered $k$-coverage protocols, called *clustered randomized connected k-coverage* (CRACC$_k$), in which part of the duties of the sink in the centralized protocol $\mathrm{CERACC}_k$ is delegated to a subset of sensors, called *cluster-heads*. These protocols differ by their degree of granularity of network clustering, and hence produce clusters of different shapes and sizes.

### 5.4.1 Cluster-Head Selection and Attributed Roles

As in the centralized protocol $\mathrm{CERACC}_k$, the family of protocols $\mathrm{CRACC}_k$ allows the sink to randomly generate in each round a slicing grid, which consists of adjacent overlapping Reuleaux triangles. In contrast with $\mathrm{CERACC}_k$, the sink randomly designates for each cluster a particular sensor, called *cluster-head*, which is responsible for $k$-coverage of its assigned cluster during a given round. Precisely, each cluster-head is located within the cluster it is in charge of selecting some of its sensing neighbours to $k$-cover it. The sink advertises a packet, called *ClusterHeadList*, including all sensors' ids that have been selected as cluster-heads. When a sensor receives *ClusterHeadList*, it checks whether its id is included. If so, it removes its id from the list and forwards the updated *ClusterHeadList* packet. Otherwise, it just forwards the original packet it has received. The $\mathrm{CRACC}_k$ family of protocols requires that each cluster-head coordinates its activity with its adjacent cluster-heads to $k$-cover its cluster, and hence select a total minimum number of sensors to $k$-cover a field. To achieve this goal, sensors that would remain active in each round to $k$-cover a cluster should be selected from lenses (i.e., intersection areas of adjacent Reuleaux triangles). For a better balance, a cluster-head attempts to select $k/3$ sensors from each lens so any Reuleaux triangle in a cluster contains exactly $k$ sensors as per Theorem 5.2. The size of a cluster and the number of its adjacent ones depend on the type of clustering being used. As can be seen, the protocols of the $\mathrm{CRACC}_k$ family are pseudo-distributed in the sense that the selection of active sensors for complete $k$-coverage of a field is not under the control of the sink. Next, we describe two protocols of the $\mathrm{CRACC}_k$ family, namely T-$\mathrm{CRACC}_k$ and D-$\mathrm{CRACC}_k$, for connected $k$-coverage in wireless sensor networks based on their degree of clustering.

### 5.4.2 The T-$\mathrm{CRACC}_k$ Protocol

In T-$\mathrm{CRACC}_k$ ("T" for Reuleaux *triangle*), a cluster is a *slice* in the obtained slicing grid and a cluster-head is called *slice-head*. In each round, the sink is responsible for randomly generating a slicing grid of a field. Given that each slice has at most three adjacent slices (Fig. 5.3), the T-$\mathrm{CRACC}_k$ protocol requires that each slice-head coordinates its activity with its adjacent slice-heads to select a minimum total number of sensors to $k$-cover a field. Figure 5.8 shows slice-head $sh_0$ sharing three lenses with slice-heads $sh_1$, $sh_2$, and $sh_3$. For instance, $sh_0$ could $k$-cover its slice by selecting sensors located in its three lenses. Then, it communicates the numbers $n_1$, $n_2$, and $n_3$ of sensors selected from lenses $Lens\,1$, $Lens\,2$, and $Lens\,3$, respectively, to its adjacent slice-heads $sh_1$, $sh_2$, and $sh_3$, respectively. Slice-head $sh_1$ would need to select $k - n_1$ more sensors from its lenses to $k$-cover its slice. It would definitely coordinate with its adjacent slice-heads to $k$-cover its slice and so does each slice-head.

Theorem 5.7 states that T-CRACC$_k$ yields minimum-energy connected $k$-coverage.

*Theorem 5.7:* The T-CRACC$_k$ protocol fully $k$-covers a field. It also is a minimum-energy connected $k$-coverage protocol.

*Proof:* Each slice-head ensures that each slice of a field is $k$-covered by exactly $k$ sensors by coordinating with each of its three adjacent slice-heads. Also, the sink assigns a slice-head to each slice in a field. Thus, by Theorem 5.2, the entire field is fully $k$-covered. Moreover, active sensors are selected only from lenses. Therefore, by Theorem 5.4, T-CRACC$_k$ guarantees that a field is $k$-covered with a minimum number of active sensors, and hence consumes a minimum amount of energy in each round. By Theorem 5.5, a $k$-covered wireless sensor network is connected, assuming that $R \geq r$. Given that T-CRACC$_k$ favours sensors with highest remaining energy in each round and benefits from slicing grid dynamics, all sensors are equally likely to be selected for $k$-coverage of a field in each round. Thus, all sensors deplete their energy slowly and uniformly, thus leading to a maximum network lifetime. ■

### 5.4.3 The D-CRACC$_k$ Protocol

The D-CRACC$_k$ ("D" for Disk) protocol has a higher network clustering granularity compared to T-CRACC$_k$. Precisely, each cluster consists of six adjacent slices forming a disk (Fig. 5.9). Notice that for ease of representation, Fig. 5.9 represents each cluster by six regular triangles instead of six Reuleaux triangles. Next, we describe all the phases of the D-CRACC$_k$ protocol in detail.

#### 5.4.3.1 Deployment Field Clustering

In addition to slicing a field, we assume that in each round, the sink is also responsible for forming *clusters of slices* from the randomly obtained slicing grid. Precisely, each cluster consists of *at most* six adjacent slices forming a *disk*. Because of the random generation of slicing grids and the geometry of a field, some clusters consist of an entire disk, and hence called *interior clusters*, while others are formed by a portion of a disk, and hence called *boundary clusters*. Figure 5.9 shows a clustered deployment field. Moreover, for each cluster, the sink selects a sensor, called *cluster-head*, which is located as near as possible to the centre of its cluster. The random generation of slicing grid ensures that *all* sensors are *equally likely* to act as cluster-heads in each round. Each cluster is defined by one point, i.e., $(x,y)$ coordinates, representing its centre and at most six other points defining its slices (or slice portions for a non-complete cluster). These seven points define the *slicing information* of a cluster, which the sink would broadcast to its corresponding cluster-head. Next, we define *interior* and *boundary lenses*.

*Definition 5.1:* An *interior lens* of a disk is a lens that is shared with no other adjacent disk and a *boundary lens* is shared by two adjacent disks. ■

Notice that each cluster overlaps with at most six others as shown in Fig. 5.9. By Lemma 5.3 and Theorem 5.4, sensors located in the boundary lenses of a given cluster should be selected first in order to minimize the necessary total number of active sensors to achieve full $k$-coverage of a field. However, this requires certain *coordination* between cluster-heads.

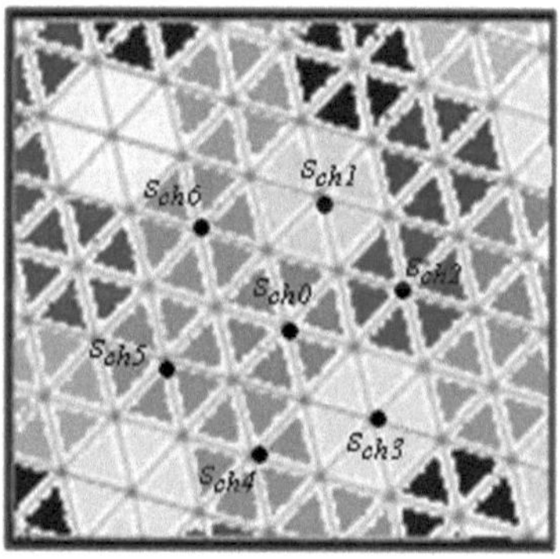


**Fig. 5.10** Clustering for D-CRACC$_k$

#### 5.4.3.2 Cluster-Heads Coordination and Sensor Selection

Each cluster-head is in charge of selecting some of its sensing neighbours to $k$-cover its cluster based on its *slicing information*. Precisely, each cluster-head exploits the overlap between the slices of its cluster as well as the overlap between its slices and those of its adjacent cluster-heads to select a minimum number of its sensing neighbours to $k$-cover its cluster. We assume that each sensor advertises its remaining energy to its sensing neighbours at the start of a round when it turns itself *on*. Each cluster-head $s_{ch_i}$ maintains a list $E_{rem}\_List(s_{ch_i})=\{E_{rem}(j):s_j\in SN(s_{ch_i})\}$ of remaining energy of its sensing neighbours, where $E_{rem}(j)$ is the remaining energy of $s_j$. It uses this list to select the ones with high remaining energy to stay active by sending a *SELECT* message including the cluster-head's *id* as well as the *id*s of all selected sensors. This would avoid those ones with low remaining energy and help the sensors deplete their energy as slowly and uniformly as possible. We assume that at the beginning of each round, *all* the sensors are active. Those ones which are selected by their corresponding cluster-heads would remain active during the underlying round, while the others turn themselves *off* (or go to sleep). For the sensor selection, each cluster-head assigns *priorities* to sensors located in boundary lenses, interior lenses, and middle of slices in descending order. That is, sensors located in boundary lenses have high priority to be selected based on Theorems 2 and 4. Given that each cluster has at most six slices, each cluster-head manages at most six interior lenses and at most six boundary lenses. On the one hand, each cluster-head is responsible for selecting sensors from its interior lenses without any coordination with its adjacent cluster-heads. On the other hand, each cluster-head coordinates with at most six adjacent cluster-heads to select sensors from its boundary lenses in order to $k$-cover its cluster with a minimum number of sensors. For instance, in the case of a disk, its cluster-head, say $s_{ch0}$ (Fig. 5.10), would advertize the subsets $S_1$, $S_2$, $S_3$, $S_4$, $S_5$, $S_6$ of sensors selected from its six boundary lenses to its adjacent cluster-heads $s_{ch1}$, $s_{ch2}$, $s_{ch3}$, $s_{ch4}$, $s_{ch5}$, $s_{ch6}$, respectively.

Theorem 5.8 states that D-CRACC$_k$ yields minimum-energy connected $k$-coverage.

*Theorem 5.8:* The D-CRACC$_k$ protocol fully $k$-covers a field. It also is a minimum-energy connected $k$-coverage protocol.

*Proof:* In each round, each disk-head uses exactly $k$ sensors to $k$-cover each of its six slices by coordinating with its six adjacent disk-heads. Also, the sink guarantees that each disk is under the responsibility of a disk-head. Thus, by Theorem 5.2, the whole field is $k$-covered. Moreover, each disk-head $k$-covers its disk with sensors selected only from interior and boundary lenses. Therefore, by Theorem 4.5, D-CRACC$_k$ uses a minimum total number of active sensors in each round such that a field is guaranteed to be $k$-covered. Thus, D-CRACC$_k$ consumes a minimum amount of energy in each round. By Theorem 5.5, a $k$-covered wireless sensor network is connected, assuming that $R \geq r$. Moreover, D-CRACC$_k$ selects sensors with maximum remaining energy, thus helping all sensors deplete their energy slowly and uniformly. Hence, D-CRACC$_k$ guarantees maximum network lifetime. ■

Notice that T-CRACC$_k$ requires more coordination between cluster-heads than D-CRACC$_k$ and hence has more overhead to $k$-cover a field. This is due to the difference of their cluster sizes. Thus, D-CRACC$_k$ is more energy-efficient than T-CRACC$_k$.

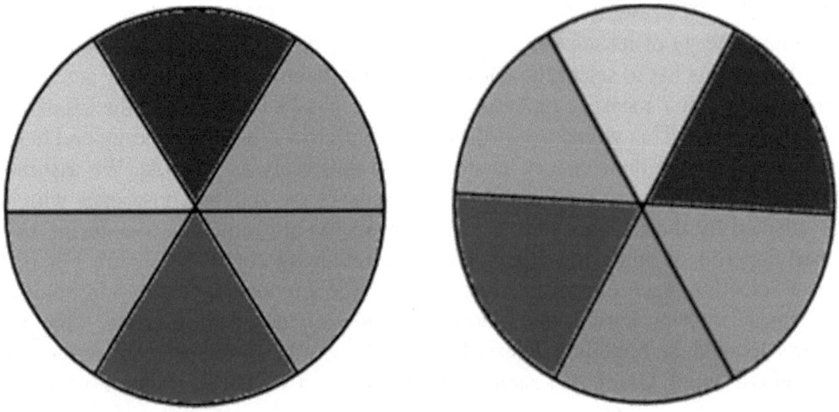

**Fig. 5.11** Slicing grids of the sensing disk of a sensor

## 5.5 Distributed *k*-Coverage Protocol

In this section, we propose a fully distributed *k*-coverage protocol, called *distributed randomized connected k-coverage* (Trig-DIRACC$_k$). The centralized protocol CERACC$_k$ presented in Sect. 5.3 does not rely heavily on global information. Thus, it can be redesigned in a fully distributed fashion based on the local information sensors have about their one-hop neighbours with regard to their physical locations and remaining energy. Also, Trig-DIRACC$_k$ design requires coordination among sensors to achieve *k*-coverage of a field. Indeed, Trig-DIRACC$_k$ is a scheduling scheme in which the sensors can be *triggered* by other sensors to become active. Clearly, this scheduling scheme assumes a high cooperation between sensors in the sense that sensors satisfy the requests of their sensing neighbours when they are solicited to turn *on*. Next, we describe Trig-DIRACC$_k$.

First, we describe our algorithm that enables a sensor to check whether it should turn active and/or select some of its sensing neighbours to turn active in order to *k*-cover its sensing range. Then, we present the state diagram of Trig-DIRACC$_k$.

```
ALGORITHM 2: k-COVERAGE-SELECTION(r, k)
(* This code is run by each sensor *)
Begin
/* Sensing disk slicing */
   Randomly decompose a sensing disk into six
      overlapping Reuleaux triangles RT_i(r), i=1..6
/* Sensors selection */
   n_k_covered_RT = 0
   For each Reuleaux triangle RT_i(r) Do
     If RT_i(r) contain k active sensors Then
       n_k_covered_RT = n_k_covered_RT + 1
     Else
       Select sensors with high remaining energy
       and in the lenses such that RT_i(r) is k-covered
       Activate the selected sensors.
     End
   End
   If n_k_covered_RT = 6 Then /* k-covered disk */
     Return("non-candidate")
   End
   Return("candidate")
End
```

**Fig. 5.12** *k*-Coverage checking algorithm of Trig-DIRACC$_k$

### 5.5.1 k-Coverage Checking Algorithm and Sensor Selection

A sensor runs a *k-coverage checking* algorithm, which is given in Fig. 5.12, in order to find out whether its sensing disk is $k$-covered. To do so, each sensor slices it sensing disk into six overlapping slices as shown in Fig. 5.11 such that two adjacent slices intersect in a lens. Thus, the *slicing grid* in Trig-DIRACC$_k$ consists of exactly six complete slices. Similarly, the $k$-coverage checking algorithm exploits the overlap between adjacent slices as in the case of centralized protocol CERACC$_k$.

Each sensor guarantees that each of the six Reuleaux triangles forming its sensing range (Fig. 5.11) is $k$-covered. To do so, it randomly decomposes its sensing disk into six Reuleaux triangles and checks whether each one of them is $k$-covered based on Theorem 5.4. A sensor starts first by selecting the sensors located in the three lenses (Fig. 5.3) to $k$-cover a given Reuleaux triangle. The selection algorithm exploits the overlap between Reuleaux triangles to select a minimum number of sensors to remain active. It selects its sensing neighbours to become active based on their remaining energy and locations in its sensing disk. Then, it activates them by sending AWAKE messages. When a sensor receives an AWAKE message, it becomes active and broadcasts a NOTIFICATION message to inform all its neighbours that it has become active. Every sensor keeps track of its active neighbours. Moreover, a sensor will turn itself active if one of the Reuleaux triangles is not $k$-covered.

Using Lemma 5.3, each sensor checks whether each of the six slices forming its sensing disk is $k$-covered. For each slice, a sensor checks whether the number of active sensors in the three lenses (Fig. 5.3) including itself is equal to $k$. Otherwise, it checks whether the number of active sensors located the entire slice, i.e., three lenses and middle of the slice (Fig. 5.3), is equal to $k$. For each slice, a sensor

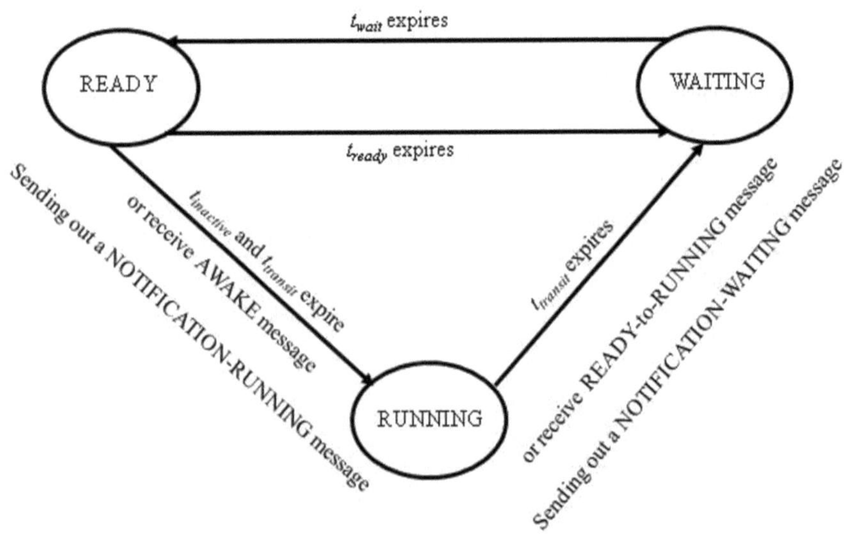


**Fig. 5.13** State diagram of Trig-DIRACC$_k$

computes its degree of coverage and save the result in an array variable *CovDegSlices*. Based on the content of *CovDegSlices*, a sensor activates a necessary number of its sensing neighbours to $k$-cover its sensing disk.

If a sensor is unable to $k$-cover its sensing area when it runs the *k-coverage checking algorithm*, this means that the minimum sensor density required for $k$-coverage is not satisfied, and hence the field cannot be fully $k$-covered.

### 5.5.2 State Transition Diagram of Trig-DIRACC$_k$

At any time, a sensor can be in one of the three states, namely READY, WAITING, or RUNNING. A *state transition diagram* associated with Trig-DIRACC$_k$ and indicating the three possible states of a sensor and transitions between them is shown in Fig. 5.13.

- READY: A sensor is listening to AWAKE messages and thus is ready to switch to the RUNNING state.
- WAITING: A sensor is neither communicating with other sensors nor sensing a field, and thus its radio is turned off. However, after some fixed time interval, it switches to the READY state to receive AWAKE messages if its neighbours decide to do so for achieving $k$-coverage.
- RUNNING: A sensor can communicate with other sensors and sense the environment.
- At the start of the monitoring task, all sensors are in the READY state except one that is in the RUNNING state. The single sensor in the RUNNING state is one of the communication neighbours of the sink that is chosen randomly to activate some of its sensing neighbours to achieve $k$-coverage of its sensing disk. Those selected sensors will in turn run their selection algorithm to $k$-cover their sensing disks. This chain of sensor activations continues until the entire field is $k$-covered. As mentioned earlier, when a given sensor is selected by any other sensor, it will send out a NOTIFICATION message to inform all its neighbours.
- While in the READY state, a sensor keeps track of its sensing neighbours that are in the RUNNING state. If it finds out that its sensing area is $k$-covered, it will switch to the WAITING state. It is not cost-effective to guarantee that a sensor is not selected more than once during one round. Indeed, guaranteeing disjoint subsets of selected sensors requires much coordination between the sensors, thus introducing unnecessary overhead.
- For energy savings, a sensor may wish to switch from the RUNNING state to the WAITING state. For this purpose, a sensor broadcasts a RUNNING-to-WAITING message and waits for some transit time $t_{transit}$. If $t_{transit}$ expires and it has not received any Running-to-Waiting message, it switches to the WAITING state and sends a NOTIFICATION-WAITING message, where it can stay there for $t_{wait}$ time. When $t_{wait}$ expires, it switches to the READY state, where it can stay in this state for $t_{ready}$ time.
- If a sensor finds out that its sensing area is not $k$-covered, it will broadcast a READY-to-RUNNING message and wait for some transit time $t_{transit}$ before

switching to the RUNNING state. If after $t_{transit}$ it has not received any other READY-to-RUNNING message, it will switch to the RUNNING state and sends a NOTIFICATION-RUNNING message; otherwise, it will switch to the WAITING state.

### 5.5.3 Ensuring Network Connectivity

Theorem 5.9 states the necessary relationship between the sensing and communication ranges of the sensors so that k-coverage implies network connectivity.

*Theorem 5.9:* Let $k \geq 3$. Under the assumption of Trig-DIRACC$_k$ protocol, a k-covered wireless sensor network is connected if $R \geq \sqrt{3}\, r$, where $r$ and $R$ are the radii of the sensing and communication ranges of the sensors, respectively.

*Proof:* Consider either configuration in Fig. 5.14. To compute the minimum communication range for the network to be connected, assume all sensors are located at one of the extreme points of a lens. With a little algebra, it is easy to check that the distance between two extreme points, say $p$ and $q$, of two adjacent slices is equal to $\delta(p,q) = \sqrt{3}\, r$. Thus, connectivity of k-covered wireless sensor networks requires that $R \geq \sqrt{3}\, r$. ■

Our result in Theorem 5.9 improves on the one given in [205], which requires $R \geq 2r$ for connectivity in k-covered WSNs.

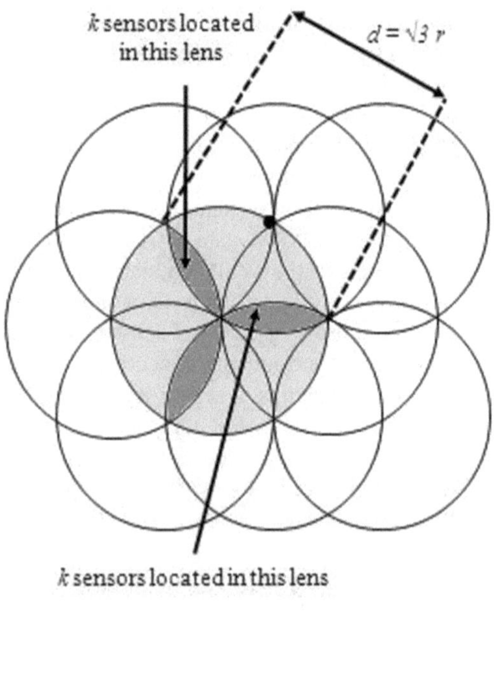


**Fig. 5.14** Maximum distance between sensors in two lenses

Theorem 5.10 states that Trig-DIRACC$_k$ is a minimum-energy distributed connected $k$-coverage protocol.

*Theorem 5.10:* The Trig-DIRACC$_k$ protocol fully $k$-covers a field. It also is as minimum-energy protocol as D-CRACC$_k$.

*Proof:* We proceed by contradiction. Assume that the total area $A$ of a field is not fully $k$-covered by active sensors. Hence, $A$ can be decomposed into a $k$-covered area $A_c$ and a non-$k$-covered area $A_{nc}$, i.e., $A = A_c \cup A_{nc}$. Thus, there is at least one sensor $s_i$ whose sensing disk intersects the area $A_{nc}$, i.e., $SD(\xi_i, r_i) \cap A_{nc} \neq \varnothing$. In particular, the sensing disk of $s_i$ is not fully $k$-covered. Hence, sensor $s_i$ is not active. Precisely, $s_i$ is in the WAITING state. According to Trig-DIRACC$_k$, however, sensor $s_i$ must be in the RUNNING state (i.e., active) given that its sensing disk is not fully $k$-covered. This contradicts our assumption. Thus, the total area $A$ of a field is fully $k$-covered by active sensors. Let us now show that Trig-DIRACC$_k$ uses as minimum number of active sensors as D-CRACC$_k$ to $k$-cover a field. Using Trig-DIRACC$_k$, each sensor checks whether its sensing area is $k$-covered. In order to $k$-cover each of its six slices, each sensor makes sure that there are exactly $k$ active within its sensing disk. Each sensor favours active sensors that are located in the interior and boundary lenses of its sensing disk. Given that there is no pre-slicing of the entire field into adjacent slices as is the case with the three other protocols, there is no guarantee that all active sensors belong to the lenses of each sensor. Therefore, it may happen that active sensors located in some lenses of the sensing disk of a sensor are located in the lenses and/or the middle of slices of other sensors. Definitely, these sensors will consider those

```
ALGORITHM 1: k-COVERAGE-CANDIDACY(r, k)
(* This code is run by each sensor *)
Begin
/* Sensing disk decomposition */
   Randomly slice a sensing disk into six
   Reuleaux triangles, RT_i(r) 1 ≤ i ≤ 6
/* Localized k-coverage candidacy checking */
   For each Reuleaux triangle RT_i(r) Do
      If RT_i(r) contains k active sensors Then
         Skip /* i.e., do nothing */
      Else
         Return ("candidate")
      End
   End
   Return ("non-candidate")
End
```

**Fig. 5.15** $k$-Coverage candidacy algorithm of Self-DIRACC$_k$

active sensors located in the middle of their slices so they $k$-cover their sensing disks with as minimum number of active sensors as possible. By Theorems 5.2 and 5.4, it follows that Trig-DIRACC$_k$ could use a little more sensors than D-CRACC$_k$ for full $k$-coverage of a field. Hence, Thus, Trig-DIRACC$_k$ consumes as minimum amount of energy as D-CRACC$_k$ during the operational network lifetime. Similarly, Trig-DIRACC$_k$ selects sensors with highest remaining energy to remain active, thus helping all sensors deplete their energy as slowly and uniformly as possible, thus extending the network lifetime. By Theorem 5.9, a $k$-covered wireless sensor network is connected if $R \geq \sqrt{3}\,r$. ■

## 5.6 Self-scheduling Based $k$-Coverage

In Sect. 5.5, we proposed a sensor scheduling protocol, called Trig-DIRACC$_k$, which allows a sensor to trigger a necessary number of its sensing neighbours to become active in order to achieve $k$-coverage of its sensing range. This protocol can be classified as a *triggered-scheduling driven k-coverage*. In this section, we suggest another connected $k$-coverage protocol using a different scheduling approaches. In this protocol, each sensor turns itself *on* based on the local information it has about its sensing neighbours in order to $k$-cover its sensing range. This protocol can be classified as a *self-scheduling driven k-coverage* and is denoted by Self-DIRACC$_k$, In Sect. 5.8, we evaluate the performance of Trig-DIRACC$_k$ and Self-DIRACC$_k$ and compare between them.

First, we present our *k-coverage candidacy* algorithm, which is shown in Fig. 5.15 and enables a sensor to check its *candidacy* to become active. Then, we present a *state transition diagram* describing the possible states a sensor can be in and transitions between those states.

### *5.6.1 k-Coverage Candidacy Algorithm*

Using a self-scheduling scheme, each sensor decides *by itself* whether to turn *on* or *off*. This decision is based solely on the local information it has about the status (*on* vs. *off*) of its sensing neighbours, i.e., sensors located in its sensing range.

A sensor turns active if its sensing disk is not $k$-covered. Based on Theorem 5.2, a sensor randomly decomposes its sensing disk into six overlapping Reuleaux triangles of width $r$ (or slices) and checks whether each one of them contains at least $k$ sensors. Each sensor should know the status of its sensing neighbours only to decide whether it is candidate to turn active. If any of the six slices does not have $k$ active sensors, a sensor is a candidate to become active. Otherwise, it is not.

### *5.6.2 State Transition Diagram of Self-DIRACC$_k$*

Figure 5.16 shows a state transition diagram for Self-DIRACC$_k$. At any time, a sensor can be in one of READY, WAITING, and RUNNING states:

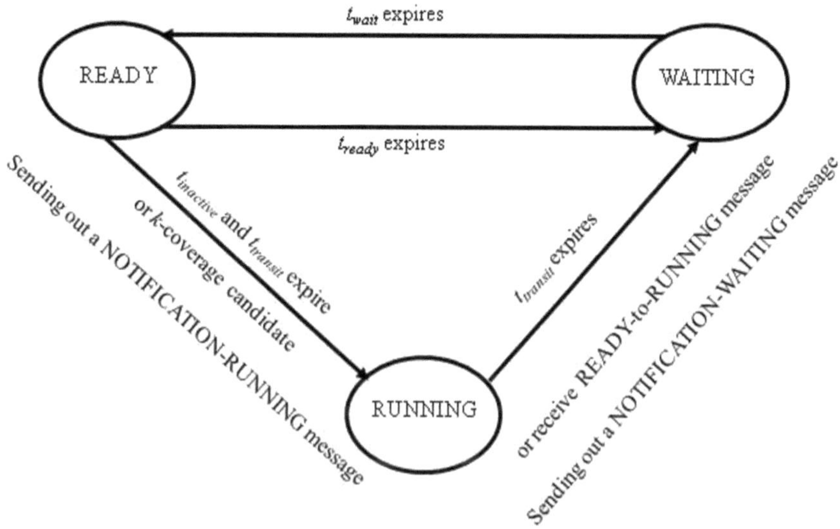


**Fig. 5.16** State transition diagram of Self-DIRACC$_k$

- READY: A sensor listens to messages and checks its $k$-coverage candidacy to switch to the RUNNING state.
- RUNNING: A sensor is active and can communicate with other sensors and sense a field.
- WAITING: A sensor is neither communicating with other sensors nor sensing a field, and thus its radio is turned *off*. However, after some fixed time interval, it switches to the READY state to check its candidacy for $k$-coverage (Fig. 5.15) and receive messages.
- At the beginning of the monitoring task, all sensors are in the RUNNING state. Moreover, each sensor chooses randomly and independently of all other sensors a value $t_{check}$ between 0 and $t_{check\text{-}max}$ after which it runs the *k-coverage candidacy* algorithm (Fig. 5.15) to check whether it stays active or switches to the WAITING state. Our intuition behind this random selection of $t_{check}$ is to avoid higher or lower coverage of any region in a field.
- When a sensor runs the $k$-coverage candidacy algorithm and finds out that it is a candidate, it sends out a NOTIFICATION-RUNNING message to inform all its neighbours. While in the READY state, a sensor keeps track of its sensing neighbours that are in the RUNNING state. If it finds out that its sensing area is $k$-covered, it will switch to the WAITING state.
- To save energy, a sensor may wish to switch from RUNNING state to WAITING state. It broadcasts a RUNNING-to-WAITING message and waits for $t_{transit}$ transit time. If $t_{transit}$ expires and it has not received any RUNNING-to-WAITING message, it switches to the WAITING state to stay $t_{wait}$ time, and sends out a NOTIFICATION-WAITING message. When $t_{wait}$ expires, it switches to the READY state, where it can stay $t_{ready}$ time. When a sensor in the READY state receives a RUNNING-to-WAITING message from its sensing

neighbour, it runs the $k$-coverage candidacy algorithm to check whether it should be active.

- If a sensor in the READY state finds out that it has not been active for some $t_{in\text{-}active}$ time, it will broadcast a READY-to-RUNNING message and wait for some $t_{transit}$ time. If $t_{transit}$ expires without receiving any READY-to-RUNNING message, it will send out a NOTIFICATION-RUNNING message and switch to the RUNNING state. Else, it stays in the READY state. A sensor in the READY state would also apply the same process if it finds out that it has not heard from one of its sensing neighbours within some $t_{active}$ time, i.e., this neighbour depleted its entire energy. Thus, each sensor in the RUNNING state should broadcast an ALIVE message after each $t_{active}$ time. We assume that each sensor remains active for at least $t_{active}$ time.

### 5.6.3 *Tri-DIRACC$_k$ Versus Self-DIRACC$_k$*

As can be seen, Tri-DIRACC$_k$ benefits well from the result of Theorem 5.4 in the sensor selection process. Indeed, each sensor prioritizes sensors located in the lenses of its sensing disk, and hence the latter is guaranteed to be $k$-covered with a minimum number of sensors. However, Self-DIRACC$_k$ does not exploit Theorem 3 since each sensor only checks whether each of the six Reuleaux triangles forming its sensing disk is $k$-covered. Thus, we can conclude that Self-DIRACC$_k$ would require more active sensors than Trig-DIRACC$_k$ to fully $k$-cover a field. On the other hand, Tri-DIRACC$_k$ requires total coordination between sensors in the sense that when a sensor receives a message from another sensor to become active, it should satisfy its request and turn itself *on*. Therefore, Trig-DIRACC$_k$ introduces more control overhead than Self-DIRACC$_k$. We anticipate that in general Trig-DIRACC$_k$ outperforms Self-DIRACC$_k$ with respect to the total energy consumption to achieve complete $k$-coverage of a field, and hence yields higher network lifetime.

## 5.7 Relaxation of Assumptions

The design of our connected $k$-coverage protocols is based on the *unit disk model* and *homogeneous sensor model*. Although these assumptions are the basis for most of the coverage and connectivity protocols in wireless sensor networks, they may not be valid in real-world wireless sensor network platforms. In this section, we relax these assumptions to promote the use of our protocols in real-world applications.

### 5.7.1 *Relaxing the Unit Disk Model*

Cao et al. [54] found that the sensing range of the sensors is non-isotropic. Also, it was found that the communication range of MICA motes is asymmetric and depends on the environments [221] and that the communication range of radios is highly probabilistic and irregular [224]. For problem tractability, we consider a *convex model*, where the communication and sensing ranges of sensors are homogeneous and *convex* but not necessarily circular.

Lemmas 5.4 and 5.5 correspond to Theorems 5.2 and 5.3, respectively. Their proof is literally the same as that in Sect. 5.2.2 by exploiting the notion of the largest enclosed disk.

Let $r_{led}$ be the radius of the largest enclosed disk of the sensing range of the sensors.

*Lemma 5.4:* Let $k \geq 3$. A field is guaranteed to be $k$-covered if for any slice of width $r_{led}$ in a field, there is at least one adjacent slice of width $r_{led}$ such that their lens contains at least $k$ active sensors. ■

*Lemma 5.5:* Let $k \geq 3$. The *minimum sensor spatial density* required to fully $k$-cover a field by *homogeneous* sensors with *convex* sensing ranges but not necessarily circular is equal to

$$\lambda(r_{led}, k) = \frac{6k}{(4\pi - 3\sqrt{3})\, r_{led}^{\,2}}$$ ■

Now, we discuss how CERACC$_k$, T-CRACC$_k$, D-CRACC$_k$, Trig-DIRACC$_k$, and Self-DIRACC$_k$ can be implemented using the convex sensing model. The unit of slicing, i.e., Reuleaux triangle, has a width equal to $r_{led}$. Any other processing remains the same for each of those four protocols. Hence, the assumption of the unit disk model can be easily relaxed with the help of the notion of the largest enclosed disk of the sensing range of the sensors.

## 5.7.2 *Relaxing the Sensor Homogeneity Model*

Real-world sensing applications [107] may require heterogeneous sensors in terms of their sensing and communication capabilities to enhance the reliability of the network and extend its lifetime [209]. Even sensors equipped with identical hardware may not always have the same sensing model. In this section, we consider heterogeneous sensors with different yet convex sensing and communication ranges.

Lemmas 5.6 and 5.7 correspond to Theorems 5.2 and 5.3, respectively.

*Lemma 5.6:* Let $k \geq 3$. A field is guaranteed to be $k$-covered if for any slice of width $r_{led}^{\min}$ in a field, there is at least one adjacent slice of width $r_{led}^{\min}$ such that their lens contains at least $k$-active sensors. ■

*Lemma 5.7:* Let $k \geq 3$. The *sensor spatial density* that is necessary for $k$-coverage of a field by *heterogeneous* sensors with *convex* sensing ranges is given by

$$\lambda(r_{led}^{\min}, k) = \frac{6k}{(4\pi - 3\sqrt{3})\, {r_{led}^{\min}}^{2}}$$

where $r_{led}^{\min}$ is the minimum radius of the largest enclosed disks of the sensing ranges of the sensors. ■

In the case of CERACC$_k$, T-CRACC$_k$, and D-CRACC$_k$, the sink has to slice a field into slices of width $r_{led}^{\min}$ and apply the same processing as in Sect. 5.2.2. For Trig-DIRACC$_k$ and Self-DIRACC$_k$, each sensor needs to consider its largest enclosed disk and run the same steps as in Sect. 5.2.2. Thus, the assumption of homogeneous sensors can also be relaxed with slight updates to our protocols.

Both protocols should consider $r_{led}^{\min}$ instead of $r$ when the sensors decompose their sensing ranges to $k$-cover them. Notice that if a single sensor with a very small sensing range is deployed, the network would have a large number of active sensors. Thus, it is necessary that Trig-DIRACC$_k$ and Self-DIRACC$_k$ adapt the sensor density to the sensing ranges of sensors in the area. Due to space limitations, we will address this issue in our future work.

## 5.8 Performance Evaluation

In this section, we present the simulation results of our protocols using a high-level simulator written in the C programming language. First, we specify the simulation environment as well as the energy consumption model that we use. Then, we present the simulation results with respect to several parameters.

### 5.8.1 *Simulation Settings*

We consider a square field of side length 1000 m. We use the energy model given in [211], where the sensor energy consumption in transmission, reception, idle, and sleep modes are 60 mW, 12 mW, 12 mW, and 0.03 mW, respectively. Following [218], the energy required for a sensor to stay idle for 1 s is equivalent to *one unit of energy*. We assume that the initial energy of each sensor is 60 J enabling a sensor to operate about 5000 s in reception/idle modes [211]. All simulations are repeated 200 times and the results are averaged.

### 5.8.2 *Simulation Results*

In this section, we present the simulation results of our protocols. Figure 5.17 shows the sensor density versus the coverage degree $k$, where the radius $r$ of the sensing range of sensors is fixed to $r = 25$ m. The sensor density increases with $k$ for a fixed $r$, as expected. As can be seen, the four protocols yield a sensor density closer to the one given in Theorem 5.3. Also, CERACC$_k$ outperforms all other protocols while Trig-DIRACC$_k$ uses more sensors than these protocols due to its distributed nature. Figure 5.18 plots the sensor density versus $r$ with $k = 3$. We

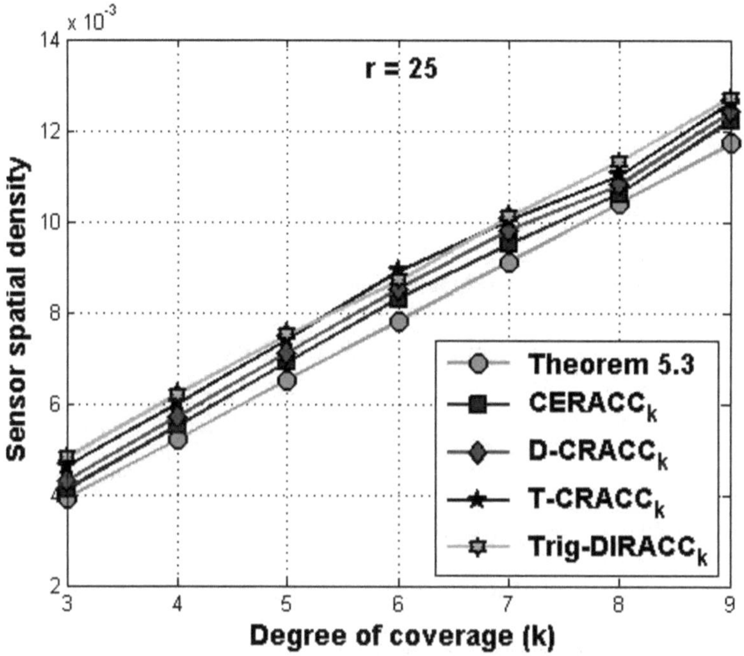


**Fig. 5.17** $\lambda(r,k)$ vs. $k$

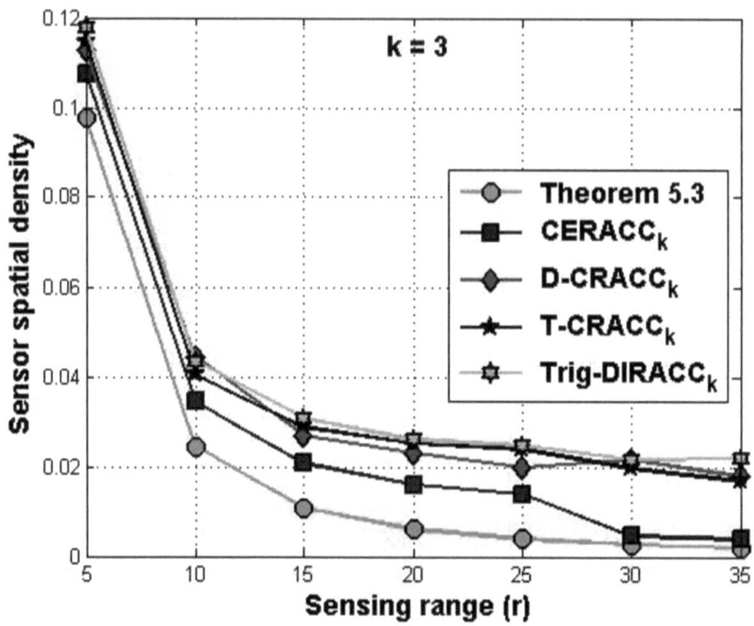


**Fig. 5.18** $\lambda(r,k)$ vs. $r$

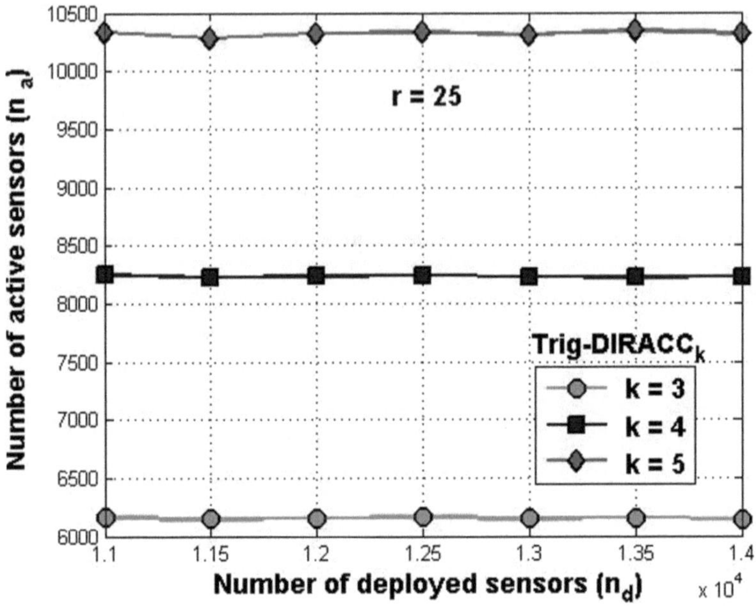


**Fig. 5.19** Number of active sensors vs. number of deployed sensors (*k* variable)

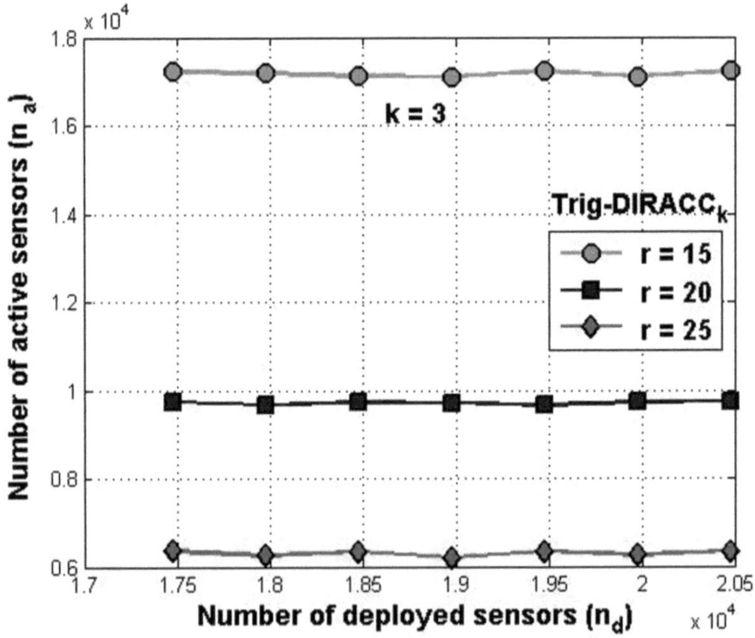


**Fig. 5.20** Number of active sensors vs. number of deployed sensors (*r* variable)

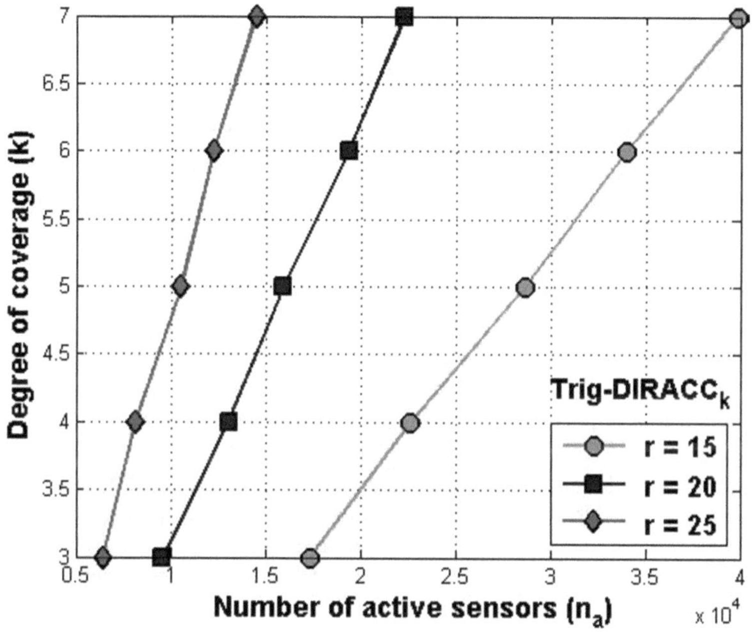


**Fig. 5.21** $k$ vs. number $n_a$ of active sensors

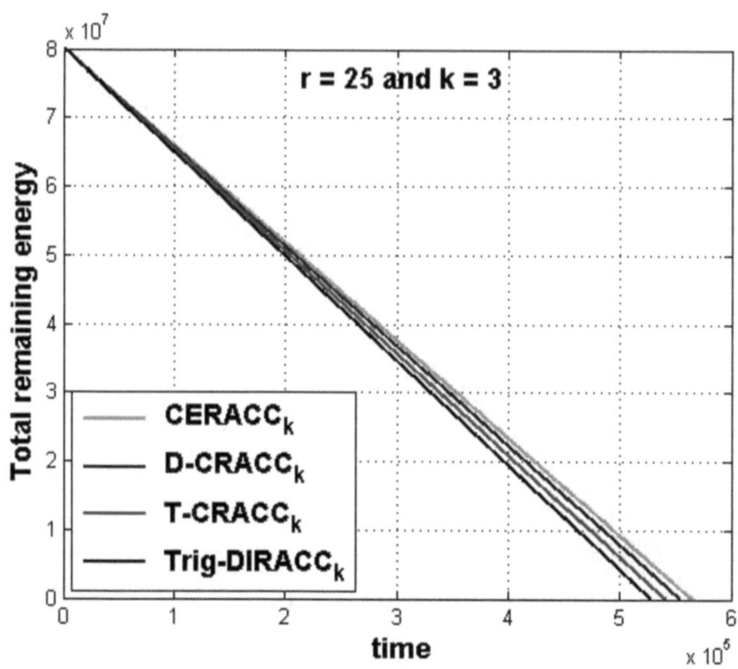


**Fig. 5.22** Remaining energy vs. time

observe that the sensor density decreases with $r$ for a fixed $k$, Likewise, the four protocols require a sensor density closer to the one computed in Theorem 5.3.

Figures 5.19 and 5.20 show the number of active sensors versus the total number of deployed sensors for Trig-DIRACC$_k$. In Fig. 5.19, we consider different values of $k$, while in Fig. 5.20, we consider different values of $r$. For higher values of $k$, more sensors need to be active to achieve the required coverage. However, for higher values of $r$, less number of sensors is needed to $k$-cover a field. In both experiments, the number of active sensors does not depend on the number of deployed sensors but only on $k$ and $r$.

Figure 5.21 shows the degree $k$ of coverage versus the total number $n_a$ of active sensors for Trig-DIRACC$_k$. Notice that $k$ increases with $n_a$. Also, for the same $n_a$, $k$ increases quickly as $r$ increases as a larger region of the field would be covered.

Figure 5.22 shows that the total remaining energy of the sensors in the four protocols decreases smoothly (in this experiment, the number of deployed sensors is 16,000). Notice that the centralized protocol CERACC$_k$ consumes less energy than all other protocols while the distributed protocol for Trig-DIRACC$_k$ consumes the highest amount of energy. Thus, CERACC$_k$ yields longer network lifetime than Trig-DIRACC$_k$. This shows the advantage of our centralized protocol (CERACC$_k$) over the distributed one (Trig-DIRACC$_k$). Indeed, the number of messages needed by CERACC$_k$ to distribute the optimal schedule to the selected sensors may be less than that required by Trig-DIRACC$_k$ due to the periodic messages exchanged by sensors to coordinate their mission for $k$-coverage of a field.

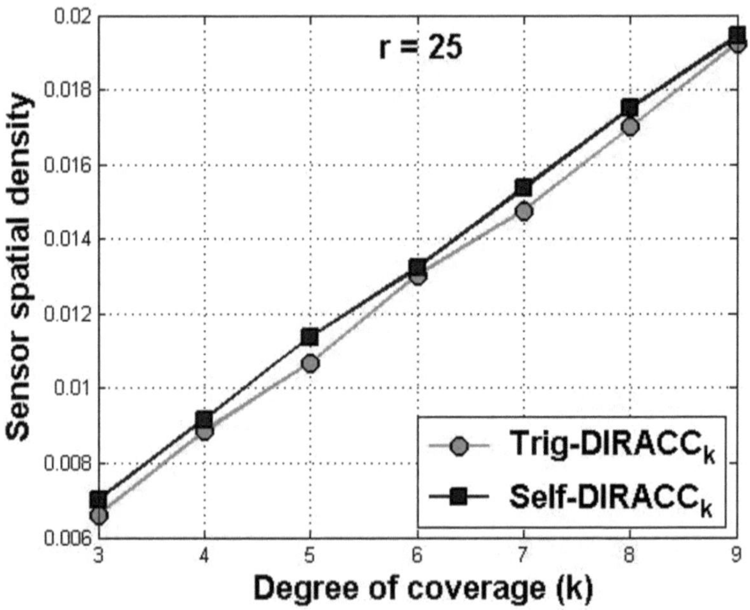


**Fig. 5.23** Trig-DIRACC$_k$ vs. Self-DIRACC$_k$ (Sensor density vs. coverage degree)

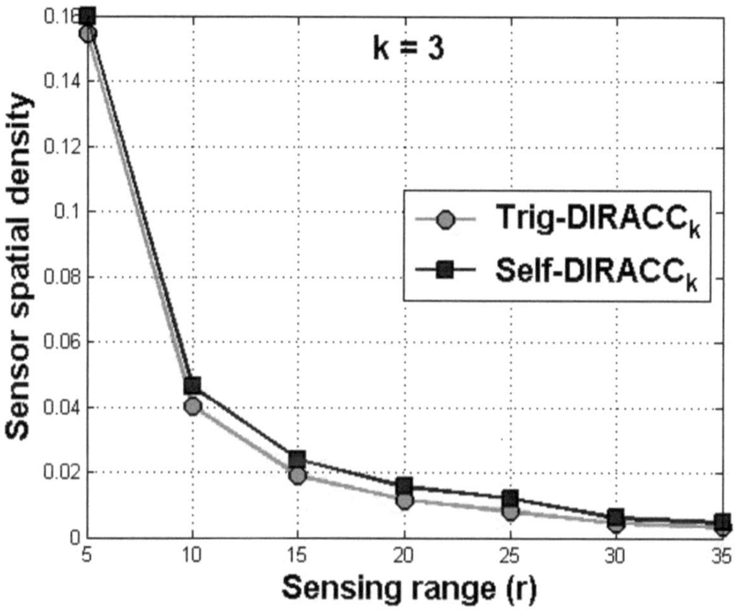


**Fig. 5.24** Trig-DIRACC$_k$ vs. Self-DIRACC$_k$ (Sensor density vs. sensing range)

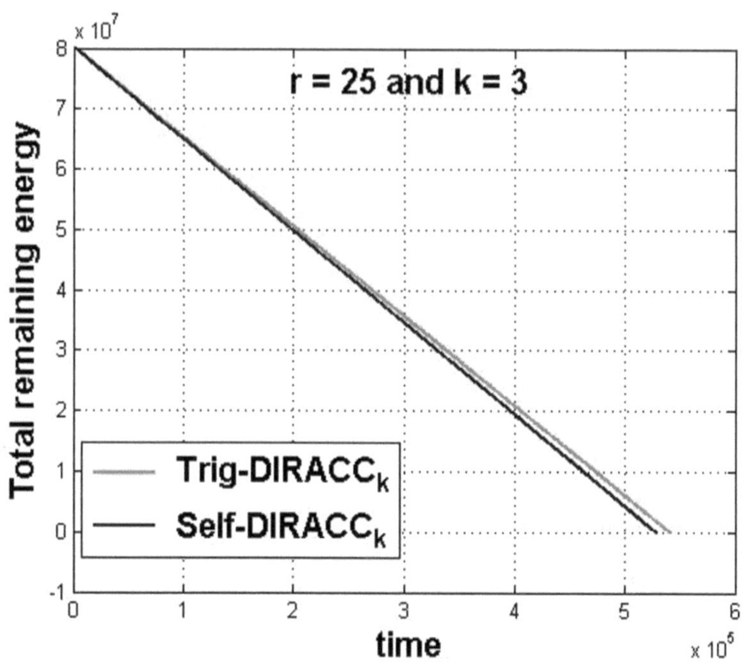


**Fig. 5.25** Trig-DIRACC$_k$ vs. Self-DIRACC$_k$ (total remaining energy vs. time)

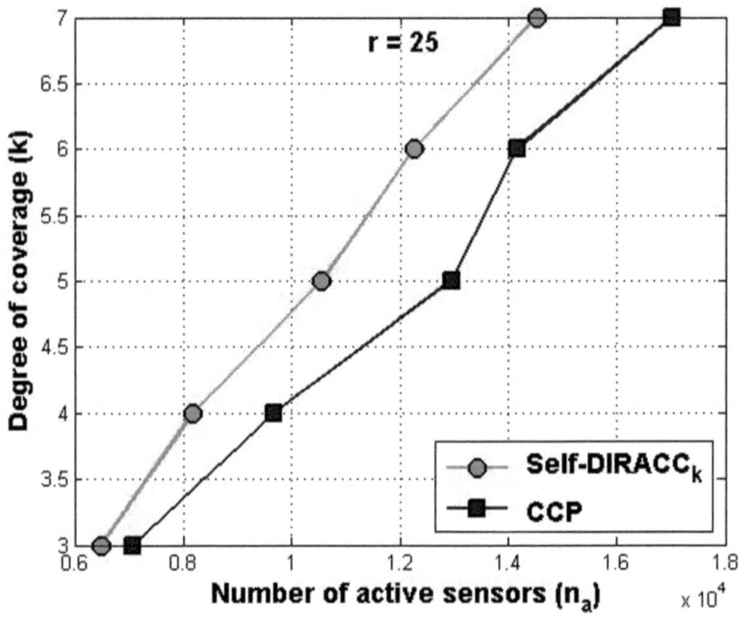


**Fig. 5.26** Self-DIRACC$_k$ compared to CCP ($k$ vs. $n_a$)

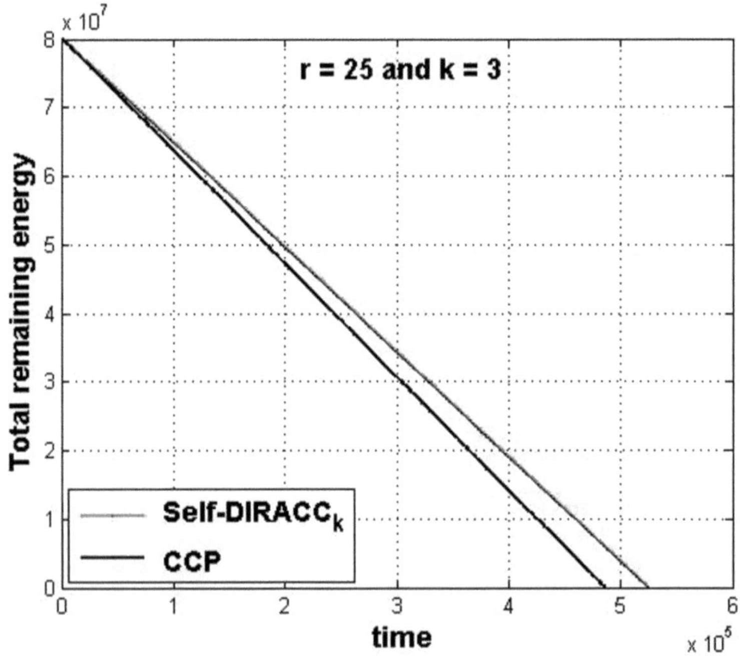


**Fig. 5.27** Self-DIRACC$_k$ compared to CCP (remaining energy vs. time)

As discussed earlier in Sect. 5.6.3, all the plots in Figs. 5.23–5.25 match our expectation. They show that Trig-DIRACC$_k$ is slightly more energy-efficient than Self-DIRACC$_k$ in terms of the necessary total number of active sensors to fully $k$-cover a field (Figs. 5.23 and 5.24) as well as the network lifetime (Fig. 5.25).

### 5.8.3 Comparison of Self-DIRACC$_k$ with CCP

In this section, for fair comparison with the CCP protocol [205], we compare Self-DIRACC$_k$. Indeed, both CCP and Self-DIRACC$_k$ share the same idea in the sense that the sensors decide whether they turn active based on the result of their coverage eligibility algorithm (for CCP) and $k$-coverage candidacy algorithm (for Self-DIRACC$_k$). The CCP protocol provides different degrees of full coverage of a convex region. CCP was the first protocol that discussed $k$-coverage and connectivity within a unified framework. It was proved that coverage implies connectivity when $R \geq 2r$ [205]. Hence, no other mechanism would be necessary to guarantee connectivity. However, CCP was integrated with a topology maintenance protocol, called SPAN [59], to provide both coverage and connectivity guarantees when $R < 2r$. Recall that a convex region $A$ is $k$-covered in CCP if all intersection points between sensing disks of sensors and between sensing disks of sensors and $A$'s boundary are at least $k$-covered.

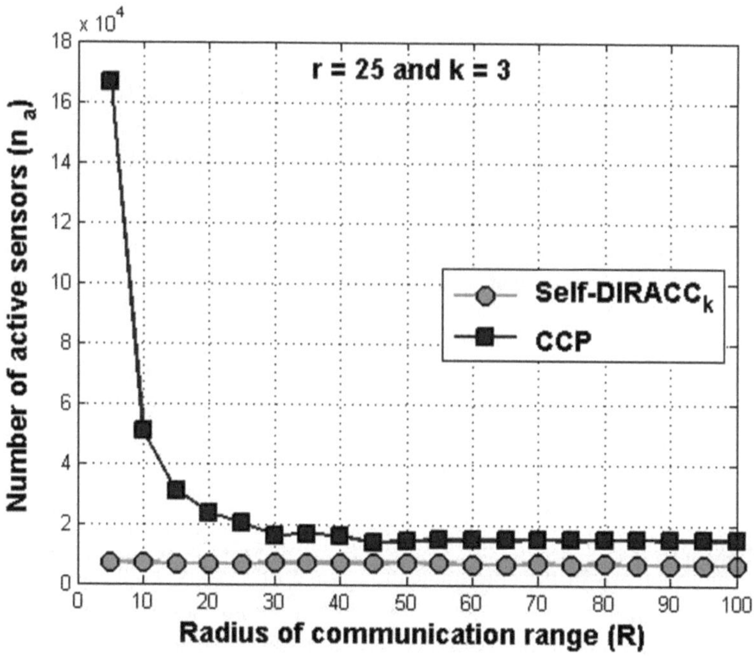


**Fig. 5.28** Self-DIRACC$_k$ compared to CCP ( $n_a$ vs. $R$ )

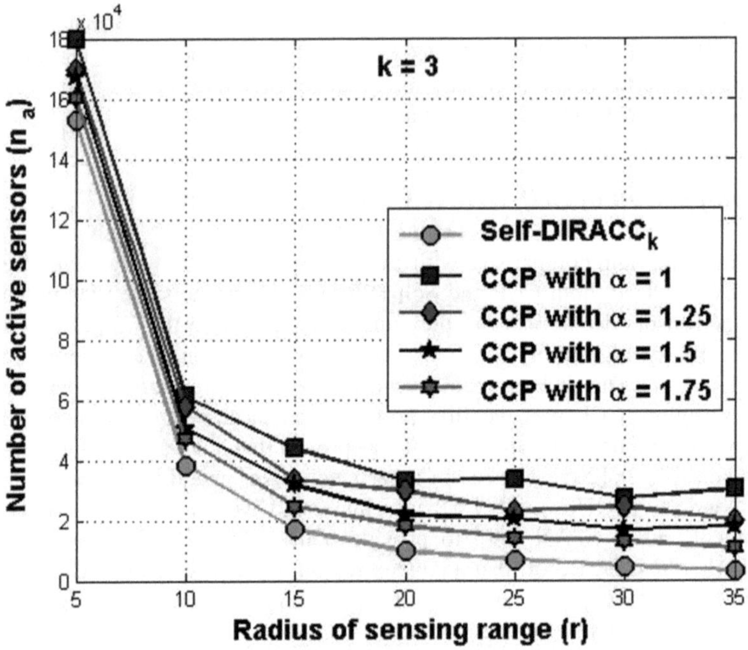


**Fig. 5.29** Self-DIRACC$_k$ compared to CCP ( $n_a$ vs. $r$)

Figure 5.26 plots the degree $k$ of coverage versus the number $n_a$ of active sensors for Self-DIRACC$_k$ as compared to CCP. It shows that Self-DIRACC$_k$ requires less active sensors than CCP to achieve the same degree of coverage, thus yielding significant energy savings. This is due not only to a higher number of active sensors required by CCP, and hence an additional energy spent in sensing, but also to the communication overhead caused by the exchange of messages between active sensors running CCP to coordinate among themselves and provide the requested $k$-coverage service. Thus, CCP consumes more energy than Self-DIRACC$_k$ as shown in Fig. 5.27. Note that while CCP requires SPAN to provide connectivity between active sensors when $R < 2r$, Self-DIRACC$_k$ does not need such a topology maintenance protocol as all it requires is $R \geq r$, thus providing connectivity when $k$-coverage is guaranteed. Indeed, Self-DIRACC$_k$ is based on the analysis of sensors' sensing range to provide $k$-coverage.

Figure 5.28 plots $n_a$ versus $R$ with $r = 30$ m, while Fig. 5.29 plots $n_a$ versus $r$ for different ratios $\alpha = R / r$, where $\alpha \geq 1$. In both cases, we fix $k = 3$. Given that $\alpha \geq 1$, any increase in the communication range of sensors would not have any impact on the performance of Self-DIRACC$_k$. It would, however, affect the performance of CCP. As can be observed, $n_a$ decreases as $R$ increases. Indeed, SPAN would require less number of sensors to maintain connectivity between active sensors as $R$ increases. However, at some point (surprisingly enough, this point corresponds to $R \geq 2r$), the number $n_a$ of active sensors required for $k$-coverage does not decrease any further. Indeed, when $R \geq 2r$, SPAN is not needed at all as both $k$-coverage and $R \geq 2r$ would guarantee connectivity. Similarly, the performance of CCP improves as the ratio $\alpha$ increases (Fig. 5.29), i.e., $R$ increases. That is, less number of active sensors is needed to provide $k$-coverage and connectivity.

## 5.9 Related Work

A variety of configuration protocols for coverage and connectivity in wireless sensor networks have been proposed in the literature with a goal to extend the network lifetime. In this section, we review a sample of these configuration protocols.

Adlakha and Srivastava [2] addressed the issue of determining the required number of sensors to achieve full coverage of a desired region. Precisely, they proposed an exposure-based model to find the sensor density based on the physical characteristics of the sensors and the properties of the target. Kumar et al. [128] showed that the minimum number of sensors needed to achieve $k$-coverage with high probability is approximately the same regardless of whether the sensors are deployed deterministically or randomly, if the sensors fail or sleep independently with equal probability. Shakkottai et al. [177, 178] gave necessary and sufficient conditions for 1-covered, 1-connected wireless sensor grid network. Also, they proposed a variety of algorithms to maintain connectivity and coverage in large wireless sensor networks. Sohrabi et al. [180, 181] proposed two self-assembly mechanisms for WSNs based on clustering, *dual network clustering* and *Rendezvous clustering algorithm*, at the link layer for efficient interconnection of clusters and, hence, end-to-end connectivity.

Kumar et al. [127] established an optimal deployment pattern for achieving $k$-barrier coverage, developed efficient global algorithms for checking $k$-barrier coverage of a given region, and showed the non-existence of localized algorithms for testing the existence of global barrier coverage [127]. To address this limitation, Chen et al. [61] proposed localized algorithms so sensors can locally determine the existence of local barrier coverage. Moreover, Kumar et al. [129] proposed optimal polynomial-time algorithms to solve the sleep-wakeup problem for the barrier coverage model using sensors with equal and unequal lifetimes.

Ai and Abouzeid [3] proposed a directional sensors-based approach for network coverage, where the coverage region of a sensor depends on its location and orientation. Li et al. [139] proposed efficient distributed algorithms to optimally solve the best-coverage problem with the least energy consumption. Megerian et al. [156, 158] proposed optimal polynomial time worst-and average-case algorithms for coverage calculation based on the Voronoi diagram and graph search algorithms. Huang and Tseng [108] presented polynomial-time algorithms, in terms of the number of sensors, for the coverage problem formulated as a decision problem. Abrams et al. [1] proposed a distributed algorithm to partition a wireless sensor network into $k$ covers, each of which contains a subset of sensors that is activated in a round-robin fashion such that as many areas are monitored as frequently as possible. Comprehensive surveys of a variety of approaches on energy-efficient coverage problems can be found in [55, 90].

Xing et al. [205] proposed the first combined study on $k$-coverage and connectivity and proved that if the radius of the communication ranges of sensors is double the radius of their sensing ranges, the network is connected provided that sensing coverage is guaranteed. Also, they computed the network connectivity based on whether the disconnected node is boundary or interior, and proposed a network configuration protocol based on the degree of coverage of the sensing application. Yang et al. [207] formalized the $k$-coverage set and the $k$-connected coverage set problems in terms of linear programming and proposed two non-global solutions for them. Bai et al. [37] proposed an optimal deployment strategy to achieve both full coverage and 2-connectivity regardless of the relationship between communication and sensing radii of the sensors. Huang et al. [110] studied the relationship between coverage and connectivity of wireless sensor networks and proposed distributed protocols to guarantee both their coverage and connectivity. Zhang and Hou [218] proposed a distributed algorithm to keep a small number of active sensors in a network regardless of the relationship between sensing and communication ranges. Tian and Georganas [185] proved that if the original network is connected and the identified active nodes can cover the same region as all the original nodes, then the network formed by the active nodes is connected when the communication range is at least twice the sensing range. Yener et al. [212] proposed a probabilistic Markov model to solve the problem of minimizing power consumption in each sensor while ensuring coverage and connectivity.

Gupta et al. [96] proposed centralized and distributed algorithms for connected sensor cover so the network can self-organize its topology in response to a query and activate the necessary sensors to process the query. Datta et al. [71] proposed two self-stabilizing algorithms to the problem of minimal connected sensor cover [96]. Zhou et al. [222] proposed a distributed and localized algorithm using the concept of the $k^{th}$-order Voronoi diagram to provide fault tolerance and extend the network lifetime, while maintaining a required degree of coverage. Cortes et al. [67] designed control and coordination algorithms for a multi-vehicle network with limited sensing and communication capabilities. Also, Liu et al. [144] proposed adaptive, distributed, and asynchronous coverage algorithms for mobile sensing networks. Moreover, they proved that mobility can be used to improve coverage in wireless sensor networks.

## 5.10 Summary

In this chapter, we studied the problem of connected $k$-coverage in wireless sensor networks, where each location in a field is *covered* by at least $k$-active sensors while *all* active sensors are being *connected* [15, 22]. First, we characterized $k$-coverage of a field based on a geometric analysis of the intersection of sensing disks of $k$ sensors. We proved that $k$-coverage implies connectivity between active sensors when the communication range of the sensors is at least equal to their sensing range. By looking at real sensor node platforms, it is always the case that the communication range of the sensors is higher than their sensing range, and hence our argument is always valid. Indeed, Tables I and II given in [218] show the communication range of Berkeley motes is much higher than the sensing range of several typical sensors, and hence support our argument. We also proved that a sufficient condition of $k$-coverage of a field is that the minimum sensor spatial density depends only on $k$ and the sensing range of the sensors. Moreover, we proposed centralized ($CERACC_k$), pseudo-distributed (T-$CRACC_k$ and D-$CRACC_k$), and fully distributed (Trig-$DIRACC_k$) protocols to solve the connected $k$-coverage problem in wireless sensor networks. Also, we proposed a fully distributed, self-scheduling protocol, called Self-$DIRACC_k$, where each sensor turns itself *on* if its sensing range is not $k$-covered. This approach is different from that of Trig-$DIRACC_k$, where each sensor selects a necessary number of sensors to become active in order to $k$-cover its sensing range. Despite the fact that Trig-$DIRACC_k$ requires coordination between the sensors to $k$-cover their sensing ranges, thus introducing additional control overhead, we found through simulations that Trig-$DIRACC_k$ outperforms Self-$DIRACC_k$ with regard to the necessary number of active sensors for full $k$-coverage of a field and network lifetime. We also extended our analysis by relaxing several widely used assumptions in $k$-coverage configuration in wireless sensor networks. We also extended our analysis by relaxing the assumptions of the unit sensing disk model and homogeneous sensor. These relaxations have helped us handle the convex sensing model and heterogeneous wireless sensor networks, and hence promote the use of our protocols in real-world applications. Our simulation results show that Self-$DIRACC_k$ is more energy-efficient than CCP [205], with respect to the number of active sensors required for $k$-coverage and network lifetime.

# Chapter 6
# Heterogeneous and Mobile Connected $k$-Coverage in Two-Dimensional Deployment Fields

This chapter introduces our solution to the problem of connected $k$-coverage in heterogeneous two-dimensional wireless sensor networks, where the sensors do not necessarily have the same capabilities. Also, it discusses our solution to the same problem in mission-oriented mobile wireless sensor networks, which are deployed in two-dimensional fields that could have more than one monitoring task, i.e. mission, to be accomplished. Specifically, it focuses on how to fully $k$-cover a region of interest in a field using a minimum number of mobile sensors while minimizing the total energy consumption, which is due to the mobility of the sensors as well as their communication in order to successfully accomplish their specific mission.

## 6.1 Introduction

In real-world applications, wireless sensor networks are composed of *heterogeneous sensors* which do not necessarily have the same capabilities in terms of their sensing range, communication range, initial energy, computation, storage, etc. These networks have a potential to increase the network lifetime and reliability without causing significant increase in their cost [209]. For instance, Intel deployed two types of sensors, namely *line-powered* and *battery-powered* sensors, in the design of a pilot application of sensor nets in order to monitor the health of mechanical equipment in its fabrication plants [126]. While line-powered sensors can be attached to pumps and motors in the fabrication plant, battery-power sensors can be used to reduce installation cost and complexity. Indeed, Yarvis et al. [209] presented several analytical, simulation, and real testbed results showing the potential benefit and impact of energy and link heterogeneity on sensor nets, where all the sensors report their sensed data to a single sink.

In addition to sensor heterogeneity, sensor mobility is another interesting feature that enables the design of mission-oriented mobile wireless sensor networks. These types of networks can be viewed as distributed and dynamic systems, where the sensors are mobile, autonomous, and interacting with each other to accomplish a specific mission in a region of interest in a field. In this type of network, these distributed sensors should be continuously self-organizing, especially for connected coverage of a region of interest and sensed data gathering, so they can achieve the goals of the mission in a dynamic and collaborative manner while minimizing their energy consumption. Also, sensor mobility should be purposeful

H.M. Ammari: Challenges and Opportunities of Connected $k$-Covered WSNs, SCI 215, pp. 111–131.
springerlink.com 

and traded off against the goals of the mission. In particular, this mobility should be energy-aware given the scarce energy resources of the individual sensors.

In this chapter, we investigate the problem of connected $k$-coverage in heterogeneous wireless sensor networks [28], where the sensors that mainly differ by their sensing and communication ranges as well as their initial energy. Although the connected $k$-coverage problem in sensor nets has been studied extensively in the literature [37, 74, 96, 110, 184], most existing studies focused on *homogeneous sensor nets*, where all the sensors have the same capabilities with regards to their storage, computational power, sensing range, communication range, initial energy, etc. Similarly, the problem of $k$-coverage-preserving *scheduling* (or *sensor duty-cycling*) in homogeneous sensor nets has gained considerable attention [22, 196, 205, 207, 218, 228]. However, this type of sensors poses a severe restriction on the design of real-life sensing applications as all the sensors are required to have the same power with respect to all of their features, including sensing, communication, and energy.

Furthermore, we address the problem of mobile connected $k$-coverage in mission-oriented wireless sensor networks under the following requirements [27]:

- *On-demand k-coverage*: A region of interest in a field should be $k$-covered, where $k \geq 3$, whenever needed. Consequently, a region of interest to be $k$-covered does not have to be the same all the time, and hence may change.
- *Connectivity*: The sensors should be maintained connected for the correct network operation.
- *Mobility*: The sensors should be able to move to designated locations in a region of interest to ensure its $k$-coverage whenever necessary.

The remainder of this chapter is organized as follows. Section 6.2 solves the problem of connected $k$-coverage in heterogeneous two-dimensional wireless sensor networks. Section 6.3 solves the problem of mobile connected $k$-coverage in mission-oriented wireless sensor networks. Section 6.4 reviews related work. Section 6.5 summarizes the chapter.

## 6.2 Heterogeneous Connected $k$-Coverage

In real-world applications, sensor nets may have sensors with different capabilities, thus increasing the network reliability and lifetime [209]. In this section, we focus on heterogeneous $k$-covered sensor nets. More precisely, we exploit the results of Chap. 5 for generating energy-efficient connected $k$-coverage configurations using *heterogeneous sensors*. Using a geometric approach, we characterize $k$-coverage of a field with homogeneous sensor nets and derive a tight condition to ensure $k$-coverage. Then, we use this characterization for heterogeneous sensor nets. Our sensor net could have more than two types of sensors, which could be either randomly or pseudo-randomly deployed in a circular field. Also, we propose centralized (whenever possible) and distributed connected $k$-coverage protocols for heterogeneous sensor nets. To the best of our knowledge, this chapter provides the first analysis and design of centralized and distributed protocols for connected $k$-coverage in heterogeneous sensor nets. More importantly, our proposed framework is more general in the sense that it can be applied to both homogeneous and heterogeneous sensor nets.

## 6.2.1 Random Deployment Approach

In this section, we assume that all the sensors are *randomly deployed* in a circular field. Next, we discuss the feasibility of centralized and distributed connected $k$-coverage protocols.

### 6.2.1.1 Centralized Connected $k$-Coverage Protocol

The design of centralized protocols for $k$-coverage in heterogeneous sensor nets, where the sensors are randomly deployed, poses a major challenge. Indeed, a central gathering node, such as the *sink*, could be responsible for designating a subset of sensors to remain active in a given *scheduling round* (or simply *round*) to achieve $k$-coverage of a field. However, it is almost impossible to select a small number of sensors to $k$-cover a field. In fact, there is no unique solution to obtain a *slicing grid* of a field (Chap. 5, Fig. 5.4), i.e., *slice* a field into overlapping Reuleaux triangles, given that the sensors are heterogeneous. In particular, there are two extreme schemes for slicing a field:

*Scheme 1:* It accounts for the least powerful sensors in terms of their sensing range. Thus, a field would be sliced into overlapping Reuleaux triangles of width $r_{min}$, the smallest radius of the sensing ranges of the sensors. This solution, however, would lead to *over k-coverage* situation, where some regions in the field are denser with sensors than others, and hence more than $k$-covered, i.e., using more than enough sensors. This is highly likely to occur when most of the selected sensors have sensing ranges whose radii are greater than $r_{min}$.

*Scheme 2:* It accounts for the most powerful sensors, thus slicing a field into overlapping Reuleaux triangles of width $r_{max}$, the largest radius of the sensing ranges of the sensors. Unfortunately, this scheme would lead to *under k-coverage* situation, where some regions in the field are less than $k$-covered. Likewise, this is highly likely to arise when most of the selected sensors have sensing ranges whose radii are less than $r_{max}$.

Both problems in the design of *centralized* connected $k$-coverage protocols are due to *pure deployment randomness* and *sensor heterogeneity*. Thus, we focus on the design of a *distributed* protocol for heterogeneous sensors (R-Het-DCC$_k$).

### 6.2.1.2 Distributed Connected $k$-Coverage Protocol (R-Het-DCC$_k$)

**Overview:** Each sensor $s_i$ randomly slices its sensing range into six overlapping Reuleaux triangles of width equal to $r_i$, the radius of $s_i$'s sensing range. Based on the result of Theorem 5.2 in Chap. 5, a sensor $s_i$ randomly picks one of the

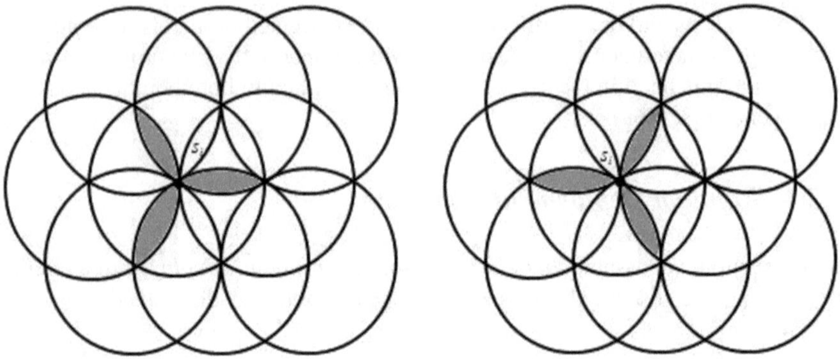


**Fig. 6.1** Three-lens flowers of $s_i$

*three-lens flowers* (as shown in Fig. 6.1) and checks whether its sensing range is $k$-covered. A sensor starts first by choosing the sensors located in the three lenses of the selected three-lens flower to remain active and $k$-cover its sensing range based on their remaining energy and the radii of their sensing ranges. Specifically, a sensor $s_i$ selects $k$ sensors from each of the lenses of the three-lens flower whose radii of their sensing ranges are at least equal to $r_i$, the radius of the sensing range of $s_i$. Then, it activates them by sending AWAKE messages. When a sensor receives an AWAKE message, it becomes active and broadcasts a NOTIFICATION message to inform all its neighbours. Every sensor keeps track of its active neighbours. Also, a sensor will turn itself active if its sensing range is not $k$-covered.

Next, we propose an efficient deployment strategy, where the sensors are *pseudo-randomly deployed* in the field, in order to exploit the *benefits* of heterogeneity.

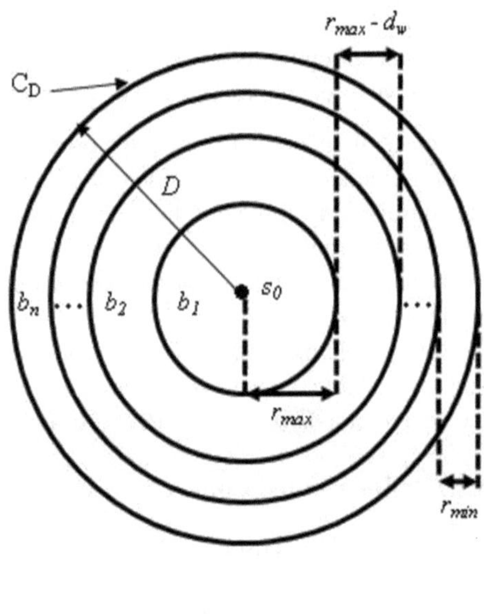


**Fig. 6.2** A field decomposed into circular bands

### 6.2.2 Pseudo-random Deployment Approach

In [17], we investigated the *energy sink-hole* problem in static sensor nets, where sensors located nearer a static sink are heavily used in forwarding data to it, thus suffering from severe energy depletion. To alleviate this problem, we considered a circular field and proposed a deployment approach using heterogeneous sensors with regard to their initial energy *only*. These sensors are assigned to the bands of the deployment field in such a way that they all deplete their initial energy at the same time. Each of these bands has a width equal to $R$, the radius of the communication range of the sensors, since we focused on data forwarding. Also, each band has homogeneous sensors. But any two bands in the field have heterogeneous sensors. Moreover, the difference between the amounts of energy in two adjacent bands should verify certain ratio so all the sensors in the network deplete their initial energy at the same time. A fine description of our multi-tier architecture can be found in [17].

In this chapter, we focus on the problem of connected $k$-coverage, and we further assume that the sensors have different radii of sensing and communication ranges. As demonstrated in [17], the most powerful sensors should be located nearer the sink. Thus, we slice a circular field ($C_D$) into $n$ bands of strictly decreasing widths starting from the inmost band. Assume that the minimum and maximum radii of the sensing range of the sensors are $r_{min}$ and $r_{max}$, respectively, and that the radius of the field is $D$. We also assume that the difference in width between any two adjacent bands is $d_w$. More specifically, the *inmost* band, denoted by $b_1$, has a width equal to $r_{max}$ while the *outmost* band, denoted by $b_n$, has a width equal to $r_{min}$ as shown in Fig. 6.2. It is important to design energy-efficient connected $k$-coverage protocols that help extend the network lifetime when sensed data routing is considered. That is, these protocols should be designed to be used later on when a network designer wishes to build a uniform framework that jointly considers sensor duty-cycling for connected $k$-coverage, and data forwarding on duty-cycled sensors. In fact, the ultimate goal of the design of sensor nets is to monitor an environment and generate sensed data to be forwarded to the sink for further analysis and processing. As stated in the beginning of this section, to alleviate the problem of the energy sink-hole problem in static sensor nets, it is necessary that the most powerful sensors should be deployed near the sink as they are heavily used in data forwarding to the sink.

The motivation behind decomposing a field into bands with such widths is the fundamental result stated by Xing et al. [205]. Precisely, they proved that a sensor net is surely connected when the radius of the communication range of the sensors is at least equal to double the radius of their sensing range provided that coverage is guaranteed. In our heterogeneous sensor deployment strategy, the width $w_j$ of a given band $b_j$ is computed such that the radius $R_j$ of the communication range of any sensor $s_j$ placed in $b_j$ should be at least equal to the sum of the width of $b_j$ and its adjacent predecessor band bj-1. That is, Rj ≥ wj + wj-1, where j ≥ 2, wj = rj and $w_{j-1} = r_{j-1}, = r_j + d_w$, with $r_{j-1}$ being the radius of the sensing range of any sensor $s_{j-1}$ located in band $b_{j-1}$ and $d_w$ is the difference in width between bands $b_{j-1}$ and $b_j$ as discussed earlier. We should recall that the width of the inmost band $b_1$ is $r_{max}$ and that of the outmost band $b_n$ is $r_{min}$. Figure 6.2 illustrates this discussion. This design ensures network connectivity to enable communication between the sensors and facilitate forwarding of the sensed data generated by the sensors towards the sink. This issue becomes clearer when a geographic forwarding protocol is built on top of our connected $k$-coverage protocols. Our sensor deployment approach is *hierarchical* due to the presence of multiple bands (or layers) forming the deployment field. Moreover, it is *pseudo-random* in the sense that the sensors are supposed to be deployed densely and randomly within each band while ensuring that the sensors in any pair of bands in the circular field are heterogeneous. Our sensor deployment approach is designed in a way that all the sensors placed in the band $b_j$ are homogeneous and hence have the same sensing range whose radius is equal to $r_{max} - (j - 1) \times d_w$, where $d_w > 0$ and $r_{max} > r_{min}$, such that the following system of equations is satisfied, where $n$ designates the number of bands forming the circular field, and hence the number of types of sensors in the heterogeneous sensor net:

$$r_{min} + (n-1) \times d_w = r_{max}$$
$$n \times r_{min} + n \times (n-1) \times d_w / 2 = D$$

with $n$ and $d_w$ being the only *unknowns* in the above equations. This is a valid assumption since we know only the capabilities of the sensors in terms of their sensing range, namely $r_{max}$ and $r_{min}$, and the radius of the circular field $D$. Therefore, knowing $r_{max}$, $r_{min}$, and $D$, it is easy to check that there is a unique solution $(n,d_w)$ to the above equations. For instance, assume that we have $r_{min} = 1$, $r_{max} = 3$, and $D = 16$, we get $n = 8$ and $d_w = 2/7$. In the following sections, we describe in detail both of our centralized and distributed connected $k$-coverage configuration protocols which exploit the characteristics of the sensors in their respective bands.

#### 6.2.2.1 Centralized Connected $k$-Coverage Protocol (PR-Het-CCC$_k$)

Generally, the sink is attached to an infinite source of energy, such as a wall outlet, and thus has no energy limitation. Thus, the single-point failure problem does not exist when the sink plays any particular role as the battery depletion problem for the sink cannot arise. In our centralized protocol, we assume that the sink is responsible for randomly decomposing each of these bands into overlapping Reuleaux triangles. This implies that the sink is aware of the locations of all the sensors in the network. Moreover, we assume that every sensor knows the *id* of the band it belongs to. Then, the sink applies the result of Theorem 5.2 in Chap. 5 to select a subset of sensors to $k$-cover each of these bands with a small number of sensors. The sensor will have to send a request to each band, where it specifies a band *id* and the *id*s of a subset of sensors located in that band so they remain active to $k$-cover the underlying band. We assume that each sensor is uniquely identified by an *id*, i.e., an integer.

The sink performs these actions at the beginning of each round under the assumption that all the sensors are awake at the beginning of a round and during some time interval $t_{awake}$. This time $t_{awake}$ should be large enough so that any sensor in the outmost band would be able to receive any message (or request) sent by the sink and destined to this band. This would help each of the sensors in the outmost band to check whether it has been designated by the sink to participate in the $k$-coverage of its band.

#### 6.2.2.2 Distributed Connected $k$-Coverage Protocol (PR-Het-DCC$_k$)

We reuse the same distributed connected $k$-coverage protocol described in Sect. 5.5 in Chap. 5. Recall that all the sensors located in the same band have the same capabilities, including their sensing range. Therefore, a sensor $s_j$ located in a band $b_j$ randomly decomposes its sensing range into six overlapping Reuleaux triangle of width $r_j = r_{max} - (j - 1) \times d_w$, the radius of $s_j$'s sensing range (i.e., the width of the band $b_j$). The only difference with the protocol proposed in Sect. 5.5 in Chap. 5 is that a sensor $s_j$ would select from its three-lens flower *only* the sensors that are located either in its band $b_j$ or in its adjacent band $b_{j-1}$ to $k$-cover its sensing range. Indeed, the sensors located in $b_{j-1}$ have higher sensing range than $s_j$ and thus will be able to participate in the $k$-coverage of $s_j$'s sensing range when they are selected. This means that a sensor $s_j$ would select sensors from its band $b_j$ which have the same power as $s_j$ (especially their sensing range) or more powerful than $s_j$ (i.e., from the band $b_{j-1}$) to ensure $k$-coverage of its sensing range.

### 6.2.3 *Performance Evaluation*

In this section, we present the simulation results of our protocols using a high-level simulator written in C. We consider a circular field of radius $D = 1000$ m. We use the energy model in [211], where the sensor energy consumption in transmission, reception, idle, and sleep modes are 60 mW, 12 mW, 12 mW, and 0.03 mW, respectively. Following [218], the energy required for a sensor to stay idle for 1 s is equivalent to *one unit of energy*. We assume that the initial energy of each sensor is 60 J enabling it to operate about 5000 s in reception/idle modes [211]. All simulations are repeated 100 times and the results are averaged.

First, we compare both of our distributed connected $k$-coverage protocols for *homogeneous sensor nets*, denoted by R-Hom-DCC$_k$, where the sensors are homogeneous and randomly deployed, and R-Het-DCC$_k$ (where the sensors are heterogeneous and randomly deployed). Indeed, R-Hom-DCC$_k$ is similar to R-Het-DCC$_k$ except that the sensors use the same sensing range $r$. For R-Hom-DCC$_k$, we assume that the sensing range of the sensors is $r = 25$ m. On the other hand, for R-Het-DCC$_k$, the sensing range of the sensors is between $r_{min} = 25$ m and $r_{max} = 50$ m. As can be seen from Fig. 6.3, R-Het-DCC$_k$ outperforms R-Hom-DCC$_k$ as it requires less number of sensors for any coverage degree $k$. Indeed, the presence of more powerful sensors helps ensure $k$-coverage of a field with less number of active sensors. We obtained the same result with regard to energy expenditure, i.e., R-Het-DCC$_k$ consumes less energy than R-Hom-DCC$_k$. Due to space limitation, we have not presented it here.

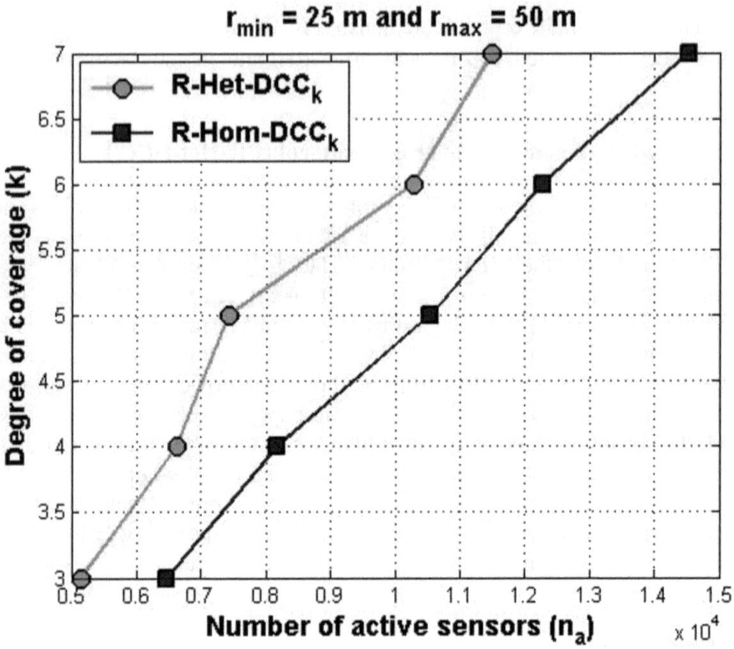


**Fig. 6.3** R-Het-DCC$_k$ vs. R-Hom-DCC$_k$

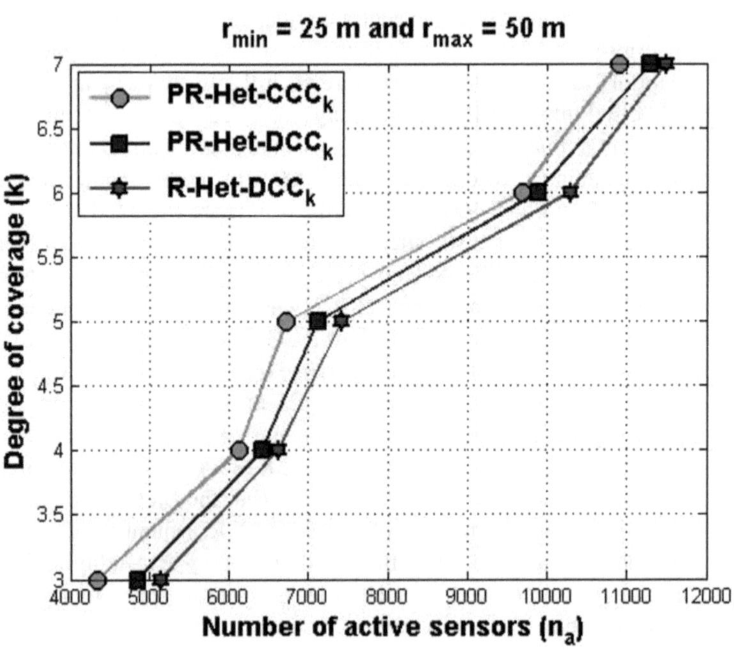


**Fig. 6.4** Comparison of PR-Het-CCC$_k$, PR-Het-DCC$_k$, and R-Het-DCC$_k$

Figure 6.4 shows that our centralized protocol for pseudo-random deployment of heterogeneous sensors (PR-Het-CCC$_k$) outperforms our distributed protocol for pseudo-random deployment of heterogeneous sensors (PR-Het-DCC$_k$). Also, PR-Het-DCCk presents better results than our distributed protocol for random deployment of heterogeneous sensors (R-Het-DCC$_k$). This shows the impact of slicing the field into concentric circular bands and deploying sensors in those bands based on their capabilities. Indeed, it is expected that our centralized protocol PR-Het-CCC$_k$ has the best performance compared to all other protocols, namely PR-Het-DCC$_k$ and R-Het-DCC$_k$. This is mainly due to the absence of coordination between adjacent sensors to achieve $k$-coverage of their sensing ranges. The sink only needs to keep track of the locations of all the sensors as well as their remaining energy. Although there is some energy consumed to forward all requests of the sink to their destination sensors, it is less than the energy needed for the coordination between the sensors to select a small number of sensors to $k$-cover the field.

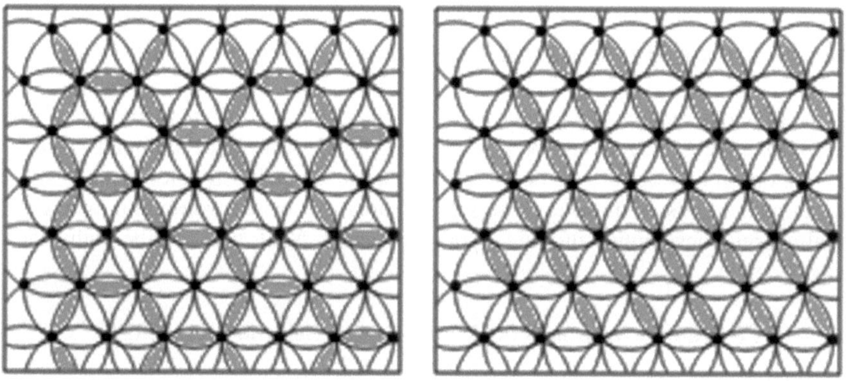

**Fig. 6.5** Decomposition of a square region of interest into Reuleaux triangles, where mobile sensors should be located in lenses to $k$-cover a region of interest

## 6.3 Mobile Connected $k$-Coverage

In this section, we focus on connected $k$-coverage in mission-oriented mobile wireless sensor networks, where a region of interest needs to be fully $k$-covered while maintaining network connectivity with the goal to minimize the total number of sensors necessary to achieve $k$-coverage and minimize the total energy consumption due to sensor mobility and communication between all the sensors in the network. Indeed, the closest work to ours is the one proposed by Wang and Tseng [196], and hence we will compare our protocols to theirs. First, we present our pseudo-random placement strategy of the mobile sensors to $k$-cover a region of interest.

### *6.3.1 Pseudo-random Sensor Placement*

In this type of wireless sensor network, there must be a node that is aware of the mission objectives of the network in the monitored field. We assume that the sink is aware of any region of interest in the field that needs to be $k$-covered and hence is responsible for computing the locations that should be occupied by the sensors in order to $k$-cover a region of interest. Moreover, it has been proved that the optimum location of the sink in terms of energy-efficient data gathering is the centre of the field [148]. Based on Theorem 5.2 in Chap. 5, the sink randomly decomposes a region of interest into overlapping Reuleaux triangles of width $r$ (or *slices*) such that two adjacent slices intersect in a region shaped as a *lens* as shown in Fig. 5.3 Chap. 5. Thus, each region of interest to be $k$-covered is sliced into regular triangles of side $r$ as shown in Fig. 6.5. The sink exploits the result of Theorem 5.2 in Chap. 5 so that a region of interest is $k$-covered by a small number of active sensors. Hence, it identifies the necessary lenses where the active sensors should be located and broadcast this information into the network. Precisely, each lens is uniquely identified by a pair of its extreme points. Figure 6.5a, b shows the lenses that should be occupied by mobile sensors to achieve energy-efficient $k$-coverage of the region of interest.

## 6.3.2 Sensor Mobility for k-Coverage of a Region of Interest

In the previous section, we showed how the sink computes the areas (or lenses) of a region of interest that should be occupied by mobile sensors to minimally $k$-cover the region. In this section, we describe two centralized and distributed protocols that decide which sensors move to these selected areas (or lenses) while minimizing the total energy consumption due to sensors movement. Next, we describe both protocols in details.

### 6.3.2.1 Centralized Approach for Mobile Sensor Selection (CAMSEL)

In addition to slicing the region of interest and determining its lenses where sensors should be placed, the sink selects the sensors that should move to occupy these lenses. To do so, the sink needs to be aware of the current locations of all the sensors. Also, the sink has to keep track of the remaining energy of each sensor in the network. Thus, the sink is required to maintain a database for all the sensors where each entry contains a sensor's *id*, its current location, and remaining energy. This information can be gathered by the sink under the assumption that the sensors move only when they are requested by the sink. Indeed, knowing the current positions of the sensors and their closest target locations in the lenses, the sink can compute the energy consumption required to reach these target locations and update their remaining energy accordingly based on the energy model [196, 211] that will be stated in Sect. 6.3.3. Also, the sink can estimate the energy consumed by each sensor while monitoring a region of interest based on the same energy model [196, 211].

Using this approach, the sink would be able to select a small number of sensors to fully $k$-cover a region of interest. Moreover, the sink would be able to minimize the total energy consumption of the sensors introduced by their mobility by choosing the sensors closer to the target region of interest to $k$-cover it. However, the network should guarantee that all the sensors selected by the sink to move to their target lenses receive the queries originated from the sink. For energy savings purposes, all queries are sent individually to all selected sensors. The sink will broadcast as many queries as the number of sensors necessary to $k$-cover a region of interest. Each query includes an *id* of a selected sensor and its target location $(x,y)_{target}$ in one of the lenses of a region of interest. Precisely, a query has the following structure: $query = <id,(x,y)_{target}>$. When a sensor receives a query, it will decide whether to forward the query towards the selected sensor based on the location of the sink and the target location $(x,y)_{target}$ in the query. Our goal is to reduce the amount of unnecessary data transmission that would flood the network to save the energy of individual sensors, thus extending the network lifetime.

#### 6.3.2.2 Distributed Approach for Mobile Sensor Selection (DAMSEL)

In the centralized approach, the sink is supposed to be aware of the status of *all* the sensors in the network with regard to their location and energy consumption. Although our centralized approach helps obtain the best schedule in terms of energy-efficient $k$-coverage and minimum energy consumption due to sensor movements, it would incur delay in the sensor selection phase especially for a large network that require a huge database for maintaining the current status of all the sensors in the network. In this section, we propose a distributed approach to remedy to this shortcoming.

In this approach, the sensors will cooperate with each other to move to the region of interest to $k$-cover it with a small number of sensors while minimizing the energy consumption introduced by their mobility. The sink will only specify the region of interest to be $k$-covered, which is supposed to be a square that is characterized by its centre $(x_0,y_0)$ and side length $a$. Thus, the sink will broadcast a *unique* query that has the following structure: $query = <(x_0,y_0),a>$. Moreover, we assume that all the sensors generate the same slicing grid of the region of interest. This means that all the sensors deterministically generate the same reference triangle whose centre coincides with that of the region of interest. This would enable all the sensors to have the same set of lenses, each of which should be occupied with at least $k$ sensors. Also, each sensor is supposed to be moving at a constant speed until it reaches its destination in the lens it selected in the region.

**State Transition Diagram of DAMSEL.** At any time, a sensor can be in one of the three states: WAITING, WILLING, MOVING, SENSING/COMMUNICATING, or SLEEPING. A *state transition diagram* associated with DAMSEL and indicating the five possible states of a sensor and transitions between them is shown in Fig. 6.6. Next, we describe each of these five states of a sensor using the protocol DAMSEL:

- *Waiting*: At the beginning of each round for mobile sensor selection, all the sensors are waiting for a query for $k$-coverage of a region of interest. Upon receipt of a $k$-coverage query from the sink, a sensor switches to the *Willing* state provided that its remaining energy would allow it to move. If a sensor does not receive any query within time $t_{waiting}$, it will switch to the *Sleeping* state.

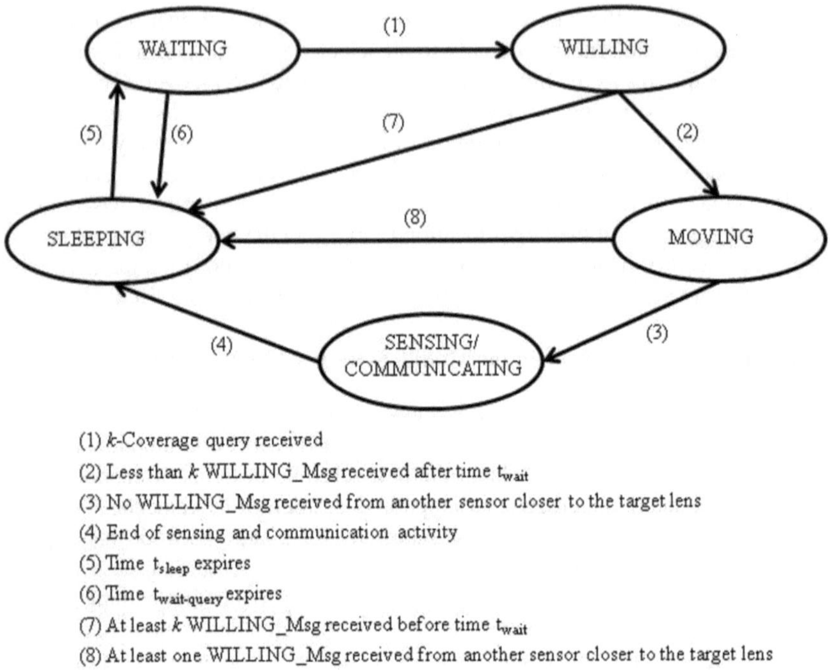


**Fig. 6.6** State diagram of DAMSEL

- *Willing*: When a sensor receives a query, it computes the slicing grid for the region of interest and chooses the closest lens to which it intends to move. Then, it broadcasts a data packet, called WILLING_Msg, including its *id*, current location $(x,y)_{current}$, and target location $(x,y)_{target}$ in the selected lens in the region of interest, thus expressing its willingness to move to this lens. Precisely, this message has the following structure: WILLING_Msg = $<id,(x,y)_{current},(x,y)_{target}>$. While in the Willing state, if a sensor $s_i$ receives at least $k$ WILLING _Msg originated from $k$ distinct sensors that are closer to the target lens than $s_i$, the sensor $s_i$ will simply discard its request and selects another target lens. If after some time $t_{willing}$ a sensor finds itself unable to move to any of the first three closest lenses, it will simply give up and automatically switch to the *Sleeping* state.
- *Moving*: If after some waiting time $t_{wait}$ the sensor that broadcasted a WILLING_Msg does not receive at least $k$ MOVING_Msg originated from $k$ distinct sensors that would surely move to the underlying lens, it will decide to move to its selected lens and broadcast a MOVING_Msg, thus expressing its final decision to move to this lens. While moving, if a sensor $s_i$ receives a WILLING_Msg from another sensor $s_j$ that is closer to the target lens than $s_i$, the sensor $s_i$ will simply give up and let $s_j$ move to this target lens. Then, the sensor $s_i$ switches to the *Sleeping* state. This would guarantee that each lens is $k$-covered by the closest $k$ sensors, thus avoiding wasting more energy due to sensor mobility.

- *Sensing/Communicating*: When a mobile sensor arrives at its target location in the selected lens, it will switch to the *Sensing/Communicating* state, where it would start sensing and communicating its sensed data to the sink. At the end of its sensing and communication activity in a region of interest, a sensor will move to the *Available* state for a new mobile sensor selection round.
- *Sleeping*: In this state, a sensor is neither communicating with any other sensor nor sensing a field, and thus its radio is turned off. However, after some time $t_{sleeping}$, it switches to the *Waiting* state in order to receive query messages.

#### 6.3.2.3 How to Ensure Network Connectivity?

In this type of mission-oriented wireless sensor network, sensor mobility is necessary so any region of interest could be $k$-covered with a small number of sensors, where the degree of coverage $k$ depends on the mission to be accomplished. However, this mobility may not guarantee that the network remains connected all the times. Although the sensors coordinate between themselves to $k$-cover a region of interest, there is no coordination between them with regard to their locations so they remain connected during the network operation. To alleviate this problem, the sensors select some of them, called *dMULEs*, based on their remaining energy to act as data MULEs [176] to keep the network connected. The dMULEs will transport messages on behalf of others and disseminate them to the other sensors in the network, thus enabling continuous communications between all the sensors including the sink. These messages could be either originated from the sink as queries to $k$-cover a region of interest or initiated from the sensors to coordinate their activity to achieve $k$-coverage of a region of interest. These dMULEs will be used efficiently in the centralized approach for mobile sensor selection to gather information about the current locations of the sensors and inform the sink accordingly so it selects the closest ones to the region of interest to be $k$-covered. This helps minimize the necessary energy consumption for $k$-coverage due to sensor mobility.

### 6.3.3 *Performance Evaluation*

In this section, we present the simulation results of CAMSEL and DAMSEL using a high-level simulator written in the C language. We use the same energy model given in Sect. 6.2.3. For fair comparison with Wang and Tseng's approach [196], we consider a square field of side length 600 m and a square region of interest of side length 300 m. Moreover, the *energy spent in moving* is given by $E_{mv}(d) = e^{move} \times d$, where $e^{move}$ is the energy cost for a sensor to move one unit distance, and $d$ is the total distance traveled by a sensor [196]. As stated in [196], $e^{move}$ is randomly selected in [0.8 J, 1.2 J] and the moving speed of each sensor is 1 m/s. Also, the sensing range of the sensors is equal to 20 m. All the simulations are repeated 100 times and the results are averaged.

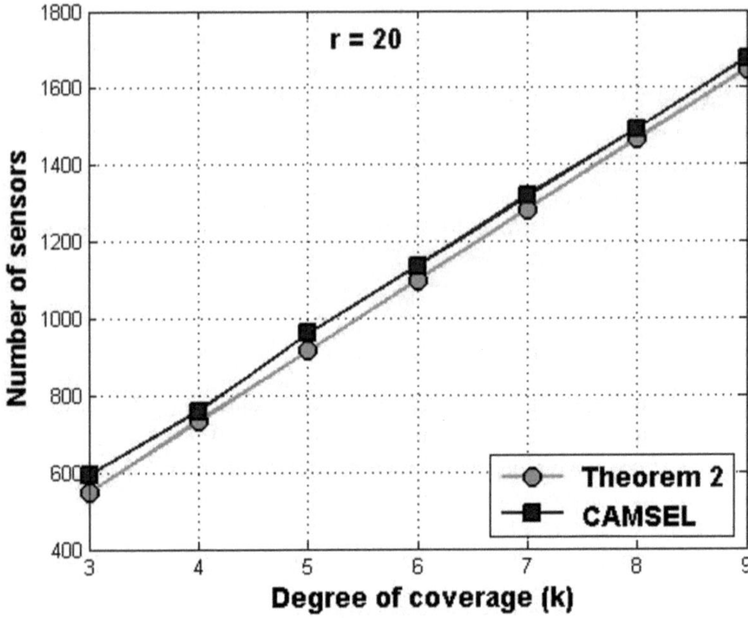


**Fig. 6.7** CAMSEL compared to the result of Theorem 5.3 in Chap. 5

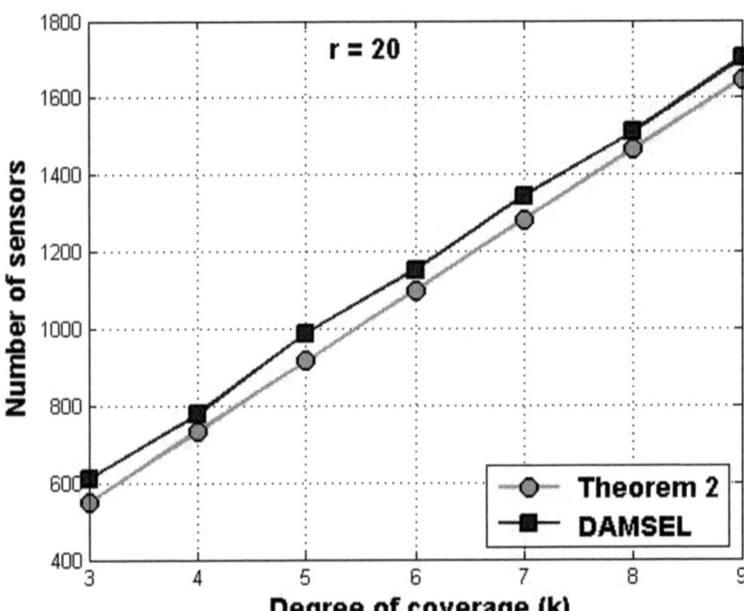


**Fig. 6.8** DAMSEL compared to the result of Theorem 5.3 in Chap. 5

Figures 6.7 and 6.8 plot the number of sensors needed to $k$-cover a region of interest, where $r = 20$ m, for both CAMSEL and DAMSEL protocols. As can be seen from Figs. 6.7 and 6.8, CAMSEL and DAMSEL require a number of sensors for $k$-coverage that is close to its corresponding theoretical value computed in Theorem 5.3 in Chap. 5. Thus, both protocols perform well compared to the theoretical result given in Theorem 5.3.

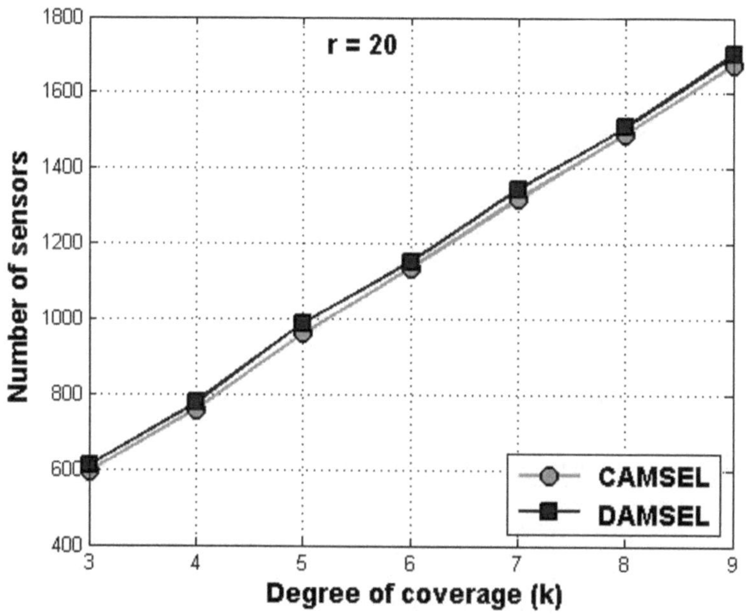


**Fig. 6.9** CAMSEL compared to DAMSEL (Number of sensors)

However, CAMSEL slightly outperforms DAMSEL as shown in Fig. 6.9. Indeed, with CAMSEL, the sink has a global view of the network and hence computes a small number of sensors to $k$-cover a region of interest, while DAMSEL helps the sensors make their decision to participate in $k$-coverage based on their local knowledge of their neighbours, which varies due to sensor mobility. Also, DAMSEL requires higher energy consumption than CAMSEL (Fig. 6.10). With DAMSEL, more sensors are willing to move to ensure $k$-coverage of a region of interest. Due to the distributed nature of DAMSEL, a moving sensor will not always be able to know that other sensors, which are not currently located in its neighborhood, decided to participate in $k$-coverage of the underlying region of interest.

Figures 6.11 and 6.12 show that our protocol DAMSEL outperforms the protocol Competition [196] with respect to the number of sensors required for $k$-coverage as well as the corresponding total moving energy. Indeed, our approach provides a fine analysis for ensuring $k$-coverage of a region of interest with a small number of sensors based on Helly's Theorem and the geometric properties of the Reuleaux

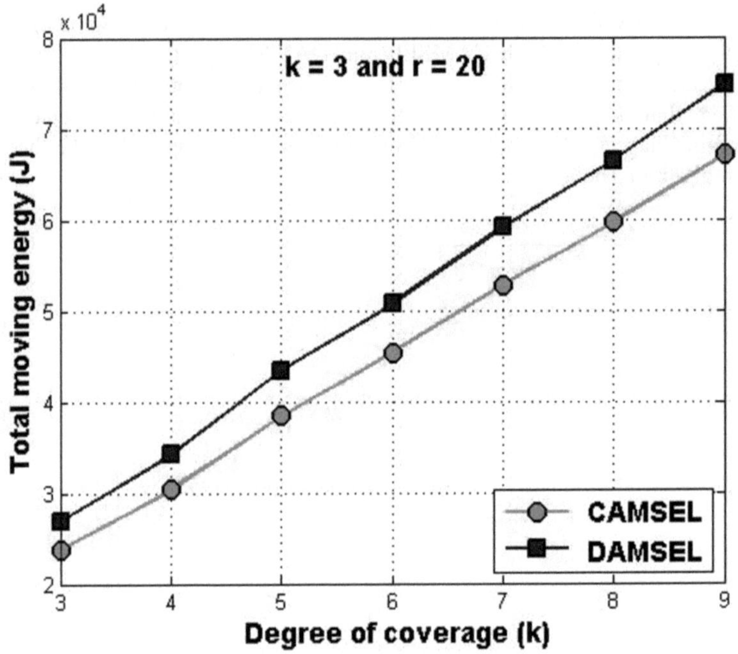


**Fig. 6.10** CAMSEL compared to DAMSEL (total remaining energy)

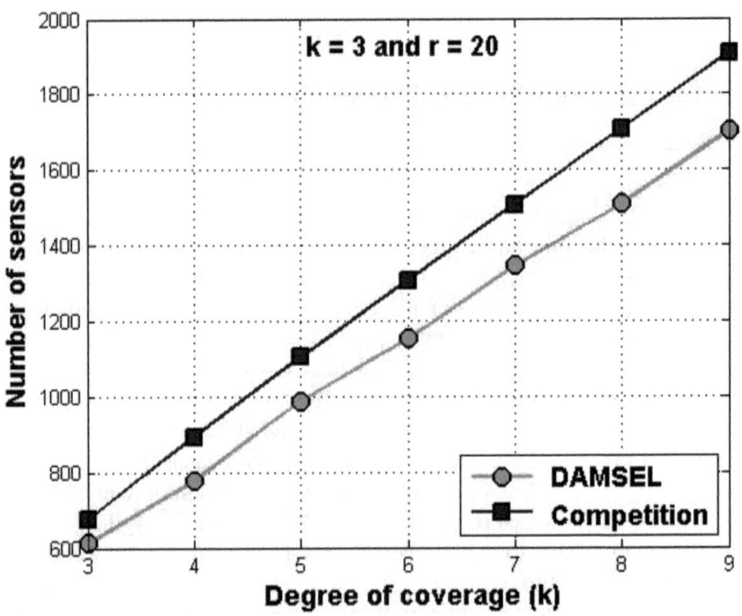


**Fig. 6.11** DAMSEL compared to Competition (number of sensors)

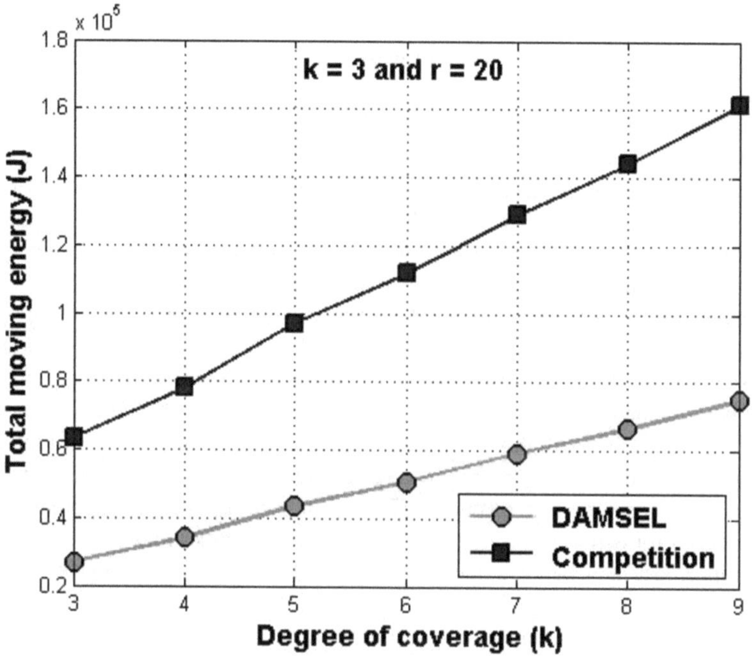


**Fig. 6.12** DAMSEL compared to Competition (total remaining energy)

triangle. In addition, the continuous collaboration between the sensors enables only a necessary number of sensors to move towards a region of interest to $k$-cover it.

## 6.4 Related Work

In this section, we describe approaches and protocols that have been proposed for heterogeneous as well as mobile wireless sensor networks. A special focus is on those on coverage and connectivity.

### *6.4.1 Sensor Heterogeneity*

While coverage and connectivity in homogeneous sensor nets has been studied well, heterogeneous sensor nets has received little attention. Wang et al. [194] proposed a fine analysis of coverage using two types of sensors with different capabilities and discussed the impact of heterogeneous sensing and communication ranges of the sensors on coverage and broadcast reachability. Duarte-Melo and Liu [78] focused on heterogeneous sensors equipped with different battery power in their analysis of a clustering approach. Precisely, they considered two sensor deployment strategies, where the first one includes type-I sensors, while the second one includes an overlay of type-II sensors, which are more powerful but fewer in number. Then, they estimated the average network lifetime and quantified the optimal number of these type-II sensors, which act as cluster-heads. Moreover, they showed how to allocate the energy between the overlay sensors and normal sensors. Lazos and Poovendran [131, 132] discussed the coverage problem in heterogeneous sensor networks. They formulated the coverage problem as a set intersection problem and derived analytical expressions, which quantify the coverage achieved by stochastic coverage. Moreover, Lazos et al. [133] addressed the problem of detecting mobile targets using sensors that have heterogeneous sensing areas of arbitrary shape with a goal to increase the robustness of target detection. Specifically, they considered a deterministic sensor deployment strategy to maximize the probability of target detection. Ma et al. [150] proposed a resource-oriented protocol that implements a network topology according to the number of heterogeneous sensors as well as their specific resources and characteristics, such as radio coverage, power capacity, processing capabilities, and mobility attributes. Du and Lin [74] proposed a differentiated coverage algorithm for heterogeneous sensor nets. The motivation behind providing different degrees of sensing coverage for different regions is that different network areas do not have the same importance and hence some areas require higher coverage degree than others. Yuce et al. [214] proposed a heterogeneous sensor network system for monitoring physiological parameters from multiple patient bodies. Their system provides patients with mobility, and medical staff with flexibility in obtaining physiological data on the patients on-demand through the Internet. Mhatre et al. [159] considered a heterogeneous sensor net with two types of nodes which differ by their intensity (or average number per unit area) $(\lambda_0,\lambda_1)$ and their battery energy $(E_0,E_1)$. While type 1 nodes perform sensing, type 2 nodes perform sensing and act as cluster heads. Their study consists of computing the optimum node intensities $(\lambda_0,\lambda_1)$ and node energies $(E_0,E_1)$ that ensure connected coverage of the surveillance area with a high probability while guaranteeing a network lifetime of at least $T$ units and minimizing the overall cost of the network.

## *6.4.2 Sensor Mobility*

Recently, significant efforts have been focused on studying coverage using mobile wireless sensor networks. Sensor mobility has been recognized as an efficient way to guarantee better sensor deployment. Indeed, Liu et al. [144] proved that mobility of the sensors can be used to improve coverage. Their study showed that dynamic aspects of the coverage using mobile sensors depend on the process of sensor movement and that new stationary configurations that help improve coverage can be obtained once the sensors move to their desired locations. They characterized area coverage as a function of time and the detection time of a randomly located target. Wang et al. [193] proposed solutions to two related deployment problems in wireless sensor networks, namely sensor placement and sensor dispatch, in the presence of obstacles. For the first problem, they proposed a solution that considers arbitrary-shaped obstacles as well as arbitrary relationship between the communication and sensing ranges of the sensors. For the second problem, they proposed centralized and distributed approaches. While the centralized one exploits the results for sensor placement and converts the dispatch problem to the maximum-weight maximum-matching problem whose objective function is minimizing the total energy consumption due to mobility or maximizing the average remaining energy of the sensors after movement, the distributed solution enables the sensors to compute their moving directions autonomously. Furthermore, Wang et al. [196] generalized their solutions to the sensor selection and dispatch problems [193] by considering multi-level coverage without obstacles. Yang and Cardei [206] dealt with the Movement-assisted Sensor Positioning (MSP) problem with a goal to increase the network lifetime. First, they proposed a solution to address the energy-hole problem caused around the sink by computing the desired non-uniform sensor density in the monitored area. Then, they proposed a centralized algorithm to relocate mobile sensors while satisfying the density requirement with minimum cost. Wu and Yang [203] proposed a method called Scan-based Movement-Assisted sensoR deploymenT (SMART) to achieve a balance state by balancing the workload of the sensors while avoiding the communication-hole problem in wireless sensor networks. SMART uses clustering, where a rectangular deployment field is partitioned into 2D mesh, which also partitioned into 1D arrays by rows and columns, and each square area is assigned a cluster-head. These rows and columns are scanned to determine the overload and underload in clusters so the load is shifted from overloaded clusters to underloaded clusters in order to achieve a balance state. Wu and Yang [202] took a step further and proposed an optimal but centralized approach based on the Hungarian method to minimize the total moving distance. They also proposed non-optimal distributed solutions for the same purpose based on the scan-based approach in [203]. Rao and Kesidis [171] investigated mobility in mission-oriented wireless sensor networks, where a sensor moves to a location so it can perform any one or all the tasks better, and hence the notion of purposeful mobility. With this type of controlled mobility, Cao et al. [53] proposed techniques for mobility-assisted sensing and routing while considering the computation complexity, network connectivity, the energy consumption due to both mobility and communication, and the network lifetime. Wang et al. [192] addressed the problem of how to meet sensing coverage requirements using mobile sensors. They proposed a Grid-Quorum solution that locate the closest redundant sensors with low message complexity and relocate them in a timely, efficient and balanced way using cascaded movement. Du and Lin [76] proposed an approach to improve the performance of wireless sensor networks in terms of coverage, connectivity, and routing by introducing a few mobile sensors in addition to the static ones, which constitute the majority of the sensors in the network. Moreover, they proposed several schemes for effective sensor mobility that helps achieve the above-mentioned goals. Wang et al. [190] proposed a proxy-based approach that allows the sensors to move directly to their target locations and not a zig-zag way with a goal to provide satisfactory coverage. In harsh environments, such as battlefields, sensor mobility helps ensure the required coverage, where mobile sensors can reach areas that cannot be reached by static sensors. Wang et al. [191] proposed a Voronoi diagram-based approach to detect coverage holes and three approaches, namely VEC (VECtor-based), VOR (VORnoi-based) and Minmax, each of which enables the sensors to move from densely deployed areas to sparsely deployed areas using less time, movement distance, and message complexity. Heo and Varshney [104] proposed distributed, energy-efficient deployment algorithms that employ a synergetic combination of cluster-structuring and a peer-to-peer deployment scheme. Also, they proposed an energy-efficient Voronoi diagram-based deployment algorithm. A comprehensive survey on node placement in wireless sensor networks can be found in [213].

## 6.5 Summary

In this chapter, we studied the problem of energy-efficient connected $k$-coverage in mission-oriented mobile wireless sensor networks [27]. First, we exploited the characterization of $k$-coverage of a region of interest based on the Helly's Theorem and the geometric properties of the Reuleaux triangle, which is introduced in Chap. 5. Then, we also proposed centralized and distributed approaches for achieving $k$-coverage of a region of interest using mobile sensors. In the centralized approach, the sink is responsible for slicing a region of interest into slices and designating a set of sensors to be moving to the selected lenses of that region in order to $k$-cover it. In the distributed approach, the sink sends only a query including the coordinates of a region of interest to be $k$-covered and its degree of coverage $k$ while the sensors coordinate between themselves to ensure energy-efficient $k$-coverage of the region. However, to maintain network connectivity, we used a few sensors as data MULEs. Simulation results that CAMSEL outperforms DAMSEL, which in turn outperforms Competition [196] in terms of the number of sensors and the total moving energy required to $k$-cover a region of interest.

We exploited the results of the homogeneous model, which are discussed in Chap. 5, to solve the connected $k$-coverage problem for heterogeneous sensor nets, where the sensors do not have the same sensing range, communication range, and initial energy [28]. Precisely, we proposed centralized and distributed protocols for generating energy-efficient connected $k$-coverage configurations in heterogeneous sensor nets. First, we considered a random deployment approach, where the sensors are distributed in a random fashion in the field. We showed that while it is possible to design distributed protocols to guarantee connected $k$-coverage of a field using heterogeneous sensors, it is impossible that a centralized protocol could be designed efficiently due to deployment randomness and sensor heterogeneity. Thus, we suggested a pseudo-random deployment approach, where the sensors are deployed in different layers in a circular deployment field with respect to the sink according to their strengths with regard to sensing range, communication range, and initial energy, and proposed energy-efficient centralized and distributed connected $k$-coverage protocols. Through simulations, we showed that our proposed approach for connected $k$-coverage in homogeneous sensor nets outperforms an existing one in terms of the number of sensors necessary to fully $k$-cover a field, and network lifetime. Moreover, we found that our pseudo-random sensor deployment approach outperforms our random deployment approach with respect to the above-mentioned metrics. Indeed, our multi-tier sensor deployment architecture is energy efficient and can be exploited as a solid basis for the design of a geographic forwarding protocol to solve the *energy sink-hole* problem in static sensor nets.

# Chapter 7
# Two-Dimensional Stochastic Connected $k$–Coverage and Three-Dimensional Connected $k$–Coverage

This chapter addresses the problem of stochastic connected $k$-coverage in two-dimensional wireless sensor networks using a more realistic, probabilistic sensing model instead of a deterministic sensing model. Moreover, it investigates the problem of connected $k$-coverage in homogeneous three-dimensional wireless sensor networks while considering a deterministic sensing model. Both studies are generalizations of the work described earlier in Chap. 5. Our goal is to enhance the practicality of our previous work and extend its scope of applicability.

## 7.1 Introduction

The problem of coverage (and in particular $k$-coverage) has been well studied in the literature [37, 127, 128, 139]. Also, the problem of coverage-preserving scheduling (or duty-cycling) has gained considerable attention [96, 110, 205, 218]. Several existing works on $k$-coverage in wireless sensor networks assumed a perfect sensing model (also known as *deterministic sensing model*), where a point in a field is guaranteed to be covered by a sensor provided that this point is within the sensor's sensing range [218]. While some approaches focused on coverage only [2], others considered both coverage and connectivity in an integrated framework to ensure the correct operation of the network [205, 218]. Indeed, coverage deals with all locations in a field, and hence informs how well a phenomenon in the field is monitored; whereas, connectivity is related to the locations of the sensors, and hence quantifies how well the active sensors communicate with each other and forward sensed data on behalf of each other to the sink. We addressed this problem in Chap. 5, where centralized, pseudo-distributed, and distributed protocols are proposed. A few works, however, considered a more realistic sensing model (also known as *stochastic sensing model*) in the design of sensor scheduling protocols while preserving either full coverage [123, 146, 157, 218] or $k$-coverage of a field [196, 205, 218], where a point is covered by a sensor with some probability.

As discussed in Chap. 4, three-dimensional settings reflect more accurately network design for real-world applications than their more traditional, two-dimensional counterparts. As of writing this dissertation, the assumption of two-dimensional space is well accepted by the sensor network community while that of three-dimensional space has been receiving little attention due to the challenges

H.M. Ammari: Challenges and Opportunities of Connected $k$-Covered WSNs, SCI 215, pp. 133–156.
springerlink.com

imposed by the design of three-dimensional wireless sensor networks [5]. Indeed, most (if not all) of the works on the design of protocols for wireless sensor networks, and particularly, those on deployment, have focused on two-dimensional space, where the sensors are deployed in a planar field. As mentioned by Poduri et al. [167], there is a tendency of ignoring the extension of protocols initially designed for two-dimensional to three-dimensional wireless sensor networks either because it is simple or straightforward. Furthermore, Poduri et al. [167] showed that there are a few properties in two-dimensional wireless sensor networks that cannot generalize at all to three-dimensional wireless sensor networks. In general, the design of connected *k*-coverage configuration protocols for wireless sensor networks in three-dimensional space is more challenging than their counterparts in two-dimensional space.

In this chapter, we consider two fundamental problems in two-dimensional and three-dimensional wireless sensor networks, respectively. First, we focus on the design of stochastic connected *k*-coverage configuration protocols for two-dimensional wireless sensor networks [25]. More specifically, we decompose this problem into two sub-problems: *stochastic k-coverage characterization problem* and *stochastic k-coverage-preserving scheduling problem*. Specifically, the first problem is to find a sufficient condition so that every point in the field is covered by at least $k$ sensors with a probability no less than $p_{th}$, called *threshold probability*, under our stochastic sensing model (see Chap. 2) and compute the corresponding minimum number of sensors. Second, we address the problem of connected *k*-coverage in three-dimensional wireless sensor networks using a deterministic sensing model of the sensors, and focus on the problem of sensor scheduling for *k*-coverage of a three-dimensional space [21], where $k > 1$. Moreover, we focus on the problem of data forwarding on duty-cycled sensors in three-dimensional connected *k*-covered wireless sensor networks. Not surprisingly, we show that the extension of our analysis of two-dimensional wireless sensor networks is really not straightforward for three-dimensional wireless sensor networks due to the non-preserving nature of some of the properties for two-dimensional space when we consider three-dimensional space.

The remainder of this chapter is organized as follows. Section 7.2 discusses the problem of stochastic connected *k*-coverage in two-dimensional wireless sensor networks under a more realistic stochastic sensing model, which is described in Chap. 2. It also presents simulation results of our proposed protocol. Section 7.3 presents a solution to the problem of connected *k*-coverage in three-dimensional wireless sensor networks and presents simulation results of our proposed protocols. Section 7.4 reviews related work. Section 7.5 summarizes the chapter.

## 7.2 Two-Dimensional Stochastic Connected *k*-Coverage

Although the approach on *k*-coverage in wireless sensor networks proposed in [205] is elegant and considers both deterministic and probabilistic sensing models, it does not provide any proof on whether its *k*-coverage eligibility algorithm would yield a minimum number of selected sensors to *k*-cover a field.

In this section, we propose a solution to the stochastic connected $k$-coverage problem stated earlier using our *stochastic sensing model*, which reflects the real behaviour of the sensing units of the sensors that are irregular in nature.

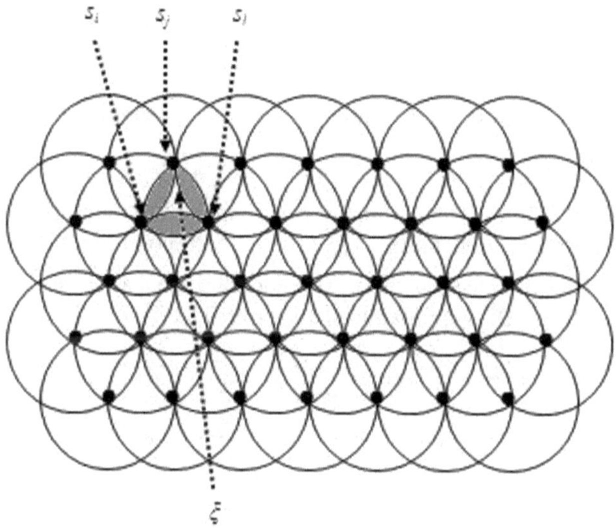


**Fig. 7.1** Reuleaux triangle

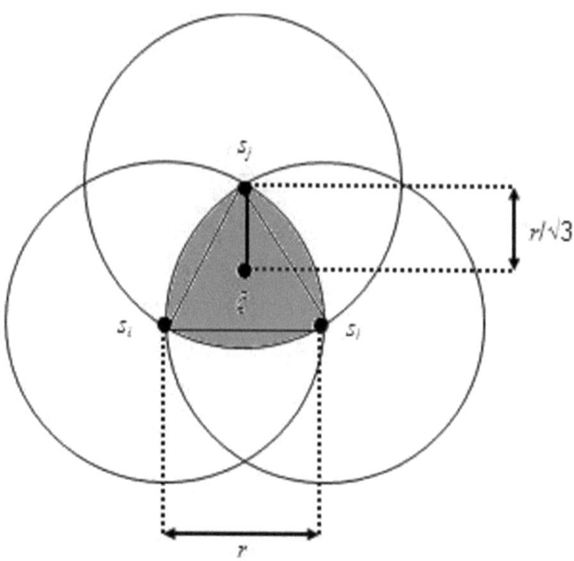


**Fig. 7.2** Location of a least $k$-covered point

### 7.2.1 Stochastic k-Coverage Characterization

In this section, we exploit the results of Sect. 5.2 of Chap. 5 to characterize $k$-coverage in wireless sensor networks based on the stochastic sensing model described earlier. It is also similar to the stochastic sensing model in [228], except that ours accounts for the type of propagation model, i.e., $2 < \alpha \leq 4$. Precisely, we use the Reuleaux triangle model to compute the *minimum k-coverage probability*, denoted by $p_{k,\min}$, such that every point in a field is $k$-covered. Theorem 7.1 computes $p_{k,\min}$.

*Theorem 7.1:* Let $r$ be the radius of the nominal sensing range of the sensors and $k \geq 3$. The minimum $k$-coverage probability $p_{k,\min}$ so that each point in a field is probabilistically $k$-covered by at least $k$ sensors under the stochastic sensing model defined in Eq. 2.2 (Chap. 2) is approximately computed as

$$p_{k,\min} \approx 1 - \left( 1 - e^{-\beta \left( \frac{r}{\sqrt{3}} \right)^{\alpha}} \right)^{k} \tag{7.1}$$

*Proof:* First, we identify the *least* $k$-covered point in a field so we can compute $p_{k,\min}$. By looking at the Reuleaux triangle corresponding to three sensors and given in Fig. 7.1, it is easy to check that the centre $\xi$ of the Reuleaux triangle is the least 3-covered point as shown in Fig. 7.2. Indeed, $\xi$ is located close to the boundaries of the sensing ranges of the sensors $s_i$, $s_j$, and $s_l$. By Lemma 5.2 (see Chap. 5), $k$ sensors should be deployed in each Reuleaux triangle regions of a field to achieve $k$-coverage with a minimum number of sensors. Thus, on the average, we can claim that the centre $\xi$ is also the least $k$-covered point in a field. Note that $\xi$ is equidistant from the sensors $s_i$, $s_j$, and $s_l$. Using the configuration in Fig. 7.2, a little algebra shows that the distance between $\xi$ and each of these three sensors is equal to $r/\sqrt{3}$. Let $S_k = \{s_1,\ldots,s_k\} \subseteq S$ be a set of sensors that $k$-cover $\xi$. As per the above observation, we can approximate the distance between any sensor in $S_k$ and $\xi$ by $r/\sqrt{3}$, i.e., $\delta(s_i,\xi) \approx r/\sqrt{3}$, for each sensor $s_i \in S_k$. Therefore, the minimum $k$-coverage probability (or detection probability) for the least $k$-covered point $\xi$ by exactly $k$ sensors $(s_1,\ldots,s_k)$ under our stochastic sensing model is given by

$$p_{k,\min} = 1 - \prod_{i=1}^{k} (1 - p(\xi, s_i)) \approx 1 - \left( 1 - e^{-\beta \left( \frac{r}{\sqrt{3}} \right)^{\alpha}} \right)^{k} \qquad \blacksquare$$

The stochastic $k$-coverage problem is to select a minimum subset $S_{\min} \subseteq S$ of sensors such that each point in a field is $k$-covered by at least $k$ sensors and that the minimum $k$-coverage probability of each point is at least equal to some given threshold probability $p_{th}$, where $0 < p_{th} < 1$. This helps us compute the sensing range $r_s$, which provide full $k$-coverage of a field with a probability no less than $p_{th}$:

$$p_{k,\min} \geq p_{th} \Rightarrow r_s \leq \sqrt{3}\left(-\frac{1}{\beta}\log\left(1-(1-p_{th})^{1/k}\right)\right)^{1/\alpha}$$

Lemma 7.1 computes the value of the *stochastic sensing range* of the sensors.

*Lemma 7.1:* Let $k \geq 3$. The *stochastic sensing range* $r_s$ of the sensors that is necessary to fully $k$-cover a field with a minimum number of sensors and with a probability not lower than $0 < p_{th} < 1$ is given by

$$r_s = \sqrt{3}\left(-\frac{1}{\beta}\log\left(1-(1-p_{th})^{1/k}\right)\right)^{1/\alpha} \tag{7.2}$$

where $\beta$ is the physical characteristic of the sensors' sensing units and $2 \leq \alpha \leq 4$ is the path-loss exponent, which depends on the propagation model (free-space model vs. multi-path model). ■

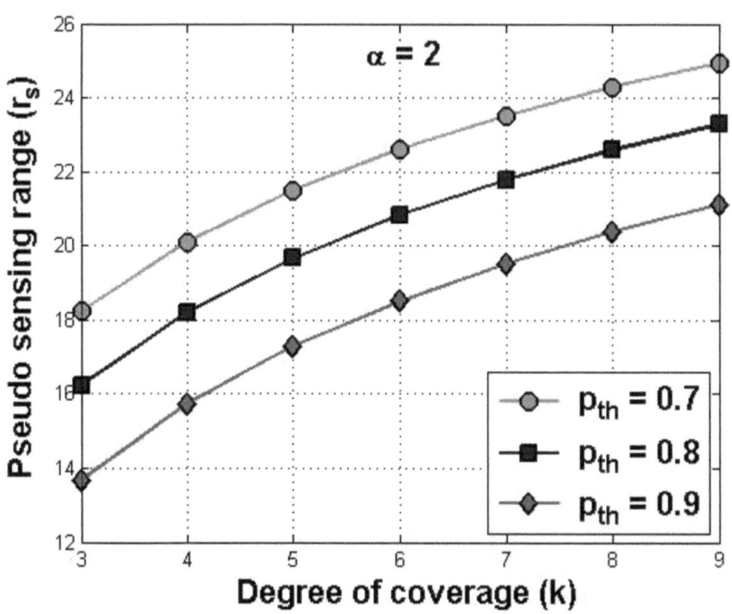


**Fig. 7.3** Upper bound of $r_s$ vs. $k$ for $\alpha = 2$

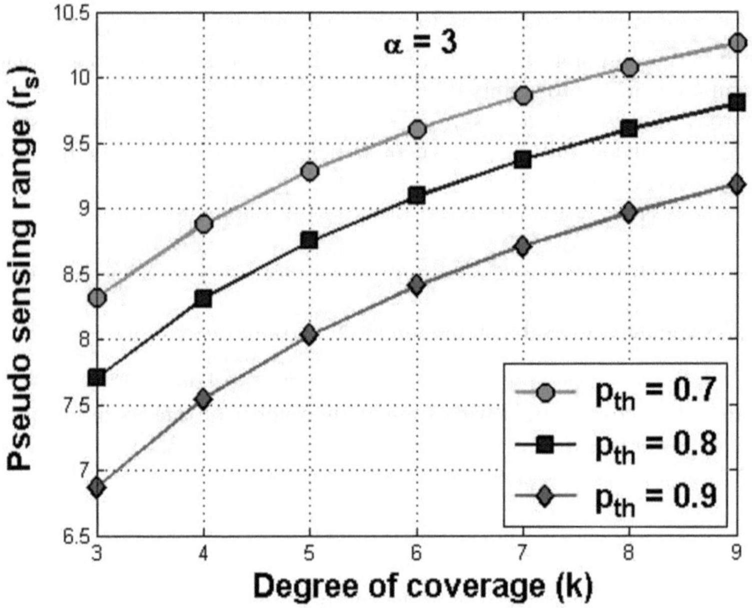


**Fig. 7.4** Upper bound of $r_s$ vs. $k$ for $\alpha = 3$

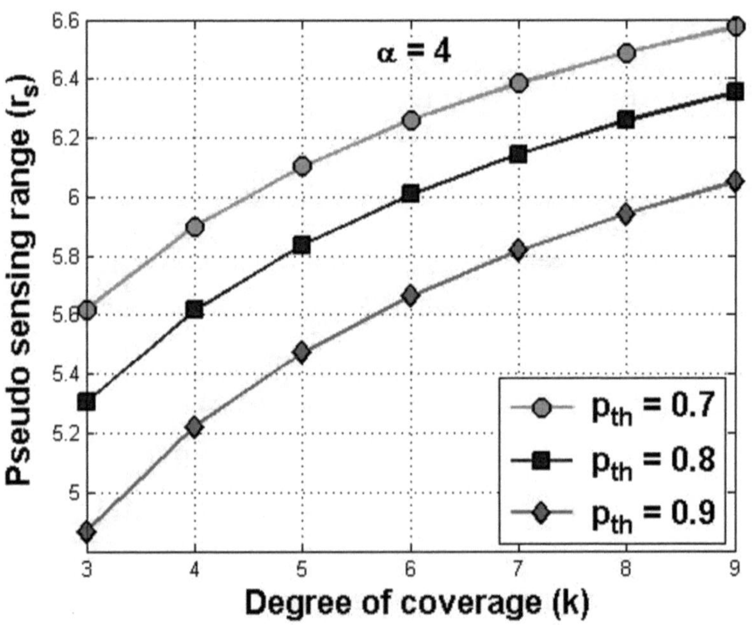


**Fig. 7.5** Upper bound of $r_s$ vs. $k$ for $\alpha = 4$

The upper bound on the stochastic sensing range $r_s$ of the sensors computed in Eq. 7.2 will be used as one of the input parameters to the *k-coverage candidacy* algorithm, which will be presented in Sect. 7.2.2. Figures 7.3–7.5 show $r_s$ for different values of $p_{th}$ and $k$ while considering the free-space model $(\alpha = 2)$ (Fig. 7.3) and the multi-path model $(\alpha = 3,4)$ (Figs. 7.4–7.5). As can be seen, $r_s$ decreases as $p_{th}$ and $\alpha$ increase; however, it increases with $k$. Indeed, to achieve higher degree of coverage, the stochastic sensing range of the sensors should increase.

Lemma 7.2 states a sufficient condition for full $k$-coverage of a field while Lemma 7.3 states a sufficient condition to guarantee connectivity between sensors under our stochastic sensing model when both $p_{th}$ and $k$ are known.

*Lemma 7.2:* Let $k \geq 3$. A field is probabilistically fully $k$-covered with a probability no lower than $0 < p_{th} < 1$ if any Reuleaux triangle region of width $r_s$ in the field contains at least $k$ sensors. ■

*Lemma 7.3:* Let $k \geq 3$. The sensors that are selected to $k$-cover a field with a probability no less than $0 < p_{th} < 1$ under our stochastic sensing model defined in Chap. 2 are guaranteed to be connected if the radius of their communication range is at least equal to $r_s$, the sensors' stochastic sensing range. ■

```
ALGORITHM 1: k-COVERAGE-CANDIDACY(r_s,k)
(* This code is run by each sensor *)
Begin
/* Sensing range slicing */
1. Randomly decompose sensing range into six
   overlapping Reuleaux triangles RT(r_s)_i, 1 ≤ i ≤ 6
/* Localized k-coverage candidacy checking */
2. For each Reuleaux triangle RT(r_s)_i Do
3.    If RT(r_s)_i contains k active sensors Then
4.       Skip /* i.e., do nothing */
5.    Else
6.       Return ("candidate")
7.    End
8. End
9. Return ("non-candidate")
End
```

**Fig. 7.6** $k$-Coverage candidacy algorithm

## 7.2.2 Stochastic k-Coverage-Preserving Scheduling

In this section, we focus on the design of a distributed sleep–wakeup scheduling protocol for stochastic $k$-coverage ($SCP_k$) of a field. The same approach could be applied for $k$-coverage under the deterministic sensing model by replacing $r_s$ by $r$. We exploit the results of Chap. 5 to present a distributed approach for the selection of a minimum number of active sensors to $k$-cover a field under the stochastic sensing model defined in Eq. 2.2 of Chap. 2. Recall the two approaches for sensor scheduling presented in Chap. 5. We use the second approach which is different from the first one in the sense that each sensor decides whether it is eligible to turn itself active. This decision is based on the degree of coverage of its sensing range. More specifically, each sensor runs our *k-coverage candidacy algorithm*, which is given in Fig. 7.6, before it takes this decision.

### 7.2.2.1 *k*-Coverage Candidacy Algorithm

A sensor turns active if its sensing disk is not $k$-covered. Precisely, a sensor randomly slices its sensing range into six overlapping Reuleaux triangle of width $r_s$ and checks whether each one of them contains at least $k$ sensors. Each sensor should know the status of its sensing neighbours only to decide whether it is a candidate to turn active or not. If any of the six overlapping Reuleaux triangles of width $r_s$ of the sensing range of a sensor $s_i$ does not have $k$ active sensors, the sensor $s_i$ is a candidate to become active. Figure 7.6 shows the pseudo-code of our *k-coverage candidacy* algorithm.

### 7.2.2.2 State Transition of $SCP_k$

The state transition diagram associated with our *stochastic* $k$-coverage *protocol* ($SCP_k$) is similar to the one given in Fig. 5.16 of Chap. 5 by replacing $r$ with $r_s$. In this case, however, each sensor decides whether to turn itself on by running the *k-coverage candidacy* algorithm given in Fig. 7.6.

## 7.2.3 Simulation Results

In this section, we present the simulation results of $SCP_k$ using a high-level simulator written in the C programming language. We consider a square field of side length 1000 m. We use the energy model given in [211], where the sensor energy consumption in transmission, reception, idle, and sleep modes are 60, 12, 12, and 0.03 mW, respectively. Following [218], the energy required for a sensor to stay idle for 1 s is equivalent to *one unit of energy*. We assume that the initial energy of each sensor is 60 J enabling a sensor to operate about 5000 s in reception/idle modes [211]. All simulations are repeated 100 times and the results are averaged.

In Figs. 7.7–7.9, we plot the sensor spatial density as a function of the degree of coverage $k$ for different values of the threshold probability $p_{th}$ and path-loss

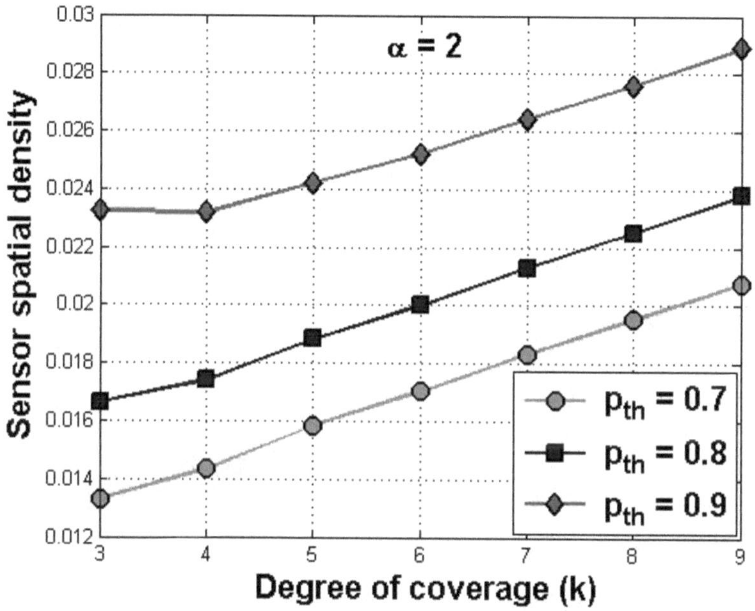


**Fig. 7.7** Sensor spatial density vs. degree of coverage $k$ for $\alpha = 2$

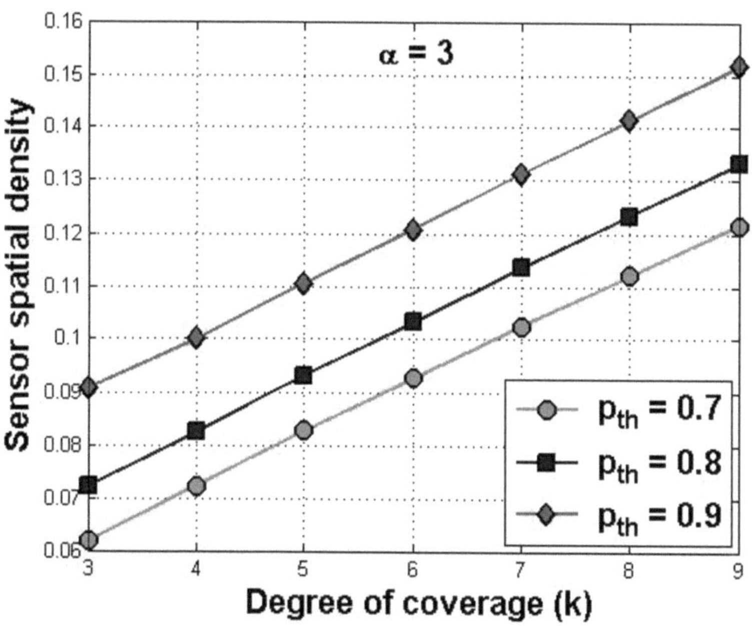


**Fig. 7.8** Sensor spatial density vs. degree of coverage $k$ for $\alpha = 3$

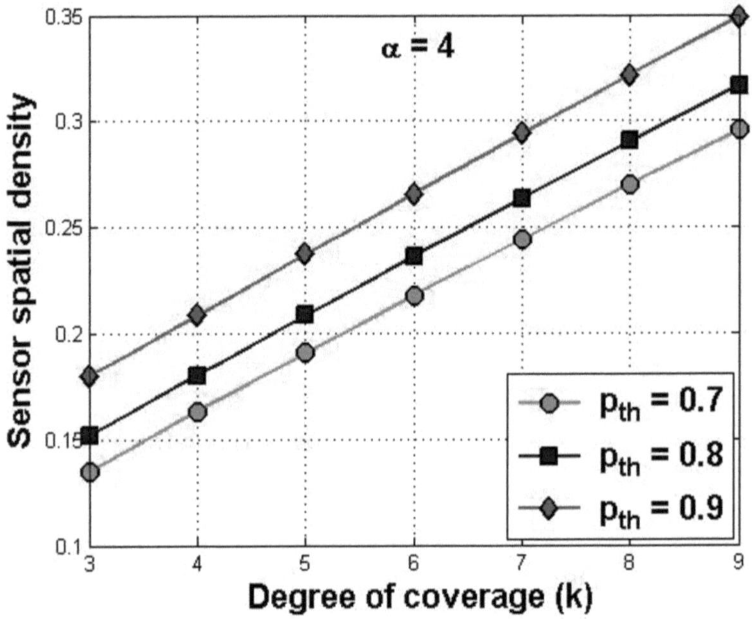


**Fig. 7.9** Sensor spatial density vs. degree of coverage $k$ for $\alpha = 4$

exponent $\alpha$. As expected, the density increases with $p_{th}$ and $\alpha$. Indeed, as we increase $p_{th}$, more sensors would be needed to achieve the same degree of coverage $k$. Recall that the width of the Reuleaux triangle that is guaranteed to be covered with exactly $k$ sensors is equal to the stochastic sensing range of the sensors given in Eq. 7.2 and hence decreases as $p_{th}$ and $\alpha$ increase. On the other hand, the sensor density required for full $k$-coverage of a field is inversely proportional to the area of this Reuleaux triangle as stated implicitly in Lemma 7.2.

Figures 7.10–7.12 plot the achieved degree of coverage $k$ versus the total number of deployed sensors. Moreover, we vary both $p_{th}$ and $\alpha$. Definitely, higher number of deployed sensors would yield higher coverage degree. Here also, any increase in $p_{th}$ and $\alpha$ would require a larger number of deployed sensors to provide the same degree of coverage. Both experiments show a good match between simulation and analytical results.

Figures 7.13–7.15 show that the number of active sensors required to provide 3-coverage increases with the characteristic of the sensors $\beta$ used in the definition of our stochastic sensing model presented in Chap. 2. Recall that $\beta$ measures the uncertainty of the sensing units of the sensors. This result is expected given the definition of the stochastic sensing range in Eq. 7.2.

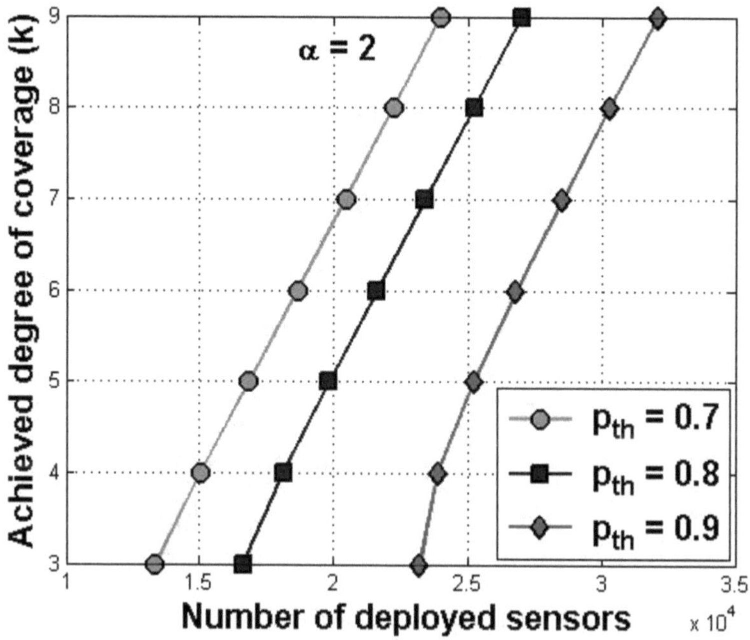


**Fig. 7.10** Degree of coverage $k$ vs. number of deployed sensors for $\alpha = 2$

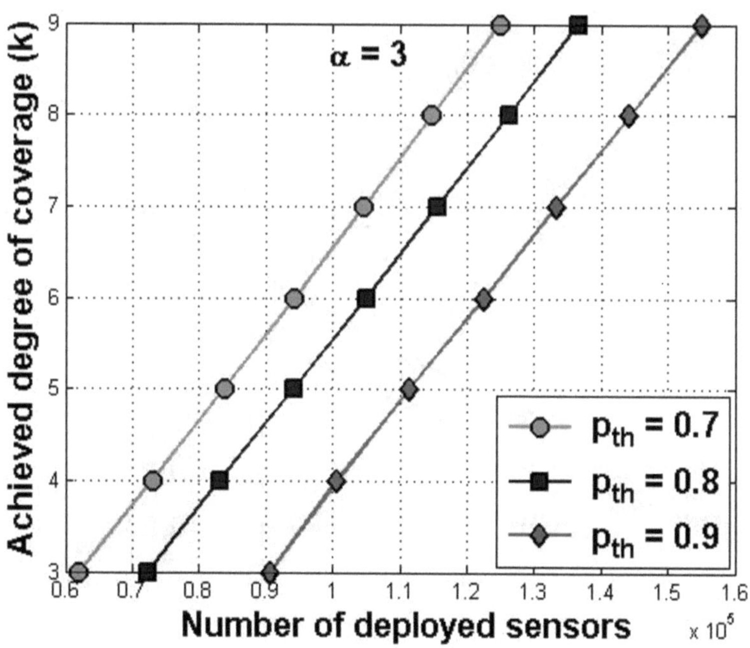


**Fig. 7.11** Degree of coverage $k$ vs. number of deployed sensors for $\alpha = 3$

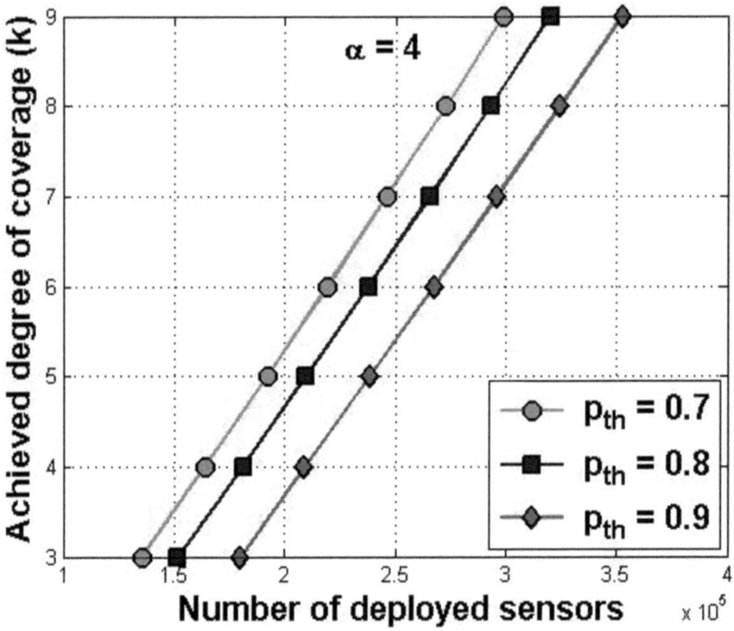


**Fig. 7.12** Degree of coverage $k$ vs. number of deployed sensors for $\alpha = 4$

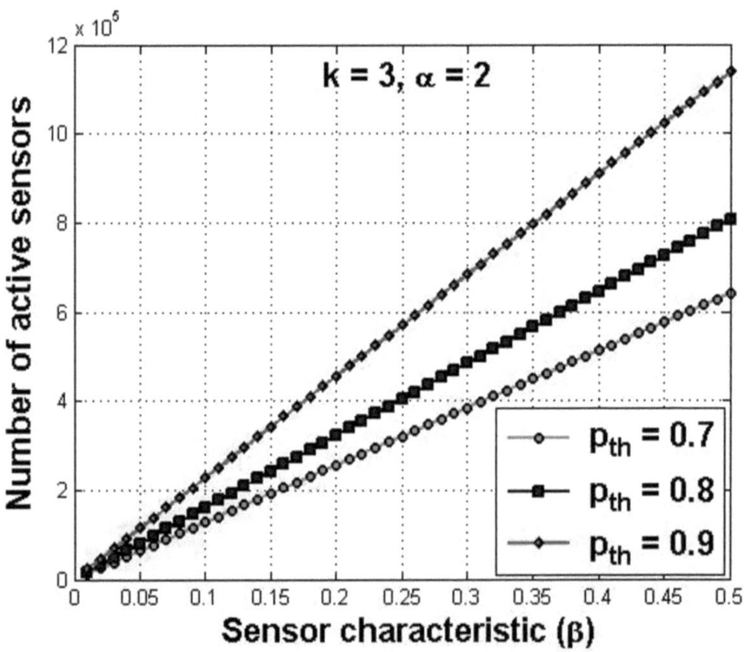


**Fig. 7.13** Number $n_a$ of active sensors vs. $\beta$ for $k = 3$ and $\alpha = 2$

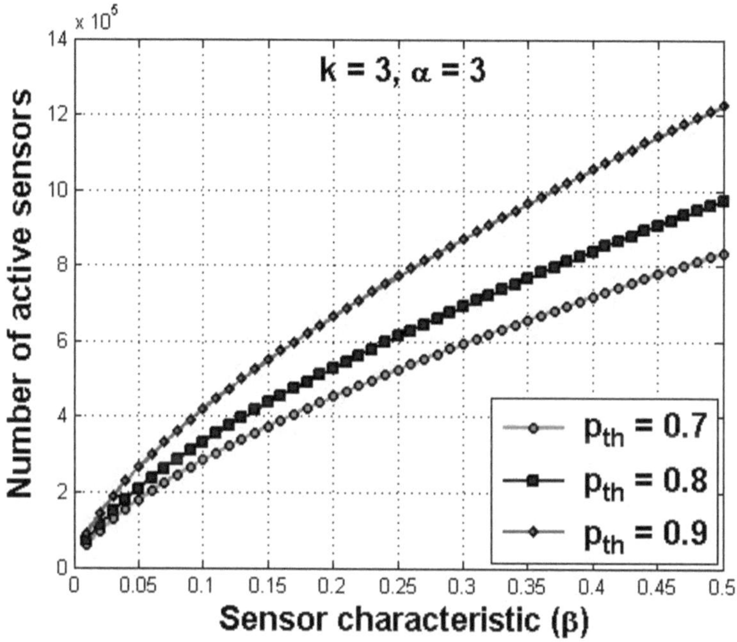


**Fig. 7.14** Number $n_a$ of active sensors vs. $\beta$ for $k = 3$ and $\alpha = 3$

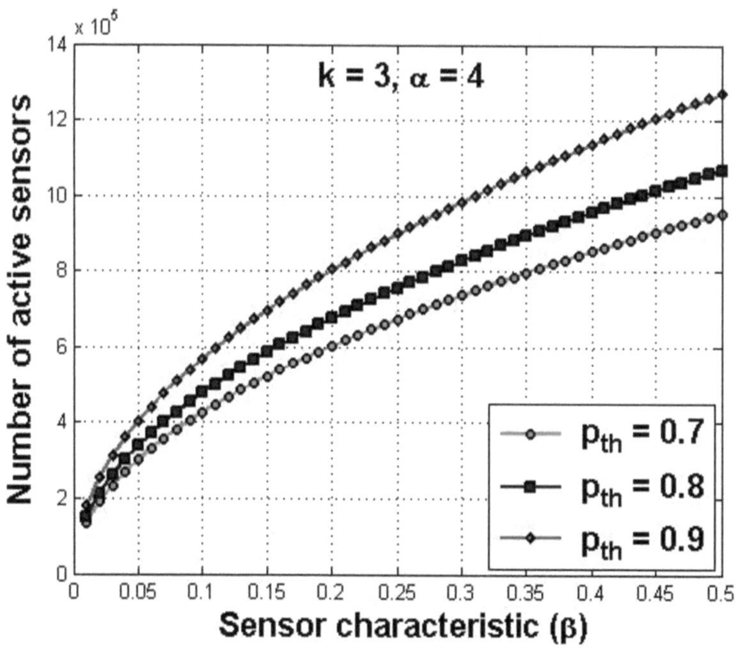


**Fig. 7.15** Number $n_a$ of active sensors vs. $\beta$ for $k = 3$ and $\alpha = 4$

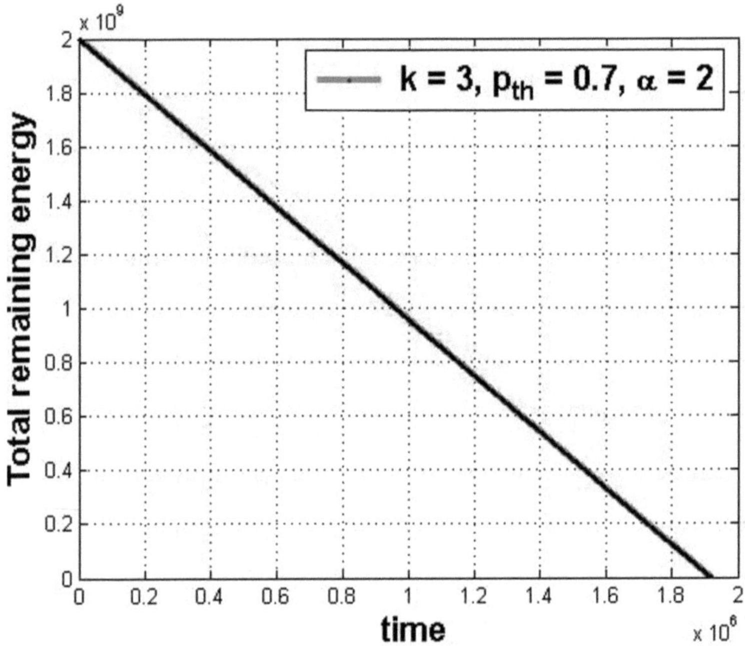


**Fig. 7.16** Total remaining energy vs. time for $k = 3$, $\alpha = 2$, and $p_{th} = 0.7$

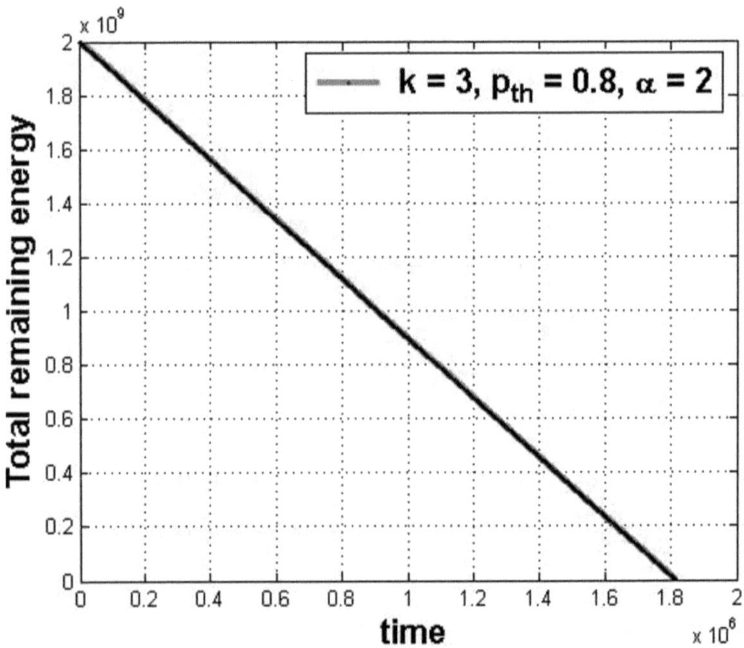


**Fig. 7.17** Total remaining energy vs. time for $k = 3$, $\alpha = 2$, and $p_{th} = 0.8$

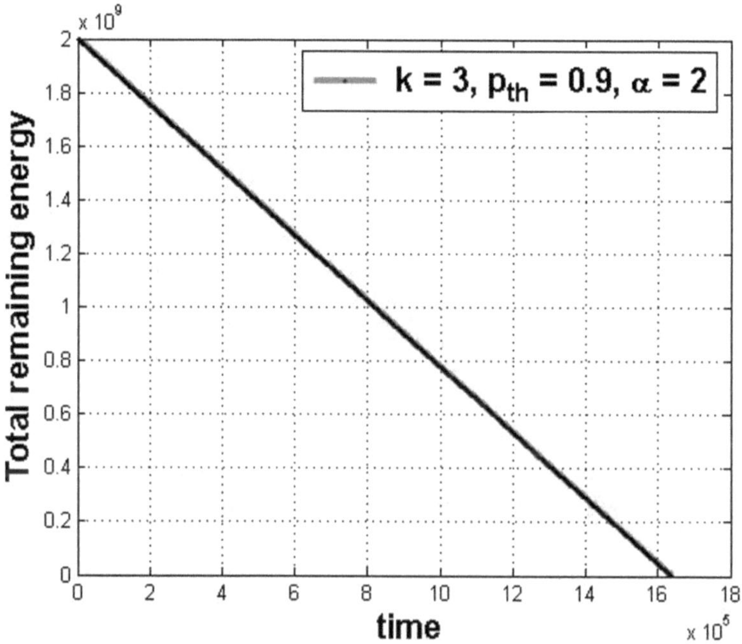


**Fig. 7.18** Total remaining energy vs. time for $k = 3$, $\alpha = 2$, and $p_{th} = 0.9$

Figures 7.16–7.18 show the impact of $p_{th}$ on the operational lifetime of the network to provide 3-coverage. As mentioned earlier, higher values of $p_{th}$ require larger numbers of active sensors, and hence more energy consumption. To the best of our knowledge, the work in [205] is the only one on probabilistic $k$-coverage. However, the probabilistic sensing model used in [205] is totally different from ours. While our stochastic sensing model quantifies the detection probability of a sensor by an exponential function, the one in [205] only assigns it a constant value. Therefore, it is impossible to provide a fair quantitative comparison between our $SCP_k$ protocol and the one in [205].

Next, we focus on the problem of connected $k$-coverage in homogeneous three-dimensional wireless sensor networks.

## 7.3 Three-Dimensional Connected $k$-Coverage

While coverage in two-dimensional wireless sensor networks have been well studied, three-dimensional wireless sensor networks have gained relatively less attention in the literature. In this section, we analyze the $k$-coverage problem in three-dimensional wireless sensor networks and propose a distributed $k$-coverage protocol. Then, we relax some widely used assumptions to promote the use of our $k$-coverage protocol in real-world scenarios. Finally, we evaluate the performance of our protocol. Our work is complementary to the existing ones, especially those few works which dealt with three-dimensional wireless sensor networks. First, we propose an energy-efficient $k$-coverage protocol for three-dimensional wireless sensor networks. Indeed, Alam and Haas [8] considered only 1-coverage and proposed deterministic sensor placement strategies to achieve full coverage of a three-dimensional space. However, 1-coverage is not always enough given that sensors are not highly reliable and some applications, such as intruder detection and tracking, require high coverage of a target field.

First, we show the problem that we encounter when we attempt to extend our analysis of connected $k$-coverage in two-dimensional to three-dimensional wireless sensor networks. We refer to this problem as the *curse of dimensionality*, which is due to the fact that some properties that are valid for two-dimensional space cannot hold for three-dimensional space. Then, we propose an energy-efficient connected $k$-coverage protocol for three-dimensional wireless sensor networks.

### 7.3.1 Problem Analysis: The Curse of Dimensionality

In this section, we analyze the above problem from the perspective of the shape of a three-dimensional region $C$ in a three-dimensional field corresponding to minimum $k$-coverage. In other words, we want to determine the shape of $C$ so that it is guaranteed to be $k$-covered when exactly $k$ sensors are deployed in it. Clearly, the *breadth* of $C$ should be less than or equal to the radius $r$ of the sensing spheres of sensors so that each location in $C$ is within the sensing spheres of these $k$ sensors. Since our goal is to achieve $k$-coverage of a three-dimensional field with a minimum number of sensors, the volume of $C$ should be maximum, and hence the breadth of $C$ must be equal to $r$. Therefore, our problem reduces to the problem of finding the shape of this three-dimensional region $C$ that has a constant breadth equal to $r$.

To solve this problem, we consider *Helly's Theorem* [44] and apply the same analysis as in the case of two-dimensional space, which was described in Chap. 5. Thus, from Helly's Theorem, we infer that a three-dimensional $(n = 3)$ convex region $C$ is $k$-covered by exactly $k$ sensors $(||E|| = k)$ if and only if $C$ is 4-covered by any four $(m = 4)$ of those $k$ sensors, where $k \geq 4$. Now, let us identify the shape of a three-dimensional convex region $C$ whose breadth is constant and equal to the radius $r$ of the sensing spheres of sensors, and hence has a maximum volume. More importantly, this region $C$ should be guaranteed to be fully $k$-covered when *exactly* $k$ sensors are deployed in it.

Let $C_k$ be the intersection of $k$ sensing spheres. Using Helly's Theorem [44], the maximum volume of the intersection of these $k$ sensing spheres is equal to that of four spheres since $k \geq 4$. However, the maximum intersection (or overlap) volume of four sensing spheres such that the maximum distance between any pair of sensors is equal to $r$, corresponds to the configuration where the centre of each sensing sphere is at distance $r$ from the centres of all other three ones. In this

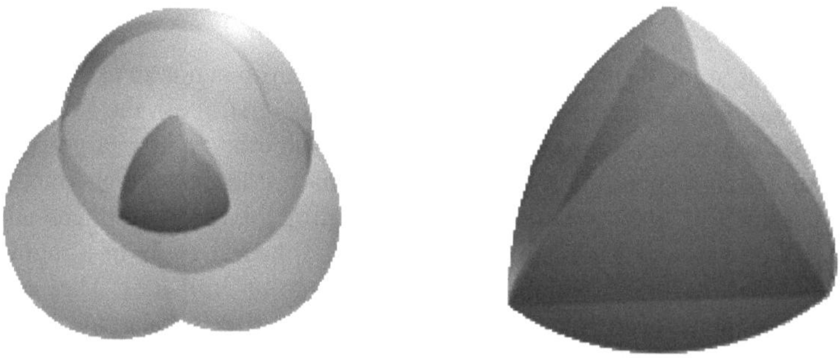

**Fig. 7.19** Intersection of four symmetric spheres and their Reuleaux tetrahedron

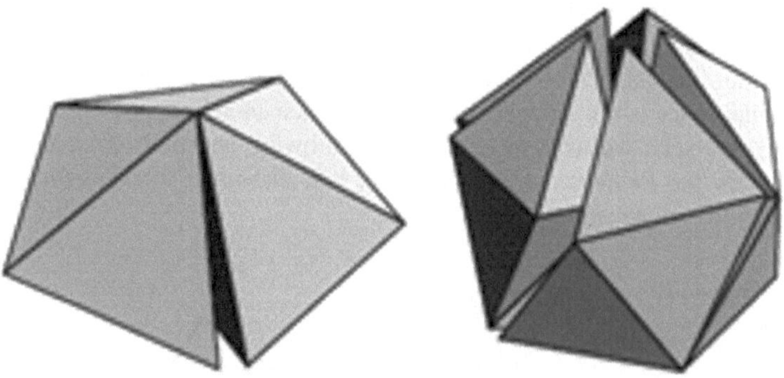

(a) (b)

**Fig. 7.20** Five regular tetrahedra about a common edge and twenty regular tetrahedra about a shared vertex [66]

configuration, the edges between the centres of these four spheres form a *regular tetrahedron* and the shape of their intersection volume is known as the *Reuleaux tetrahedron* [234] (Fig. 7.19).

In [15, 22], we used Helly's Theorem [44] in our analysis of the $k$-coverage problem and exploited the geometric properties of the Reuleaux triangle to derive a sufficient condition to fully $k$-cover a two-dimensional field. Note that the Reuleaux triangle of side $r$, which represents the intersection of three symmetric, congruent disks of radius $r$, consists of a central regular triangle of side $r$ and three curved regions. More importantly, it has a constant width equal to $r$ [233]. We found that a Reuleaux triangle region of width $r$ of a two-dimensional field is guaranteed to be $k$-covered with exactly $k$ sensors, where $r$ is the radius of the sensing range of the sensors [22]. Also, the regular triangle allows a perfect tiling of two-dimensional space. Based on this characterization, we designed an energy-efficient $k$-coverage configuration protocol for two-dimensional wireless sensor networks [15, 22].

Now, we provide some facts why the Reuleaux tetrahedron is not an appropriate solution to our *minimum connected k-coverage* problem. First of all, the Reuleaux tetrahedron does not have a constant breadth whose value is slightly larger than the radius $r$ of the corresponding spheres [234]. In contrast to the regular triangle, the regular tetrahedron does not allow a perfect tiling of a three-dimensional space. Indeed, Conway and Torquato showed that the dihedral angle of a regular tetrahedron is equal to 70.53º, which is not sub-multiple of 360º [66]. They also gave two arrangements of regular tetrahedra such that five regular tetrahedra packed around a common edge would result in a small gap of 7.36° as shown in Fig. 7.20 (left diagram), and that 20 regular tetrahedra packed around a common vertex yield gaps that amount to a solid angle of 1.54 steradians as shown in Fig. 7.20 (right diagram) [66]. This shows that some properties that hold for two-dimensional space are not valid for three-dimensional space. Thus, the extension of the analysis of $k$-coverage in two-dimensional space [22] to three-dimensional space is not straightforward, and hence another approach should be used. More precisely, we want to address the following question: *What is the "closest shape" to the Reuleaux tetrahedron that will guarantee energy-efficient k-coverage of a three-dimensional space?*

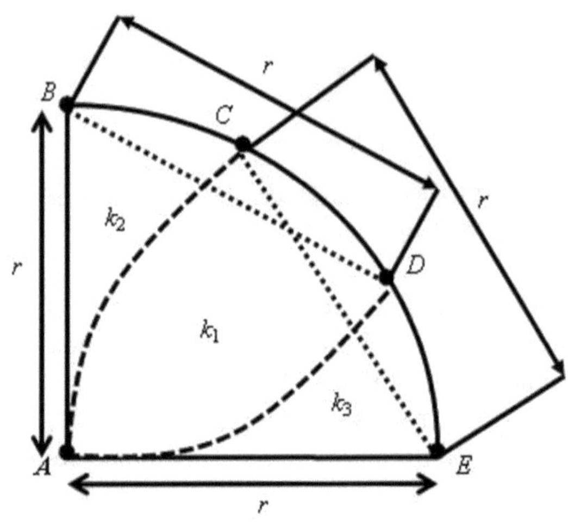


**Fig. 7.21** Two-dimensional projection of a slice

To address this question efficiently, we consider two halves of the sensing sphere of the sensors, i.e., the top half and bottom half. Note that in two-dimensional wireless sensor networks, we divide the sensing disk of the sensors into six overlapping Reuleaux triangles [22]. By analogy with the two-dimensional analysis given in [22], we divide each of the halves of a sensing sphere into six congruent three-dimensional regions, called *slices*, each of which has three flat faces and one curved face representing an *equilateral spherical triangle*. Unfortunately, the distance between the point $B$ at the top of a slice and all the points $E$ on the edge of any spherical triangle is larger than $r$. Thus, a sensor located at $B$ cannot cover any point $E$ and a sensor located at any point $E$ cannot cover $B$. Any sensor located in the region $< A,C,D >$ is able to cover the whole slice as shown in Fig. 7.21,. However, sensors located in the regions $< A,B,C >$ and $< A,D,E >$ cannot cover the entire slice. Thus, if $k_1 = k$, then $k_2 = k_3 = 0$; otherwise, $k_2 = k_3 = k - k_1$. In other words, to guarantee *k*-coverage of a slice and hence a sensing sphere with a minimum number of sensors, it is necessary that active sensors should be located in the region $< A,C,D >$, thus efficiently solving the *minimum connected k-coverage* problem.

Theorem 7.2, which follows from the above analysis, states a *tight sufficient condition* for *k*-coverage of a three-dimensional field.

*Theorem 7.2:* Let $k > 1$. A three-dimensional field is *k*-covered if any slice of the field contains at least $k$ active sensors. ■

Theorem 7.3, which follows from Theorem 7.2, computes the *minimum sensor spatial density necessary* to fully *k*-cover a three-dimensional field.

*Theorem 7.3:* Let $r$ be the radius of the sensing spheres of sensors and $k > 1$. The minimum sensor spatial density required to guarantee *k*-coverage of a three-dimensional field is computed as

$$\lambda(r,k) = \frac{9k}{\pi r^3}$$

*Proof:* The volume of a slice is $vol(slice) = \pi r^3/9$. By Theorem 7.2, each slice should contain at least *k* sensors. Thus, *k*-covering a three-dimensional field with a minimum number of sensors requires that every slice in the three-dimensional field contain exactly $k$ sensors. Thus, the minimum sensor density to *k*-cover a three-dimensional field is equal to $k/vol(slice)$. ■

Using Theorem 7.2, Theorem 7.4 states a sufficient condition to maintain connectivity in three-dimensional *k*-covered wireless sensor networks.

*Theorem 7.4:* Let $k > 1$. A three-dimensional *k*-covered wireless sensor network is connected if $R \geq r$, where $r$ and $R$ stand for the radii of the sensing and communication spheres of sensors, respectively. ■

### 7.3.2 Our Distributed k-Coverage Protocol

In this section, we describe our *distributed randomized connected k-coverage protocol* (3DIRACC$_k$) for three-dimensional field. First, we present our algorithm that enables a sensor to check its *candidacy* to turn active.

```
Algorithm 1: k-Coverage-Candidacy
(* This code is run by each sensor *)
Begin
/* Sensing sphere slicing */
1. Randomly decompose a sensing sphere
   into 12 slices
/* Localized k-coverage candidacy checking */
2. For each slice Do
3.    If it contains k awake sensors Then
4.       Skip /* i.e., do nothing */
5.    Else
6.       Return ("candidate")
7.    End
8. End
9. Return ("non-candidate")
End
```

**Fig. 7.22** *k*-Coverage-Candidacy algorithm

*k-Coverage-candidacy algorithm:* A sensor turns active if its sensing sphere is not *k*-covered. Based on Lemma 5.2 in Chap. 5, a sensor randomly decomposes its sensing sphere into 12 slices of side $r$ and checks whether each one of them contains at least $k$ sensors. Each sensor should know the status of its sensing neighbours only to decide whether it is candidate to turn active or not. If any of the 12 slices does not have $k$ active sensors, a sensor is a candidate to become active. Else, it is not. Fig. 7.22 shows the pseudo-code of our *k*-Coverage-candidacy algorithm.

*State Transition Diagram of 3DIRACC$_k$:* We use the same state transition diagram described in Sect. 7.2.2.

### 7.3.3 Performance Evaluation

We consider a cubic deployment field of side length $1000\,\mathrm{m}$ where all sensors are randomly and uniformly deployed. We use the same energy model used earlier in Sect. 7.2.3. All simulations are repeated 20 times and the results are averaged.

Figure 7.23 plots the sensor spatial density $\lambda(r,k)$ versus the coverage degree $k$, where the radius $r$ of the sensing range of sensors is equal to 30 m. We observe a close to perfect match between our simulation and analytical results. Notice that $\lambda(r,k)$ increases with $k$ for a fixed $r$. Indeed, higher coverage degree of a field would require more active sensors.

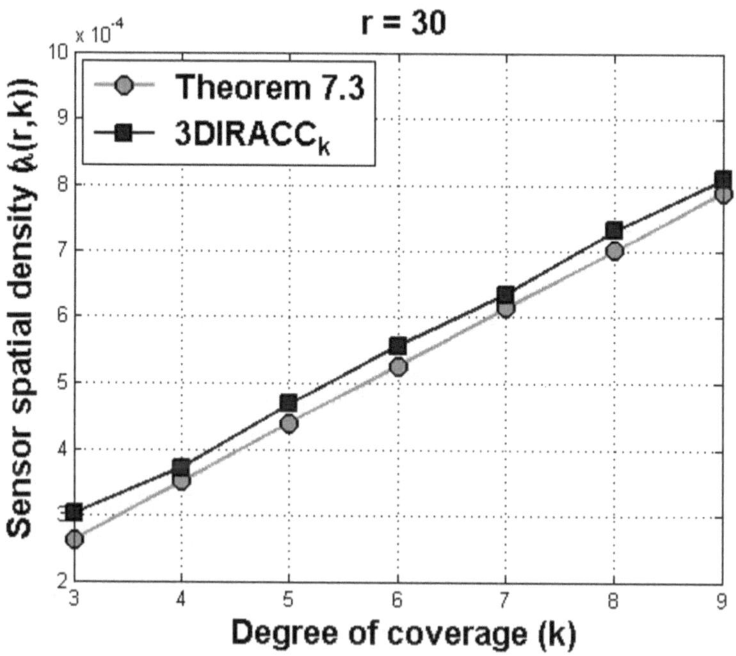


**Fig. 7.23** $\lambda(r,k)$ vs. $k$

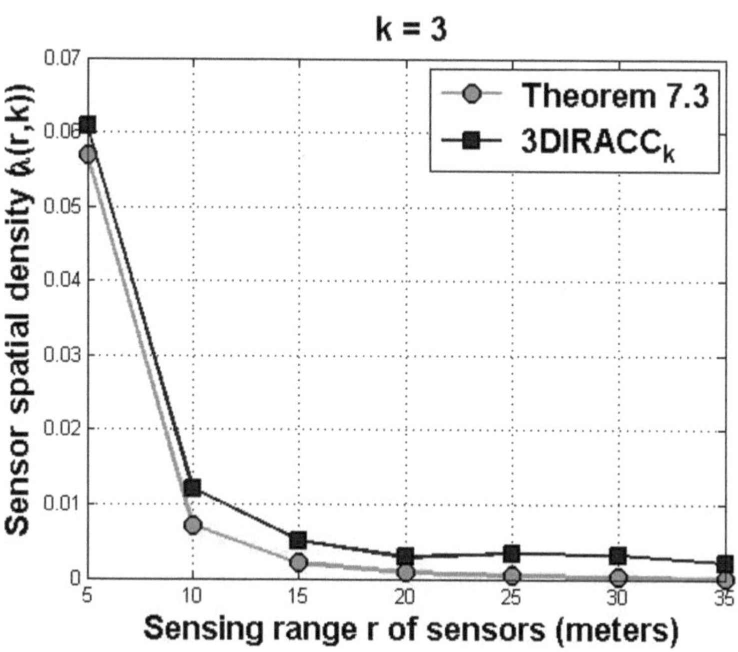


**Fig. 7.24** $\lambda(r,k)$ vs. $r$

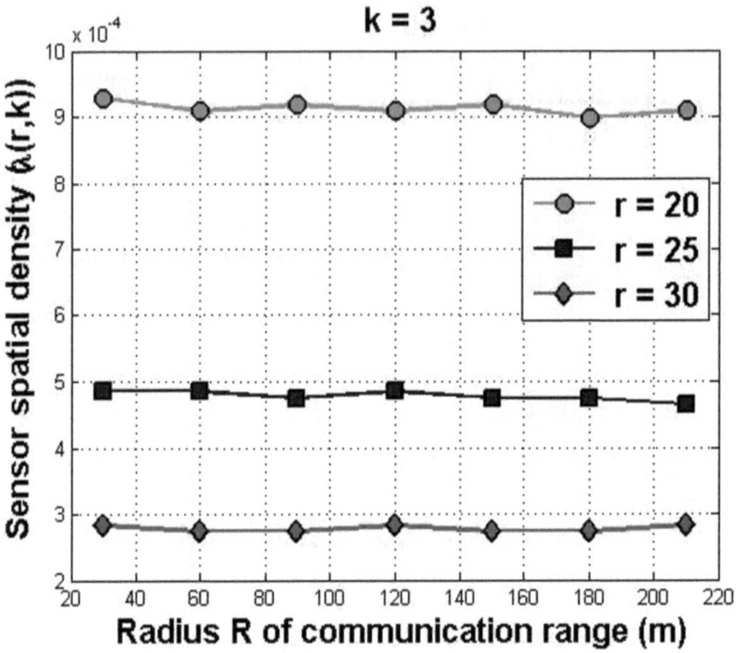


**Fig. 7.25** $\lambda(r,k)$ vs. $R$

Figure 7.24 plots $\lambda(r,k)$ versus the radius $r$ of the sensing range of sensors, where the degree of coverage $k$ is equal to 3. We observe that $\lambda(r,k)$ decreases with $r$ for a fixed $k$. In fact, sensors with larger sensing range would cover more areas and hence less number of active sensors is required to achieve a certain coverage degree $k$ of a field.

Figure 7.25 plots $\lambda(r,k)$ versus the radius $R$ of the communication range of sensors for different values of the radius $r$ of their sensing range, where the degree $k$ of coverage is equal to 3. Notice that $\lambda(r,k)$ does not increase with $R$. Indeed, as we found in Theorem 7.3, the sensor spatial density $\lambda(r,k)$ of active sensors to $k$-cover a field depends only on the radius $r$ of the sensing range of sensors. Also, our $k$-coverage protocol 3DIRACC$_k$ is based on the sensing range of sensors in the sense that each sensor guarantees that its sensing range only is $k$-covered.

## 7.4 Related Work

In this section, we present sample approaches on stochastic coverage in two-dimensional wireless sensor networks. Existing work on coverage and connectivity in three-dimensional wireless sensor networks is described in Chap. 4.

As mentioned in Chap. 6, Lazos and Poovendran [132] formulated the coverage problem in heterogeneous wireless sensor networks as a set intersection problem and derived analytical expressions, which quantify the coverage achieved by stochastic coverage. Megerian et al. [157] studied the exposure in wireless sensor networks, which is related to the quality of coverage provided by these networks, based on a general sensing model, where the sensing signal of a sensor at an arbitrary point by a function that is inversely proportional to the distance between the sensor and point. Liu and Towsley [146] studied three coverage measures, namely area coverage, node coverage, and detectability, using the general sensing model defined in [157].

Liu et al. [147] presented a joint scheduling scheme based on a randomized algorithm for providing statistical sensing coverage and guaranteed network connectivity. This scheme does not make any assumption on the relationship between sensing and transmission ranges, and works without the availability of per-node location information. Zou and Chakrabarty [228] proposed a distributed approach for the selection of active sensors to fully cover a field based on the concept of connected dominating set. This approach is based on a probabilistic sensing model, where the probability of the existence of a target is defined by an exponential function that represents the confidence level of the received sensing signal.

Wang and Tseng [196] proposed solutions to the $k$-coverage sensor deployment problem using both deterministic and probabilistic sensing models. These solutions compute the minimum number of sensors required to $k$-cover a field as well as their locations, and schedule the sensors to move to these locations. In the first solution, the sink computes those locations and the sensors bid for their closest locations. The second solution enables the sensors to derive the target locations by themselves. Xing et al. [205] extended their CCP protocol to provide probabilistic coverage guarantee based on a probabilistic coverage model, where the sensors may have non-uniform and irregular communication and sensing regions. According to this model, a point in a convex coverage area is guaranteed to be $k$-covered with a probability no lower than $\beta$. CCP provides probabilistic coverage via a mapping of the $(k,\beta)$-coverage requirement to a *pseudo coverage degree* $k'$, which is computed analytically.

## 7.5 Summary

In this chapter, we proposed a distributed approach to solve the scheduling problem in stochastic $k$-covered wireless sensor networks [25], where the sensing ability of the sensors is represented by a probability function. Indeed, stochastic sensing models are more realistic than the deterministic sensing model, which does not capture the probabilistic nature of the sensors' characteristics. Our methodology is based on a geometric analysis using the Reuleaux triangle model. For problem tractability, we considered the deterministic sensing model and then extended the analysis to a stochastic sensing model. First, we characterized $k$-coverage in wireless sensor networks and provided a necessary and sufficient condition to achieve $k$-coverage with a minimum number of sensors. Then, we presented our $k$-coverage-preserving scheduling protocol ($SCP_k$) based on this characterization. Precisely, sensors activate themselves by running a $k$-coverage candidacy algorithm to ensure that their sensing ranges are $k$-covered. We found a good match between simulation and analytical results.

We also investigated the problem of connected $k$-coverage in three-dimensional wireless sensor networks [21]. We found that our model for $k$-coverage in wireless sensor networks does not generalize to three-dimensional wireless sensor networks due to the inherent characteristics of the Reuleaux tetrahedron that totally differ from those of its two-dimensional Reuleaux triangle counterpart. Hence, we proposed a new model and derived the sensor spatial density to guarantee full $k$-coverage of a three-dimensional field. Also, we proposed a distributed randomized connected $k$-coverage protocol ($3DIRACC_k$) for three-dimensional field and evaluated its performance. We found that our simulation results match the theoretical ones.

# Chapter 8
# Network Connectivity and Fault-Tolerance Measures in Two-Dimensional Deployment Fields

This chapter gives our measures of *unconditional* (or *traditional*) and *conditional* connectivity and fault-tolerance of two-dimensional $k$-covered wireless sensor networks. The latter measures are more realistic than the former as they impose a restriction on the subsets of sensors that can fail at the same time. Precisely, conditional measures take into consideration the inherent properties of $k$-covered wireless sensor networks, such as high sensor density and sensor heterogeneity. In particular, the neighbour set of a sensor is defined as a *forbidden faulty set*, and hence cannot fail at the same time. This concept defines the new measure of connectivity, called *conditional connectivity*, which seems to be more realistic.

## 8.1 Introduction

*Connectivity*, primarily a graph-theoretic concept, helps define the *fault tolerance* of wireless sensor networks in the sense that it enables the sensors to communicate with each other so their sensed data can reach the sink. On the other hand, *sensing coverage*, an intrinsic architectural feature of wireless sensor networks, plays an important role in meeting application-specific requirements, for example, to reliably extract relevant data about a sensed field. A fundamental aspect in the design of wireless sensor networks is to keep them functional as long as possible. Because of scarce battery power (or *energy*), sensors may entirely deplete energy or have remaining energy below some threshold that is required for the sensors to function properly. Those sensors are called *faulty* as they cannot perform any monitoring task properly. A wireless sensor network is said to be *functional* if at any time there is at least one communication path between every pair of non-faulty sensors in the network. The existence of communication paths between pairs of sensors, however, is related to another fundamental property of wireless sensor networks, called *vertex-connectivity* (or simply *connectivity*). In general, sensing applications are required to be fault tolerant, where any pair of sensors is usually connected by multiple communication paths. Therefore, network functionality and hence network fault-tolerance strongly depends on connectivity. Another important issue in the design of wireless sensor networks is what is called *sensing coverage*, which is a good indicator of the quality of surveillance of a field of interest. Some sensing applications demand full coverage where every location in

H.M. Ammari: Challenges and Opportunities of Connected $k$-Covered WSNs, SCI 215, pp. 157–174.
springerlink.com 

the field is covered by at least one sensor. Moreover, to cope with the problem of faulty sensors, duplicate coverage of the same region is desirable.

Notice that sensing coverage and network connectivity are not totally orthogonal concepts. While sensing coverage depends on the *sensing range*, connectivity relates to the communication range of the sensors. Intuitively, sensing coverage becomes meaningless if the sensed data cannot be exchanged by the sensors so they reach a central gathering point, i.e., *sink*, for further analysis. Thus, for a network to function properly, sensing coverage and network connectivity should be maintained. In fact, it has been proven that connectivity strongly depends on coverage and hence considerable attention has been paid to establish tighter connection between them, although only loose lower bound on connectivity of wireless sensor networks is known. It was proved that the connectivity of a *homogeneous* $k$-covered wireless sensor network is equal to $k$ provided that the radius of the communication range of the sensors is at least double the radius of their sensing range [205]. In this chapter, we investigate connectivity based on the degree of sensing coverage by studying *k-covered* wireless sensor networks [13].

Although network connectivity can be used to measure the fault tolerance of small-scale networks, it is not appropriate for large-scale dense networks, such as $k$-covered wireless sensor networks. Traditional (or unconditional) connectivity has no restriction on the faulty sensor set and assumes that any subset of sensors can potentially fail at the same time, including all the neighbours of a given sensor. However, highly dense sensor networks, such as $k$-covered wireless sensor networks, can consist of thousands of sensors for which it is highly unlikely in this type of network that all the neighbours of a given sensor fail simultaneously. This is due to the following two reasons:

- Assuming a uniform sensor distribution, the ratio of the size of the neighbour set of a given sensor to the total number of sensors in a field of area size $A$ is given by $\pi R^2 / A$, where $R$ $(R << \sqrt{A})$ is the radius of the communication ranges of the sensors forming a homogeneous wireless sensor network. The probability of the failure of the entire neighbour set of a given sensor can be identified with this ratio and hence is very low. It is worth mentioning that there are two types of faults: random faults and arbitrary faults. Definitely, random fault can be dealt with much easier than arbitrary faults. In practice, the failure of sensors would come from the (uncontrollable) environment, and if a sensor is damaged due to a wild animal, it is very likely that the same animal will damage some neighbouring sensors as well, but it is highly unlikely that *all* the neighbouring sensors would be damaged by the same animal in highly densely deployed networks, such as $k$-covered wireless sensor networks.
- In real-world scenarios, wireless sensor networks can be *heterogeneous*, where sensors have different sensing, processing, and communication capabilities, thus increasing the network reliability and lifetime [209]. Hence, the probability that an entire neighbour set of a given sensor fail simultaneously in this type of heterogeneous network is very low.

Thus, the classical (or unconditional) connectivity may not reflect the actual fault tolerance of large-scale dense networks, such as $k$-covered wireless sensor networks, due to the above shortcomings. To alleviate this problem, we use a more general concept of connectivity, called *conditional connectivity* (or P-connectivity), which was originally introduced in [101] with respect to some property. Precisely, we exploit an instance of conditional connectivity, called *restricted connectivity*, which was proposed in [82]. Specifically, the restriction is on the faulty set (i.e., set of nodes that can fail). Restricted connectivity uses the concept of *forbidden faulty set* in which the entire neighbour set of a given node cannot be faulty at the same time. In this chapter, we propose a new measure of fault tolerance for $k$-covered wireless sensor networks, called *conditional fault-tolerance*, based on the concepts of conditional connectivity and forbidden faulty sensor set that includes all the neighbours of a given sensor. We prove that $k$-covered wireless sensor networks can sustain a large number of sensor failures provided that the faulty sensor set does not include a forbidden faulty sensor set.

The remainder of this chapter is organized as follows. Section 8.2 computes unconditional connectivity and fault-tolerance measures for two-dimensional $k$-covered wireless sensor networks while Sect. 8.3 computes their conditional connectivity and fault-tolerance measures. Section 8.4 reviews related on conditional connectivity and its variations. Section 8.5 summarizes the chapter.

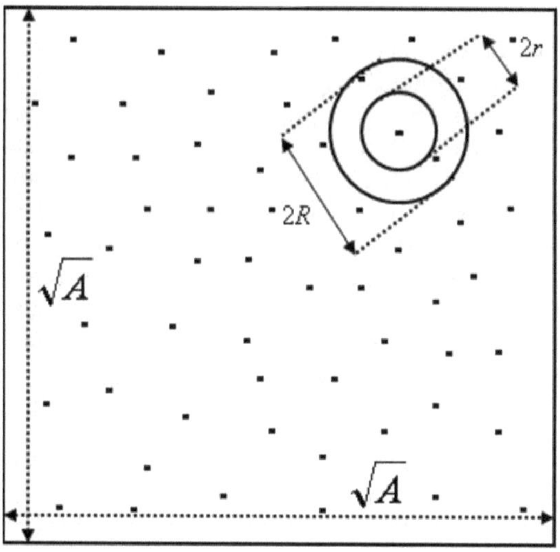


**Fig. 8.1** Sensor distribution in a square field of area $A$. The radius of the sensing ranges of the sensors is $r$ while the radius of their communication ranges is $R$

## 8.2 Unconditional Fault-Tolerance Measures

In this chapter, we compute connectivity and fault-tolerance measures of homogeneous and heterogeneous two-dimensional $k$-covered wireless sensor networks. We consider a square field of area size $A$ as shown in Fig. 8.1. We also assume that the size of the sensing and communication ranges of the sensors is much less than $A$.

### *8.2.1 Homogeneous Sensors*

We assume that the sink has the same communication range as the other sensors. In Chap. 5, we have computed the sensor density required to achieve $k$-coverage of a two-dimensional field based on the Reuleaux triangle [233] model. Also, we have proved that when a two-dimensional wireless sensor network is guaranteed to be connected when it is configured to provide $k$-coverage of a two-dimensional deployment field with $k \geq 3$ and the radius of the communication ranges of the sensors is at least equal to the radius of their sensing ranges, i.e., $R = r$. To make this chapter self-contained, we recall Lemma 5.2 given in Chap. 5.

*Lemma 5.2:* Let $r$ be the radius of the sensing disks of the sensors and $k \geq 3$. A field is $k$-covered if any Reuleaux triangle region of width $r$ in a field contains at least $k$ active sensors. ■

From Lemma 5.2, we deduce that the minimum sensor spatial density required to guarantee $k$-coverage of a field is given by

$$\lambda(r,k) = \frac{2k}{(\pi - \sqrt{3})\, r^2} \tag{8.1}$$

Indeed, since we are interested in the number of neighbours of a sensor, we need to consider this lemma instead of Theorems 5.2, which characterizes minimum $k$-coverage, and Theorem 5.3, which gives a tight bound on the sensor spatial density required for $k$-coverage.

Ammari and Das [22] proposed a sleep-wakeup scheduling protocol for full $k$-coverage of a field while using a minimum number of active sensors based on the above characterization. Theorem 8.1 computes the connectivity of homogeneous $k$-covered wireless sensor networks and derives their fault tolerance. While the works in [205] compute connectivity based on boundary and interior sensors, our study considers the sink given its critical role in data collection and analysis. Precisely, we focus on the size of the connected component containing the sink. Indeed, it is more realistic to relate network connectivity measure to the sink. For instance, we may have a giant connected component of sensors that does not include the sink, and hence their data cannot reach the sink.

*Theorem 8.1:* Let $G$ be a communication graph of a homogeneous $k$-covered wireless sensor network deployed in a square field of area $A$. The connectivity of $G$ is given by

$$\frac{2\pi\alpha^2 k}{(\pi - \sqrt{3})} \leq \kappa(G) \leq \frac{2R\sqrt{A}\,k}{(\pi - \sqrt{3})\, r^2} \tag{8.2}$$

where $\alpha = R/r$ and $k \geq 3$. Its fault tolerance, $\eta(G)$, is given by

$$\frac{2\pi\alpha^2 k}{(\pi - \sqrt{3})} - 1 \leq \eta(G) \leq \frac{2R\sqrt{A}\,k}{(\pi - \sqrt{3})\, r^2} - 1 \tag{8.3}$$

*Proof:* To compute the network connectivity of homogeneous *k*-covered wireless sensor networks, we need to consider the following three cases, which depend on the sink, denoted by $s_0$ and located at $\xi_0$. Since we assume that sensor failure is due to low battery power, we can use a powerful sink with infinite energy, thus eliminating the possibility of sink failure.

*Case 1—Isolated sink:* This situation occurs when the disconnected network has at least two connected components, one of them being the trivial component containing the sink. Given the definition of network connectivity, the number of disconnected components should be equal to two. Notice that the optimum location of the sink in terms of energy-efficient data gathering from the available sensors is the centre of the square field [148]. The sink can be isolated only when all its neighbours fail. Therefore, we compute the number of neighbours of the sink whose failure would disconnect the sink.

Let $n_0$ be the minimum number of sensor failures to isolate the sink $s_0$. Assuming that the sensors are deterministically deployed to achieve *k*-coverage of a field, the value of $n_0$ is given by

$$n_0 = \lambda(r,k)\,|D(\xi_0,R)| \tag{8.4}$$

where $|D(\xi_0,R)| = \pi R^2$ is the measure of the area of the communication disk $D(\xi_0,R)$ of the sink $s_0$ located at $\xi_0$. Hence, the network connectivity in this case is given by

$$\kappa_1(G) = n_0 = \lambda(r,k)\,\pi R^2 \tag{8.5}$$

Substituting Eq. 8.1 in Eq. 8.5 yields

$$\kappa_1(G) = \frac{2\pi\alpha^2 k}{(\pi-\sqrt{3})} \tag{8.6}$$

where $\alpha = R/r$ and $k \geq 3$. Notice that connectivity increase with the ratio $\alpha$ and the sensing coverage $k$. We observe that the network connectivity $\kappa_1(G)$ is also higher than the sensing coverage $k$.

*Case 2—Non-trivial connected components*: Similarly, the disconnected network has two connected components. We distinguish two particular network configurations that are worth of study. In the first one (Fig. 8.2), the connected component including the sink corresponds to its communication disk (whose area is $\pi R^2$) and is surrounded by a circular band that contains no sensors. Furthermore, the distance between any pair of sensors from these two components is at least $R$ in order to disable any communication between the two connected components. Thus, the width of this empty band should be at least $R$. Hence, the area of the smallest empty circular band $B(\xi_0,R)$ should be equal to

$$|B(\xi_0,R)| = \pi(2R)^2 - \pi R^2 = 3\pi R^2 \tag{8.7}$$

Thus, the minimum number $n_0$ of sensor failures to isolate the connected component of the sink is given by

$$n_0 = \lambda(r,k)\,|B(\xi_0,R)| \tag{8.8}$$

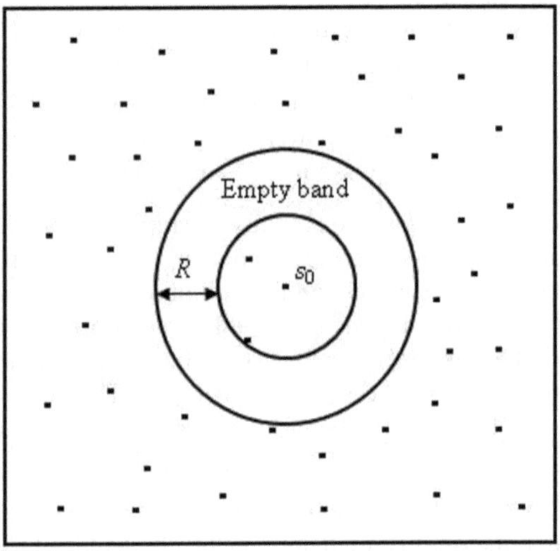


**Fig. 8.2** First non-trivial connected component of the disconnected network

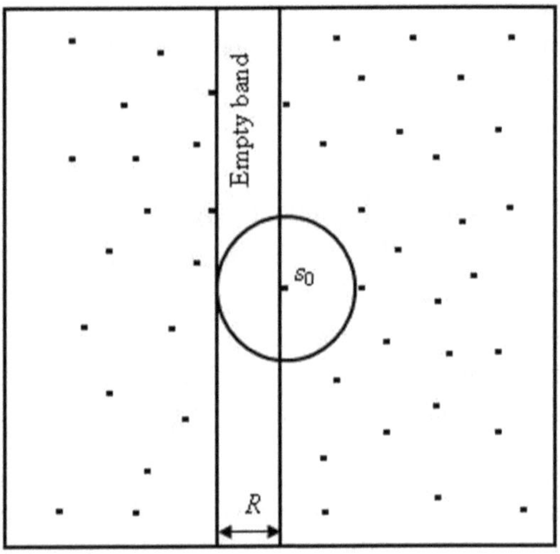


**Fig. 8.3** Second non-trivial connected component of the disconnected network

Hence, the network connectivity is equal to

$$\kappa_2(G) = n_0 = 3\,\lambda(r,k)\,\pi\,R^2 \tag{8.9}$$

If we set $\alpha = R/r$ and substitute Eq. 8.1 in Eq. 8.9, we obtain

$$\kappa_2(G) = \frac{6\,\pi\,\alpha^2\,k}{(\pi - \sqrt{3})} \tag{8.10}$$

The second configuration of the disconnected network (Fig. 8.3) corresponds to the smallest connected component containing the sink if the field has to be divided into two regions such that none of them surrounds the other. Hence, the width of the empty rectangular band, denoted by $\mathrm{B}(R,\sqrt{A})$, which splits the field vertically should be equal to $R$. The minimum number $n_0$ of sensor failures to isolate the sink is given by

$$n_0 = \lambda(r,k)\, R\sqrt{A} \tag{8.11}$$

We find that the network connectivity is equal to

$$\kappa_3(G) = n_0 = \lambda(\mathrm{r,k})\, R\sqrt{A} \tag{8.12}$$

Setting $\alpha = R/r$ and substituting Eq. 8.1 in Eq. 8.12 yields

$$\kappa_3(G) = \frac{2R\sqrt{A}\,k}{(\pi-\sqrt{3})\,r^2} \tag{8.13}$$

Notice that $\kappa_3(G) > \kappa_2(G)$ given the hypothesis $R << \sqrt{A}$.

*Case 3—Largest connected component*: One of the components of the disconnected network has only one sensor that is not the sink. This case is similar to the first case in that a single sensor becomes isolated when all of its neighbours fail. Applying the same reasoning leads to the same result found in Case 1. Thus, we have $\kappa_1(G) \le \kappa(G) \le \kappa_3(G)$ and hence the network fault tolerance, $\eta(G)$, is given by $\kappa_1(G) - 1 \le \eta(G) \le \kappa_3(G) - 1$. ■

It is easy to check that $\kappa(G) > k$. However, it was proved in [205] that the connectivity of $k$-covered wireless sensor networks is equal to $k$ provided that $R \ge 2r$.

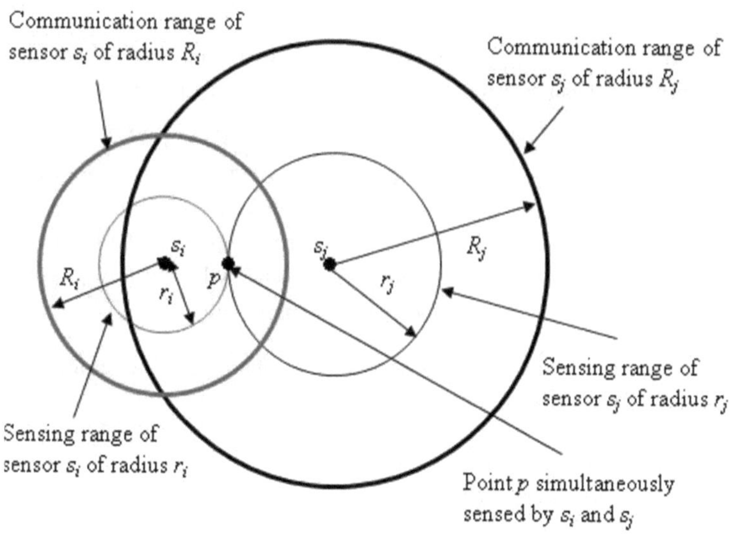


**Fig. 8.4** 1-Coverage and $R_i \ge 2r_i$ do not imply connectivity

### 8.2.2 Heterogeneous Sensors

In this section, we consider $k$-covered wireless sensor networks with heterogeneous sensors whose sensing and communications are not the same, i.e., both parameters are varying. In this case, it is easy to prove that the relationship $R \geq 2r$ cannot guarantee network connectivity even when the network is configured to provide sensing coverage. Figure 8.4 shows that the sensor $s_j$ can connect to the sensor $s_i$, but $s_i$ cannot connect to $s_j$. Lemma 8.1 establishes a necessary and sufficient condition for connectivity of heterogeneous $k$-covered wireless sensor networks.

*Lemma 8.1:* A heterogeneous $k$-covered wireless sensor networks is connected if for any sensor, the radius of its communication disk is at least equal to the sum of the radii of its own sensing disk and that of the most powerful sensor in terms of sensing capability, i.e., for all $s_i \in S$, $R_i \geq r_i + r_{\max}$, where $r_{\max} = \max\{r_j : s_j \in S\}$.

*Proof:* Consider two sensors $s_i$ and $s_j$ whose sensing disks are tangential to each other at point $p$ (Fig. 8.4). Let $R_i = r_i + r_{\max}$ and $R_j = r_j + r_{\max}$. Thus, $R_i \geq r_i + r_j$ and $R_j \geq r_i + r_j$, implying $|\xi_i - \xi_j| \leq \min\{R_i, R_j\}$. Hence, $s_i$ and $s_j$ are mutually connected. Thus, the underlying heterogeneous wireless sensor network is connected because it is $k$-covered. ■

It is worth noting that achieving $k$-coverage depends on the least powerful sensors in terms of their sensing capability. Corollaries 8.1 and 8.2 for heterogeneous $k$-covered wireless sensor networks correspond to Lemmas 5.1 and 5.2, respectively, given in Chap. 5.

*Corollary 8.1:* If the width of a closed convex area $\mathrm{A}$, $\omega(\mathrm{A})$, satisfies $\omega(\mathrm{A}) \leq r_{\min}$, where $r_{\min} = \min\{r_j : s_j \in S\}$, then $\mathrm{A}$ is guaranteed to be $k$-covered when $k$ heterogeneous sensors are deployed in it, where $k \geq 3$.

*Proof:* Our reasoning should be based on the least powerful sensors in terms of their sensing capability. In the worst case, when $k$ least powerful sensors (all of them have the smallest radius of their sensing disks) are deployed in the Reuleaux triangle of constant (maximum) width equal to $r_{\min}$, denoted by $RT(r_{\min})$, then $RT(r_{\min})$ is guaranteed to be $k$-covered, where $r_{\min} = \min\{r_j : s_j \in S\}$. ■

*Corollary 8.2:* The sensor spatial density necessary to guarantee $k$-coverage of a field sensed by spatially distributed heterogeneous sensors is given by

$$\lambda(r_{\min}, k) = \frac{2k}{(\pi - \sqrt{3})\, r_{\min}^{2}} \tag{8.14}$$

where $r_{\min} = \min\{r_j : s_j \in S\}$ and $k \geq 3$.

*Proof:* The proof is verbatim and stems from the fact that if $k$ least powerful sensors (in terms of their sensing ranges) are able to $k$-cover a region $\mathrm{A}$, then any subset of $k$ sensors deployed in $\mathrm{A}$ will be able to do so. Using the AET model, we can easily prove that the maximum size of $\mathrm{A}$ is $\mathrm{A}_{\max}(r_{\min}) = (\pi - \sqrt{3})\frac{r_{\min}^{2}}{2}$. Thus,

the required sensor spatial density is $\lambda(r_{min},k) = \dfrac{k}{A_{max}(r_{min})} = \dfrac{2k}{(\pi - \sqrt{3})\ r_{min}^2}$,
where $r_{min} = \min\{r_j : s_j \in S\}$. ■

While for homogeneous $k$-covered wireless sensor networks "$k$-connected" is the proper parameter, we believe that "$k$-node disjoint paths to the sink" is the right argument for heterogeneous $k$-covered wireless sensor networks since differences in communication ranges induce unidirectional networks. Indeed, "$k$-strong connectivity" is an over-kill in the case of heterogeneous $k$-covered wireless sensor networks networks. From now on, connectivity of heterogeneous $k$-covered wireless sensor networks should be understood in the sense of "$k$-node disjoint paths to the sink". Corollary 8.3 computes connectivity of heterogeneous $k$-covered wireless sensor networks and their fault tolerance.

*Corollary 8.3:* Let $G$ be a communication graph of a heterogeneous $k$-covered wireless sensor networks with $k \geq 3$. The connectivity of heterogeneous $k$-covered wireless sensor networks (in the sense of $k$-node disjoint paths to the sink) is given by

$$K_1 \leq \kappa(G) \leq K_2 \tag{8.15}$$

where

$$K_1 = \frac{2\pi R_{min}^2 k}{(\pi - \sqrt{3})\, r_{min}^2}$$

$$K_2 = \frac{2 R_{max} \sqrt{A}\, k}{(\pi - \sqrt{3})\, r_{min}^2}$$

$r_{min} = \min\{r_j : s_j \in S\}$ and $R_{max} = \max\{R_j : s_j \in S\}$. The fault tolerance, $\eta(G)$, is given by

$$K_1 - 1 \leq \eta(G) \leq K_2 - 1 \tag{8.16}$$

*Proof:* Similarly, we consider three cases depending on the types of components of the disconnected network that contain the sink.

*Case 1—Isolated sink:* Although there are several cases for the sink to consider in terms of its communication capability, we limit our study to the following extreme two cases.

*Case 1.1:* In this case, the sink is supposed to be the most powerful node in the network in terms of sensing, communication, computation, and storage capabilities. Hence, the radius of its communication disk is equal to $R_{max}$. Indeed, in some sensing applications, the sink would receive all sensed data from the sensors in the network which has to be analyzed and processed for a better and more accurate decision-making process. Thus, the minimum number $n_0$ of sensor failures to isolate the sink is given by

$$n_0 = \lambda(r_{min}, k)\,|\,D(\xi_0, R_{max})\,| \tag{8.17}$$

Substituting Eq. 8.14 in Eq. 8.17, we find that the network connectivity of heterogeneous $k$-covered wireless sensor networks is computed as

$$\kappa_1(G) = n_0 = \frac{2\pi {R_{\max}}^2 k}{(\pi - \sqrt{3})\, {r_{\min}}^2} \tag{8.18}$$

where $r_{\min} = \min\{r_j : s_j \in S\}$, $R_{\max} = \max\{R_j : s_j \in S\}$, and $k \geq 3$.

*Case 1.2:* In this case, the sink is supposed to be the least powerful node in the network and hence the radius of its communication disk is equal to $R_{\min}$. Indeed, in other sensing applications, in addition to traffic relaying, the sensors are also responsible for in-network data processing. In this case, the sink would receive only a small amount of data and hence does not have to possess high capabilities. Thus, the minimum number $n_0$' of sensor failures to isolate the sink is given by

$$n_0' = \lambda(r_{\min}, k)\,|\,D(\xi_0, R_{\min})\,| \tag{8.17'}$$

Substituting Eq. 8.14 in Eq. 8.17', we find that the network connectivity of heterogeneous $k$-covered wireless sensor networks is computed as

$$\kappa_1'(G) = n_0' = \frac{2\pi {R_{\min}}^2 k}{(\pi - \sqrt{3})\, {r_{\min}}^2} \tag{8.18'}$$

where $r_{\min} = \min\{r_j : s_j \in S\}$, $R_{min} = \min\{R_j : s_j \in S\}$, and $k \geq 3$.

*Case 2—Non-trivial connected components:* We consider the same two network configurations, which were studied in the case of homogeneous $k$-covered wireless sensor networks. Notice that the sensors located around the sink are heavily used in data forwarding and hence should be the most powerful ones in terms of sensing and communication capabilities. Otherwise, they will suffer severe energy depletion and die very quickly. Thus, the communication disk of the sink contains only powerful sensors. Hence, the width of the circular empty band surrounding the connected component containing the sink should be equal to $R_{\max}$. Thus, the area of this empty circular band $B(\xi_0, R_{\max})$ should be equal to

$$|\,B(\xi_0, R_{\max})\,| = \pi\,(2R_{\max})^2 - \pi\, {R_{\max}}^2 = 3\pi\, {R_{\max}}^2 \tag{8.19}$$

The minimum number $n_1$ of sensor failures to isolate the sink is given by

$$n_1 = \lambda(r_{\min}, k)\,|\,B(\xi_0, R_{\max})\,| \tag{8.20}$$

If we substitute Eq. 8.14 in Eq. 8.20, we find that connectivity is equal to

$$\kappa_2(G) = n_1 = \frac{6\pi R_{\max}^{\ 2}\, k}{(\pi - \sqrt{3})\, r_{\min}^{\ 2}} \tag{8.21}$$

Likewise, for the second configuration of the disconnected network, the width of the empty rectangular band $\mathrm{B}(R,\sqrt{A})$ should be equal to $R_{\max}$ and hence its area is equal to $|\mathrm{B}(R,\sqrt{A})| = R_{\max}\sqrt{A}$. Therefore, the minimum number $n_2$ of sensor failures to isolate the sink is given by

$$n_2 = \lambda(r_{\min}, k)\,|\,\mathrm{B}(R,\sqrt{A})\,| \tag{8.22}$$

Thus, the network connectivity is given by

$$\kappa_3(G) = n_2 = \frac{2\, R_{\max}\sqrt{A}\, k}{(\pi - \sqrt{3})\, r_{\min}^{\ 2}} \tag{8.23}$$

*Case 3—Largest connected component:* In this case, the single-node component may include the least powerful or the most power sensor in terms of its communication capability. Hence, the network connectivity will have lower and upper bounds depending on whether the isolated sensor is the least or most powerful sensor, respectively. The minimum number $n_{lb}$ of sensor failures to isolate a least powerful sensor is given by

$$n_{lb} = \lambda(r_{\min}, k)\,|\,D(\xi_0, R_{\min})\,| \tag{8.24}$$

while the minimum number $n_{ub}$ of sensor failures to isolate a most powerful sensor is given by

$$n_{ub} = \lambda(r_{\min}, k)\,|\,D(\xi_0, R_{\max})\,| \tag{8.25}$$

Let $K_{lb}$ and $K_{ub}$ be lower and upper bounds, respectively, on connectivity $\kappa_4(G)$ of heterogeneous $k$-covered wireless sensor networks. In this case, it is easy to establish that $\kappa_4(G)$ satisfies

$$K_{lb}(G) \le \kappa_4(G) \le K_{ub}(G) \tag{8.26}$$

where
$$K_{lb}(G) = n_{lb} = \frac{2\pi R_{\min}^{\ 2}\, k}{(\pi - \sqrt{3})\, r_{\min}^{\ 2}}$$

$$K_{ub}(G) = n_{ub} = \frac{2\pi R_{\max}^{\ 2}\, k}{(\pi - \sqrt{3})\, r_{\min}^{\ 2}}$$

Using the results of the above three cases, we find that connectivity of heterogeneous $k$-covered wireless sensor networks satisfies $K_1 \le \kappa(G) \le K_2$ and their fault tolerance is given by $K_1 - 1 \le \eta(G) \le K_2 - 1$, where $K_1 = K_{lb}(G) = \frac{2\pi R_{\min}^{\ 2}\, k}{(\pi - \sqrt{3})\, r_{\min}^{\ 2}}$,

$K_2 = \kappa_3(G)\dfrac{2\,R_{\max}\,\sqrt{A}\,k}{(\pi - \sqrt{3})\,{r_{\min}}^2}$, $r_{\min} = \min\{r_j : s_j \in S\}$, $R_{\min} = \min\{R_j : s_j \in S\}$, and $R_{\max} = \max\{R_j : s_j \in S\}$. Similarly, it is easy to prove that $\kappa(G) > k$. ■

### 8.2.3 Conditional Fault-Tolerance Measures

Previous works on fault tolerance in wireless sensor networks assumed that any subset of sensors and, in particular, the neighbour set of any sensor, can potentially fail at the same time. As discussed earlier, this assumption is unrealistic for large-scale dense wireless sensor networks, such as *k*-covered wireless sensor networks, which are highly dense so they can achieve desirable levels of redundant coverage. Also, deploying heterogeneous sensors for real-world applications reduces the probability of simultaneous failure of the entire neighbour set of a given sensor. In this chapter, we restrict a faulty sensor set to some subsets of sensors that do not include the forbidden faulty set.

Next, we compute the conditional connectivity and conditional fault-tolerance of homogeneous and heterogeneous *k*-covered wireless sensor networks.

### 8.2.4 Homogeneous Sensors

In this section, we consider homogeneous sensors that possess the same sensing range and the same communication range. Our results prove that *k*-covered wireless sensor networks can sustain a larger number of sensor failures under the restriction imposed on a faulty sensor set. Precisely, we show that the conditional

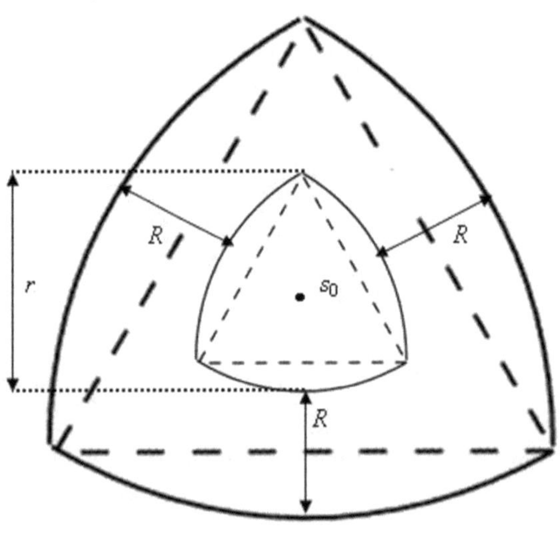


**Fig. 8.5** $RT(\xi_0, r)$ and $A(\xi_0, R)$ regions

connectivity of homogeneous $k$-covered wireless sensor networks is larger than $k$. Theorem 8.2 computes the conditional connectivity and conditional fault-tolerance of homogeneous $k$-covered wireless sensor networks.

*Theorem 8.2:* The conditional connectivity of homogeneous $k$-covered wireless sensor networks is given by

$$\kappa(G:P)=\frac{4\,R\,(R+r)\,k}{r^2} \tag{8.27}$$

where $\alpha = R/r$ and $k \geq 3$. The conditional fault-tolerance of $G$, $\eta(G:P)$, is given by $\eta(G:P)=\kappa(G:P)-1$.

*Proof:* We consider two cases based on the type of component to which the sink belongs.

*Case 1—Smallest-size component including the sink*: Under the assumption of forbidden faulty set, we assume the sink belongs to the smallest connected component that is disconnected from the rest of the network. Let $\xi_0$ be the location of the sink $s_0$. By hypothesis, any location in the field is $k$-covered with $k \geq 3$, and in particular the location $\xi_0$. Therefore, there must be a subset of sensors located at distance at most equal to $r$ from $\xi_0$. Using the Reuleaux Triangle model, the Reuleaux triangle of width $r+\varepsilon_1$ and centered at $\xi_0$, denoted by $RT(\xi_0, r+\varepsilon_1)$, where $\varepsilon_1$ is an infinitesimal value, must be not empty; otherwise, the $k$-coverage property at $\xi_0$ is not satisfied, and particularly the forbidden faulty set constraint is not met. Our goal is to compute the minimum number of sensors to fail in order to disconnect the sink under the forbidden set constraint. Notice that the smallest connected component including the sink requires $\varepsilon_1 = 0$. The region $RT(\xi_0, r)$ is a guarantee that the sink will not be isolated by itself and hence the forbidden faulty sensor set constraint with respect to the sink is not violated. Indeed, only a subset of its neighbours fails and not all of them as in the case of classical connectivity (see Sect. 8.2.1, case 1 of proof of Theorem 8.1). In this configuration, the majority of the sensors are not connected to the sink, and hence the network is dead. Now, to disconnect the sink together with its neighbours located in $RT(\xi_0, r)$, the annulus surrounding the region $RT(\xi_0, r)$ and centered at $\xi_0$ should be empty and have a width equal to $R+\varepsilon_2$; otherwise, the network remains connected. Notice that the minimum number of sensor failure requires $\varepsilon_2 = 0$. Thus, this empty annulus, denoted by $A(\xi_0, R)$ (Fig. 8.5), will guarantee that the connected component in $RT(\xi_0, r)$ is disconnected. Notice that the width of the outmost Reuleaux triangle centered at $\xi_0$ is equal to $2R+r$. Hence, the area of the annulus $A(\xi_0, R)$ is given by

$$|A(\xi_0, R)| = |RT(\xi_0, 2R+r)| - |RT(\xi_0, r)| = 2(\pi-\sqrt{3})\,R\,(R+r) \tag{8.28}$$

Thus, the conditional minimum number $n(P)$ of sensor failures to disconnect the smallest component including the sink is computed as

$$n(P) = \lambda(r,k)\,|\,A(\xi_0, R)\,| \tag{8.29}$$

Substituting Eq. 8.1 and Eq. 8.28 in Eq. 8.29, we find that the conditional connectivity is computed as

$$\kappa_1(G:P) = n(P) = \frac{4\,R\,(R+r)\,k}{r^2} \tag{8.30}$$

where $r$ and $R$ are the radii of the sensing and communication disks of the sensors, respectively, and $k$ is the degree of coverage of the field. It is easy to prove

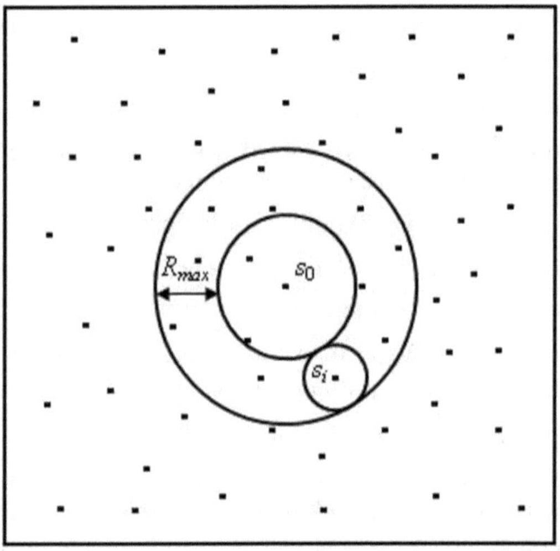


**Fig. 8.6** The forbidden fault set constraint is violated (neighbour set of $s_i$ is within the circular band of width $R_{\max}$ )

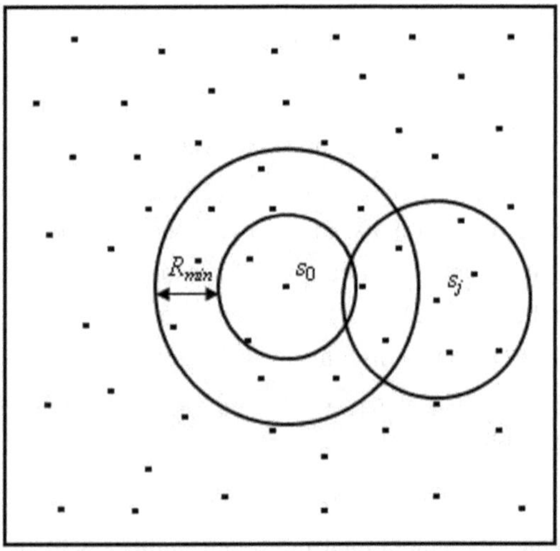


**Fig. 8.7** Connectivity is maintained (the radius of $s_j$'s communication disk is larger than $R_{\min}$ )

that the forbidden faulty sensor set constraint is satisfied for both the faulty and non-faulty sensors. Any sensor inside the region $RT(\xi_0,r)$ still has non-faulty neighbours located in $RT(\xi_0,r)$. Also, any sensor outside the region $RT(\xi_0,R+r)$ has non-faulty neighbours within $RT(\xi_0,R+r)$. Similarly, any faulty sensor within the annulus $A(\xi_0,R)$ has non-faulty neighbours located in $RT(\xi_0,r)$ and outside $RT(\xi_0,R+r)$.

*Case 2—Largest connected component*: We assume that the sensors located in the annulus $A(\xi_i,R)$ as defined earlier fail. This case is similar to the previous one except that the sink belongs to the largest connected component of the disconnected network. Using the same reasoning as in case 1, we obtain the same conditional network connectivity. Here again, to consider whether or not the resulting network is connected or not depends on the type of coverage (full coverage or partial coverage) required by the sensing application. Thus, we have $\kappa_2(G:P)=\kappa_1(G:P)$.

From both cases 1 and 2, it follows that the conditional network connectivity is $\kappa(G:P)=\kappa_1(G:P)$, and hence the conditional network fault-tolerance, $\eta(G:P)$, is given by $\eta(G:P)=\kappa(G:P)-1$. It is easy to check that $\kappa(G:P)>\kappa(G)$ and hence $\kappa(G:P)>k$. By definition (Chap. 2, Definition 2.32), the conditional network fault tolerance, $\eta(G:P)$, is given by $\eta(G:P)=\kappa(G:P)-1$. ■

### 8.2.5 Heterogeneous Sensors

Computing the conditional connectivity of heterogeneous $k$-covered wireless sensor networks is not a straightforward generalization of the process used previously for homogeneous $k$-covered wireless sensor networks. We found that disconnecting the network while satisfying the forbidden faulty set constraint is a challenging problem. If, on the one hand, we choose the width of the annulus to be $R_{max}$, then the sensors with communication range less than or equal to half of $R_{max}$ may be located in the annulus. Thus, the property P will be violated (Fig. 8.6) as the entire neighbour set of some sensors located within the annulus would fail at the same time. If, on the other hand, the width of the annulus is less than $R_{max}$, then the non-faulty sensors of one connected component might be able to connect to the non-faulty sensors of the other connected component of the disconnected network. Hence, the obtained network is not disconnected (Fig. 8.7). As can be seen, we cannot find an exact value of the conditional connectivity of heterogeneous $k$-covered wireless sensor networks in the absence of any deterministic sensor deployment strategy. Hence, we propose to compute lower and upper bounds on conditional connectivity based on particular sensor configurations in the annulus and around the annulus. While in the first scenario we assume that the annulus contains only least powerful sensors, the second scenario supposes that the annulus consists of most powerful sensors.

Corollary 8.4 computes the conditional connectivity and conditional fault-tolerance of heterogeneous $k$-covered wireless sensor networks.

*Corollary 8.4:* The conditional connectivity of heterogeneous $k$-covered wireless sensor networks is given by

$$\kappa_1(G:P) \le \kappa(G:P) \le \kappa_2(G:P) \tag{8.31}$$

where

$$\kappa_1(G:P) = \frac{4\,R_{\min}\,(R_{\min} + r_{\min})\,k}{r_{\min}^{\;2}}$$

$$\kappa_2(G:P) = \frac{4\,R_{\max}\,(R_{\max} + r_{\max})\,k}{r_{\min}^{\;2}}$$

$k \ge 3$ $r_{\min} = \min\{r_j : s_j \in S\}$, $r_{\max} = \max\{r_j : s_j \in S\}$, $R_{\min} = \min\{R_j : s_j \in S\}$, and $R_{\max} = \max\{R_j : s_j \in S\}$. The conditional fault tolerance of the network is given by

$$\kappa_1(G:P) - 1 \le \eta(G:P) \le \kappa_2(G:P) - 1 \tag{8.32}$$

*Proof:* First assume that the annulus as well as the area surrounding it contains only least powerful sensors, and hence its width is equal to $R_{\min}$. Furthermore, in order to guarantee that the sink will not be isolated, which would violate the forbidden faulty sensor set constraint, the width of the Reuleaux triangle centered at the location $\xi_0$ of the sink $s_0$ should be equal to $r_{\min}$. These two conditions yield a disconnected network that satisfies the forbidden faulty sensor set constraint. First assume that the annulus as well as the area surrounding it contains only least powerful sensors, and hence its width is equal to $R_{\min}$. The area of the annulus $A(\xi_0, R_{\min})$ is given by

$$\begin{aligned} |\,A(\xi_0, R_{\min})\,| &= |\,RT(\xi_0, 2\,R_{\min} + r_{\min})\,| - |\,RT(\xi_0, r_{\min})\,| \\ &= 2\,(\pi - \sqrt{3})\,R_{\min}\,(R_{\min} + r_{\min}) \end{aligned}$$

Hence, the conditional minimum number $n(P)$ of sensor failures to disconnect the connected component including the sink from the rest of the network is computed as

$$n_1(P) = \lambda(r_{\min}, k)\,|\,A(\xi_0, R_{\min})\,|$$

Thus, the lower bound on conditional connectivity is computed as

$$\kappa_1(G:P) = n_1(P) = \frac{4\,R_{\min}\,(R_{\min} + r_{\min})\,k}{r_{\min}^{\;2}} \tag{8.33}$$

where $r_{\min} = \min\{r_j : s_j \in S\}$ and $R_{\min} = \min\{R_j : s_j \in S\}$. To compute the upper bound on network connectivity, we assume that the sensors inside the annulus and around it are the most powerful ones. The analysis is similar to the previous one except that we just replace $r_{\min}$ by $r_{\max}$ and $R_{\min}$ by $R_{\max}$ in the denominator of the first part of Eq. 8.7 in order to disconnect the network while meeting the forbidden faulty set constraint. We found that the network connectivity is given by

$$\kappa_2(G:P) = \frac{4\,R_{\max}\,(R_{\max} + r_{\max})\,k}{r_{\min}^{\;2}} \tag{8.34}$$

where $r_{max} = \max\{r_j : s_j \in S\}$, $R_{min} = \min\{R_j : s_j \in S\}$, and $R_{max} = \max\{R_j : s_j \in S\}$. Thus, the conditional network connectivity of heterogeneous $k$-covered wireless sensor networks with respect to the forbidden faulty set constraint $P$ satisfies

$$\kappa_1(G : P) \leq \kappa(G : P) \leq \kappa_2(G : P)$$

and their conditional network fault-tolerance is given by

$$\kappa_1(G : P) - 1 \leq \eta(G : P) \leq \kappa_2(G : P) - 1$$ ■

Note that there is neither a polynomial-time algorithm for computing $\kappa(G : \mathrm{P})$ for a general graph nor any tight upper bound for $\kappa(G : \mathrm{P})$. However, our characterization *of* $k$-coverage based on the intersection of $k$ sensing disks and the Reuleaux triangle make it possible to compute the corresponding minimum sensor spatial density. This helps us derive conditional connectivity and conditional fault-tolerance of $k$-covered wireless sensor networks.

## 8.3 Related Work

Existing works on coverage and connectivity in wireless sensor networks assumed the concept of traditional connectivity. Our proposed approach, however, considers both concepts of traditional and conditional connectivity. The latter is based on the concept of forbidden faulty sensor set that includes subsets of sensors that cannot be faulty at the same time. Furthermore, our measures of connectivity and fault tolerance for $k$-covered wireless sensor networks take into consideration the morphology of wireless sensor networks, where the sink is the most crucial node. In this section, we describe existing work that exploited the concept of conditional connectivity.

The concept of conditional connectivity [101] has been investigated in several research works. Esfahanian [81] presented a new fault-tolerance analysis for the $n$-cube networks based on the concept of forbidden faulty set. Latifi et al. [130] introduced a new measure of conditional connectivity for the $n$-dimensional cube, where every node is required to have at least $g$ good neighbours. Wu and Guo [201] computed the fault tolerance of the $m$-ary $n$-dimensional hypercubes using forbidden faulty sets. Also, Chen et al. [60] proposed a probabilistic approach for computing the fault tolerance of hypercube network using forbidden faulty sets. Malde and Oellermann [151] introduced the notion of $F$-connectivity as the smallest number of vertices of $G$ whose removal produces a trivial graph or a disconnected graph with each component a subgraph of $F$, where $F$ is an induced subgraph of $G$. Oellermann [163] proposed the $\mathrm{P}$-connectivity of a graph $G$ with respect to hereditary properties, where every induced subgraph $F$ of a graph $G$ having property $\mathrm{P}$ also has property $\mathrm{P}$. Ammari and Das [29] proposed measures of conditional fault-tolerance of $k$-covered wireless sensor networks but considered connectivity from graph theory perspective, which is quite different from connectivity with the sink. Indeed, network connectivity is not necessarily a condition for the network to operate whereas connectivity to the sink is. Thus, more realistic measures of fault tolerance should be defined with respect to the sink.

## 8.4 Summary

One important issue in the design of wireless sensor networks is fault tolerance. Precisely, the network should remain functionally connected in spite of some sensor failures. In the context of graph theory, connectivity is an appropriate measure of fault tolerance. In this chapter, we have investigated coverage, connectivity, and fault tolerance in $k$-covered wireless sensor networks [13]. In order to compute their connectivity and fault tolerance, we have found that it is necessary to compute the minimum sensor spatial density to guarantee $k$-coverage of a convex field. For this purpose, we have characterized $k$-coverage based on the intersection of sensing disks of $k$ sensors and the Reuleaux Triangle model. Our measures take into account the size of the connected component that includes the sink. Indeed, the sink is a very critical node in data collection and decision making. Regardless of the degree of coverage $k$ provided by the network, we have proved that connectivity is always higher than $k$. We have derived fault tolerance of $k$-covered wireless sensor networks based on connectivity. Traditional (or unconditional) connectivity assumes that any subset of sensors can be potentially faulty at the same time. For large-scale dense wireless sensor networks, such as $k$-covered wireless sensor networks, this assumption is not realistic. To alleviate this problem, we have proposed a new measure of fault-tolerance for $k$-covered wireless sensor networks based on the concepts of conditional connectivity and forbidden faulty sensor set, where the entire neighbour set of a given sensor cannot fail at the same time. We have proved that $k$-covered wireless sensor networks can sustain a large number of sensor failures under this assumption. We believe that our results can be used in the design of wireless sensor networks with prescribed degrees of coverage, connectivity, and fault tolerance. In Appendix 1, we provide network and fault-tolerance measures for three-dimensional homogeneous and heterogeneous $k$-covered wireless sensor networks using the same approach for their counterpart in two-dimensional deployment fields.

# Chapter 9
# Geographic Forwarding on Always-On Sensors

This chapter addresses the problem of energy-efficient data forwarding in wireless sensor networks. It presents our solution to this problem for always-on wireless sensor networks, where the sensors are active all the time. Specifically, it investigates both short-range and long-range data forwarding schemes on always-on sensors and provides several theoretical and simulation results. It shows that short-range data forwarding scheme is more appropriate than long-range data forwarding scheme to save energy of the sensors and promote the longevity of wireless sensor networks with scarce energy resources. Our proposed solution is based on the geometric properties of Delaunay triangulation.

## 9.1 Introduction

In contrast to traditional wireless networks, energy efficiency is a critical determinant to extend the lifetime of wireless sensor networks. *Data forwarding* is an essential function in *wireless sensor networks*. As discussed earlier, the sensors communicate with each other via wireless, multi-hop links and have limited *battery power* (or *energy*), which is the most crucial resource. With these challenges in mind, energy efficiency is the primary *constraint* that should be met in the design and implementation of protocols for wireless sensor networks, which will be used by the sensors in their sensing, communication, and processing tasks. Because energy is the most crucial resource for the sensors to perform efficiently and correctly, the design of energy-efficient data forwarding protocols for wireless sensor networks has been receiving much attention to extend the network lifetime.

It is well-known that *geographic forwarding* is an energy-efficient and practical scheme for wireless sensor networks in the sense that the sensors are not required to maintain global and detailed information on the topology of the entire network. The sensors need only maintain local knowledge on their one-hop communication neighbours with respect to their geographic location information in order to progress data towards their final destinations, i.e., to forward data to a central gathering point, i.e., the sink, for further analysis and processing.

There is an ongoing debate on short-range versus long-range data forwarding in multi-hop wireless networks. This chapter supports the short-range data forwarding strategy for wireless sensor networks, where energy should be given the highest priority [20]. Precisely, our protocol is designed to efficiently use the limited energy of the sensors by minimizing their average energy consumption in forwarding data originated from sources to a single sink with a goal to prolong the

H.M. Ammari: Challenges and Opportunities of Connected $k$-Covered WSNs, SCI 215, pp. 175–200.
springerlink.com 

operational network lifetime. Our proposed data forwarding protocol exploits the properties of Delaunay triangulation to forward data via short range.

The remainder of this chapter is organized as follows. Section 9.2 describes our WLDT protocol. Section 9.3 analyzes WLDT while Sect. 9.4 discusses both short-range and long-range data forwarding schemes and presents several theoretical results. Section 9.5 addresses a few reliability issues of WLDT. Section 9.6 reviews related work. Section 9.7 summarizes the chapter.

## 9.2 The WLDT Protocol

In this section, we propose an energy-efficient data forwarding protocol for wireless sensor networks so they remain operational as long as possible. Our protocol, *Weighted Localized Delaunay Triangulation-based data forwarding* (WLDT), uses 1-lookahead scheme to guarantee data delivery to the sink. WLDT aims to minimize the average energy consumption of the sensors during data forwarding towards the sink. It exploits the geometric properties of the Delaunay triangulation [36] to build an energy-efficient path between the source and the sink as a sequence of sub-paths whose endpoints are called *checkpoints*. These checkpoints are selected based on their location in the field and their remaining energy. A sub-path between a pair of checkpoints consists of a *series of forwarders*, which are the endpoints of short Delaunay edges and are selected based on their location and remaining energy to forward the data between their checkpoints. Next, we describe WLDT in details.

### *9.2.1 Long-Range Versus Short-Range Forwarding*

Let $S=\{s_0,...,s_{n-1}\}$ be a set of $n$ sensor nodes and $s_m$ the single sink connected by wireless links over a wireless sensor network.

Lemma 9.1 states that, under some specific condition, data forwarding through a short-range forwarding scheme is more energy efficient than that using a long-range forwarding scheme. We assume that the sensors $s_i$ and $s_j$ are one-hop neighbours of each other. Note that the "one-hop neighbour" relationship is symmetric. In other words, we assume that all the sensors are homogeneous, i.e., have the same physical capabilities and, in particular, their transmission range.

*Lemma 9.1:* The total energy consumption for forwarding one data packet from the sensor $s_i$ to the sensor $s_j$ along a short-range path is smaller than the energy spent along a long-range path between them if there is a sensor $s_k \in NS(s_i)$ such that

$$\delta(s_k,s_{k,p}) < \sqrt{\left(\frac{d^2}{2}-\frac{E_{elec}}{\varepsilon}\right)^{2/\alpha}-\frac{d^2}{4}}$$

where $\delta(s_i,s_j)=d$ and $s_{k,p}$ is the orthogonal projection of $s_k$ on the segment $\overline{s_is_j}$ .

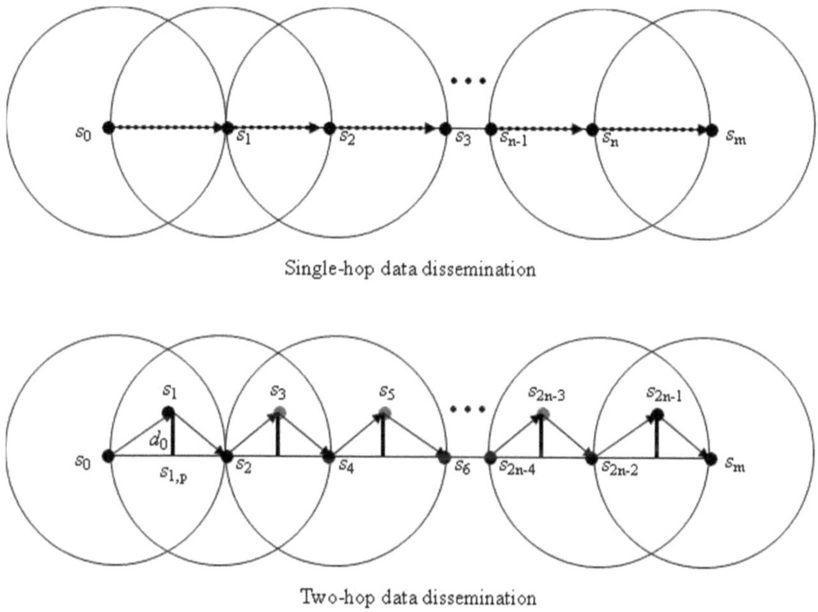


**Fig. 9.1** Single-hop vs. two-hop data forwarding

*Proof:* Let us compute the critical distance $d_0$ between $s_k$ and $s_{k,p}$ for which long-range and short-range data forwarding consume the same energy. Let

$$E_{1-hop}(s_i, s_j) = 2\,\kappa\, E_{elec} + \kappa\, \varepsilon\, \delta^{\alpha}(s_i, s_j)$$

and

$$E_{2-hop}(s_i, s_j) = 4\,\kappa\, E_{elec} + \kappa\, \varepsilon\, (\delta^{\alpha}(s_i, s_k) + \delta^{\alpha}(s_k, s_j))$$

be the energy consumption required to forward the data from $s_i$ to $s_j$ through 1-hop and 2-hop paths, respectively. It is easy to check that $E_{2-hop}(s_i, s_j)$ reaches its minimum when $\delta(s_i, s_k) = \delta(s_k, s_j)$. Thus, $E_{1-hop}(s_i, s_j) = E_{2-hop}(s_i, s_j)$ implies

$$\delta(s_i, s_k) = \left( \frac{d^2}{2} - \frac{E_{elec}}{\varepsilon} \right)^{1/\alpha}$$

where $\delta(s_i, s_j) = d$ . By Pythagorean Theorem, $\delta^2(s_i, s_k) = \delta^2(s_i, s_{k,p}) + {d_0}^2$, where $d_0 = \delta(s_k, s_{k,p})$, i.e.,

$$d_0 = \sqrt{\left( \frac{d^2}{2} - \frac{E_{elec}}{\varepsilon} \right)^{2/\alpha} - \frac{d^2}{4}}$$

Thus, a short-range path between the sensors $s_i$ and $s_j$ contains at least one forwarder between them if $\delta(s_k, s_{k,p}) < d_0$. ■

The WLDT protocol is based on Lemma 9.1 and the result in [32] which states that the total energy consumption for forwarding one data packet from the source $s_0$ to the sink $s_m$ reaches its minimum value only when all forwarders between $s_0$ and $s_m$ lie on the line segment $[s_0, s_m]$ between $s_0$ and $s_m$.

Lemma 9.2 compares the average energy consumption of the sensors in short-range and long-range data forwarding schemes.

*Lemma 9.2:* Long-range data forwarding leads to higher average energy consumption of the sensors than short-range data forwarding even when the total energy consumption in both schemes is the same.

*Proof:* According to Lemma 9.1, it is clear that when $\delta(s_k, s_{k,p}) = d_0$, the total energy consumption required for long-range and short-range data forwarding is the same. Without loss of generality, let us assume that long-range data forwarding uses the shortest path between the source $s_0$ and the sink $s_m$. Moreover, we assume that all the forwarders are located at the perimeter of the transmission ranges (within distance $r$ from each other). The average energy consumption of the sensors participating in long-range data forwarding is given by

$$E_{avg}^{sh}(s_i) = 2\,\kappa\, E_{elec} + \kappa\,\varepsilon\, r^{\alpha}$$

We assume that we deal with two-hop data forwarding, where half of the sensors lie on the shortest path $[s_0, s_m]$ and the other half lie above it as shown in Fig. 9.1, where $\delta(s_i, s_{i+1}) = \left(\frac{r^2}{2} - \frac{E_{elec}}{\varepsilon}\right)^{1/\alpha}$ (see Lemma 9.1). Hence, the average energy consumption of the sensors in two-hop data forwarding is $E_{avg}^{th}(s_i) = 2\,\kappa\, E_{elec} + \kappa\,\varepsilon\, \delta^{\alpha}(s_i, s_{i+1})$. Thus,

$$E_{avg}^{th}(s_i) = 2\,\kappa\, E_{elec} + \kappa\,\varepsilon \left(\frac{r^2}{2} - \frac{E_{elec}}{\varepsilon}\right)$$

Note that $\left(\frac{r^2}{2} - \frac{E_{elec}}{\varepsilon}\right) < r^{\alpha}$, for all $\alpha \geq 2$. Thus,

$$E_{avg}^{th}(s_i) < E_{avg}^{sh}(s_i) \qquad \blacksquare$$

*Corollary 9.1:* Assume $\alpha = 2$ (free space model) and $s_k \in NS(s_i)$ lies on $\overline{s_i s_j}$. If $\delta(s_i, s_k)\delta(s_k, s_j) > \frac{E_{elec}}{\varepsilon}$, Then

$$E_{1-hop}(s_i, s_k) + E_{1-hop}(s_k, s_j) < E_{1-hop}(s_i, s_j) \qquad \blacksquare$$

Corollary 9.1 recommends two-hop data forwarding between any pair of sensors $s_i$ to $s_j$ only if they are separated by a distance $\delta(s_i, s_j)$ such that

$$\delta(s_i, s_k)\delta(s_k, s_j) > \frac{E_{elec}}{\varepsilon}$$

where $s_k$ is a forwarder that lies on the line segment $[s_i, s_j]$.

Algorithm: WLDT
Begin
*// Actions executed by source* $s_0$
Step 1: Checkpoint selection
1. Compute the localized DT, $LDT(s_0)$
2. Find the subset of candidate checkpoints, $Clos(s_0, s_m)$
3. Select the checkpoint $s_p$ using equation (9.1)

Step 2: Checkpoint-based short-range data forwarding
4. Consider the localized DT, $LDT(s_0)$
5. Assign weights to the Delaunay edges adjacent to $s_0$
6. Sort the weighted Delaunay edges adjacent to $s_0$
7. Run the 1-lookahead scheme to find a forwarder $s_k$
8. If $s_k$ found Then
9.   Forward the sensed data to $s_k$
  Else
10.   Forward the sensed data directly to $s_p$
  EndIf

*// Actions executed by any checkpoint or forwarder* $s_i$
11. While (data has not reached the sink $s_m$) Do
12.   If $s_i$ is a forwarder Then
13.     Execute Steps 4-11 and replace $s_0$ with $s_i$
    Else
14     If $s_i$ is a checkpoint Then
15       If sink node $s_m \in NS(s_i)$ Then
16         save $temp = s_i$
17         set $s_p = s_m$
18         Execute Steps 4-11 and replace $s_0$ with *temp*
       Else
19         Execute Steps 1-11 and replace $s_0$ with $s_p$
       EndIf
     EndIf
   EndIf
 EndWhile
End

**Fig. 9.2** The WLDT protocol

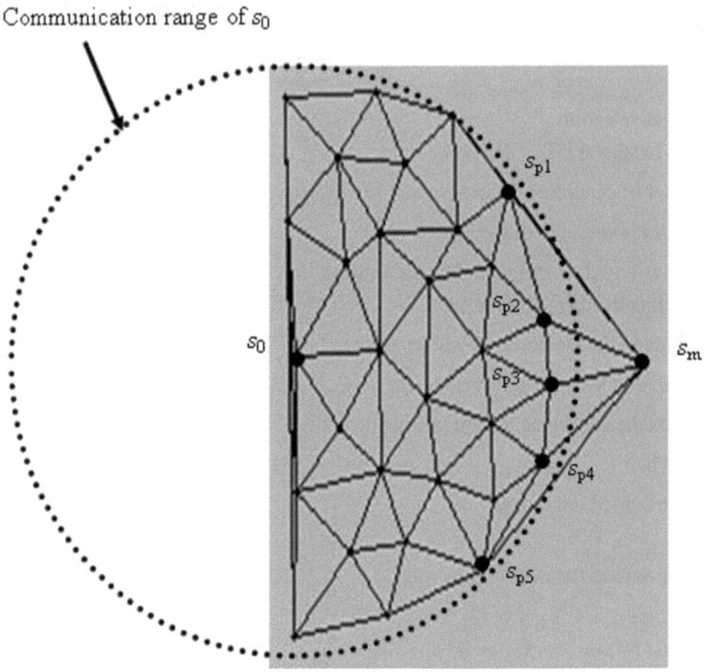


**Fig. 9.3** Localized DT and candidate checkpoints

## 9.2.2 A Two-Step Data Forwarding Protocol

We propose a data forwarding protocol (Fig. 9.2), which benefits from the energy gain introduced by short-range data forwarding as stated in Lemmas 9.1 and 9.2 and Corollary 9.1, and uses the geometric properties of the Delaunay triangulation (DT). Since we are interested in the area between the sending and the receiving sensors, the neighbour set $NS(s_i)$ of the sensor $s_i$ will contain only the sensors located between $s_i$ and the sink $s_m$. Figure 9.3 shows a localized DT as well as a set of candidate checkpoints of $s_0$ (Chap. 2, Definitions 2.17 and 2.18).

The WLDT protocol is composed of two steps: *checkpoint selection* and *checkpoint-based short-range data forwarding*.

### 9.2.2.1 Checkpoint Selection

We consider a scenario where the source $s_0$ wishes to forward its sensed data to the sink $s_m$. The goal of this step is to identify a subset of candidate checkpoints, $CCP(s_0,s_m)$, of $NS(s_0)$ that are closest to the sink $s_m$. First, the source $s_0$ constructs its localized Delaunay triangulation, $LDT(s_0)$, as shown in Fig. 9.3. Intuitively, only the sensors in $CCP(s_0,s_m)$ will be able to get the sensed data out of the transmission range of $s_0$. Figure 9.3 shows the $LDT(s_0)$ and the subset $CCP(s_0,s_m)$ that includes $s_{p1}$, $s_{p2}$, $s_{p3}$, $s_{p4}$, and $s_{p5}$, which are adjacent to the sink $s_m$ in the $LDT(s_0)$.

From the subset of candidate checkpoints, $CCP(s_0,s_m)$, the source $s_0$ selects the checkpoint sensor $s_p$ such that

$$CE_p = \max\{CE_i = c_i\, e_i : s_i \in CCP(s_0,s_m)\} \tag{9.1}$$

where

$$c_i = \frac{\delta(s_0,s_m)}{\delta(s_0,s_i)+\delta(s_i,s_m)} \tag{9.2}$$

and

$$e_i = \frac{E_{rem}(s_i)}{\sum_{s_j \in CCP(s_0,s_m)} E_{rem}(s_j)} \tag{9.3}$$

As can be seen, the weight $c_i$ measures the degree of *closeness* of $s_i$ to the shortest path $[s_0,s_m]$, while the term $e_i$ is the percentage of the remaining energy of the sensor $s_i$ with respect to the total remaining of the subset of sensors $CCP(s_0,s_m)$. Note that $c_i$ attains its maximum $(c_i = 1)$ when $s_i$ lies on $[s_0,s_m]$. Intuitively, the source $s_0$ and the sink $s_m$ are the first and last checkpoints, respectively.

#### 9.2.2.2 Checkpoint-Based Short-Range Forwarding

The objective is to forward the sensed data to the checkpoint $s_p$ that was selected in the previous step. First, the source $s_0$ assigns weights to each of the Delaunay edges adjacent to it as follows: if $s_j$ is an adjacent node to $s_0$ in the $LDT(s_0)$, then the weight placed on the edge $(s_0,s_j)$ is $CED_j(s_0) = c_j\, e_j\, d_j$, where

$$c_j = \frac{\delta(s_0,s_p)}{\delta(s_0,s_j)+\delta(s_j,s_p)}$$

$$e_j = \frac{E_{rem}(s_j)}{\sum_{s_k \in Adj(s_0)} E_{rem}(s_k)}$$

and

$$d_j = \frac{1/\delta(s_0,s_j)}{\sum_{s_k \in Adj(s_0)} 1/\delta(s_0,s_k)}$$

where $Adj(s_0)$ denotes the subset of sensors adjacent to $s_0$ in the $LDT(s_0)$. As can be observed, the term $d_j$ measures the degree of closeness of the sensor $s_j$ to the source $s_0$. This means that $s_0$ favours closer sensors so it transmits its sensed data over short distances and hence saves its energy. Then, the source $s_0$ selects its next forwarder using $1-lookahead$ scheme described as follows:

- The source $s_0$ sorts the list of Delaunay edges adjacent to it, *sorted-list*, based on their weights in decreasing order.
- The source $s_0$ considers the node, $s_h$, with the highest weight and examines its adjacent neighbours in the $LDT(s_0)$. If $s_h$ has at least one path that originates from one of its adjacent sensors and leads to the checkpoint $s_p$ using only positive progress (condition COND), the node $s_h$ is selected as the next forwarder. Otherwise, the source $s_0$ repeatedly picks the next node in *sorted-list* and checks if it satisfies COND.

The WLDT protocol is said to be $1-lookahead$ because any forwarding sensor uses the information about the adjacent of its adjacent sensors so that it can make appropriate forwarding decision. At the end of this phase, the source sensor identifies its appropriate forwarder and forwards the sensed data to it. By Lemma 9.1, the source $s_0$ prefers to forward its data to its corresponding checkpoint (located at the perimeter of its transmission range) through a series of forwarders, hence favouring short-range over long-range data forwarding. This will enable all the sensors deplete their energy slowly, thus extending the network lifetime. From the function $CED_j$, it is clear that when the sensor $s_0$ selects a forwarder $s_k$, it takes into consideration three metrics: remaining energy of the sensor $s_k$, position of the sensor $s_k$ with respect to the shortest path $[s_0,s_p]$, and the Euclidean distance, $\delta(s_0,s_k)$, between the sensors $s_k$ and $s_j$; recall that $E_{tr}(s_0,s_k) \propto \delta(s_0,s_k)$. This means that WLDT attempts to build energy-efficient sub-paths between the source and its checkpoint or between any pair of consecutive checkpoints, which include a series of forwarders linked by short Delaunay edges. Then, the source $s_0$ fills in two fields in the data packet, namely *Checkpoint*, which contains the checkpoint $s_p$, and *Forwarder*, which contains the forwarder $s_h$, and forwards the data to $s_h$. When a node $s_k$ receives the sensed data, it will examine both fields to check whether it is a checkpoint or a forwarder. If $s_k$ is a checkpoint, it will act like the source $s_0$ by running steps 1 and 2. Otherwise, it will run only step 2. The pseudo-code of WLDT is given in Fig. 9.2.

### 9.2.3 Illustrative Example

Figure 9.4 shows a path marked by arrows between the source $s_0$ and its checkpoint $s_{p2}$. Data are forwarded by $s_1$, $s_2$, $s_3$, and $s_4$ along the corresponding Delaunay edges before it reaches $s_{p2}$. In order to identify its checkpoint, $s_0$

**Fig. 9.4** Data forwarding path between $s_0$ and $s_{p2}$

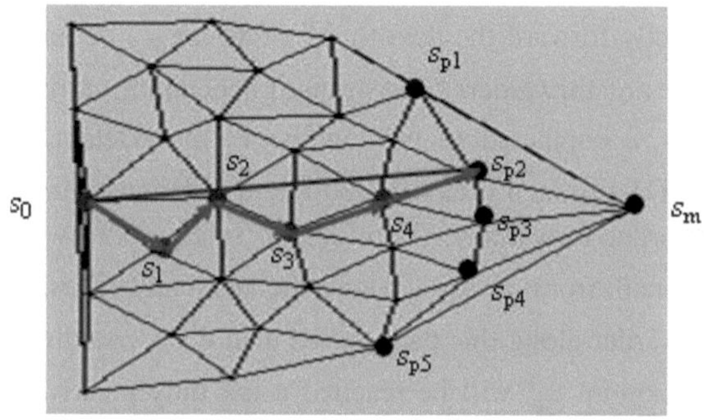


constructs $LDT(s_0)$. Then, it computes its forwarder $s_1$, sets up fields *Checkpoint* $= s_{p2}$ and *Forwarder* $= s_1$, and forwards its data to $s_1$, which in turn forwards it to $s_2$. The same forwarding process repeats until $s_{p2}$ receives the data. When $s_{p2}$ gets the data, it will act as $s_0$ in order to forward the data to its next checkpoint. The entire process of determining checkpoints and series of forwarders between any pair of consecutive checkpoints repeats until the sink $s_m$ receives the data. At each forwarding step, the fields *Checkpoint* and *Forwarder* are updated accordingly.

## 9.3 Analysis of WLDT

In this section, we compute a lower bound on the energy consumption in short-range data forwarding and the corresponding optimum number of forwarders between a source and the sink. This bound helps extend the battery lifetime of the sensors, thus prolonging the operational network lifetime. Specifically, we prove that any checkpoint on the data forwarding path between the source $s_0$ and the sink $s_m$ is reachable (Lemma 9.3), and hence the sink itself (Corollary 9.2). Next, we approximate the length of any edge in the transmission graph $G = (S,T)$ (Lemma 9.4) and deduce the length of any Delaunay edge in $LDT(s_i)$ (Lemma 9.5). Then, we analyze the energy consumption in short-range data forwarding (Theorems 9.1 and 9.2) and compare WLDT with BVGF and GPSR, which implement a long-range data forwarding scheme (Theorems 9.3 and 9.4). Finally, we show that checkpoints have a positive impact on data forwarding in terms of energy savings (Theorem 9.5).

*Lemma 9.3:* Any checkpoint between the source $s_0$ and the sink $s_m$ is reachable.

*Proof:* Let us prove that the checkpoint of the source $s_0$, say $s_p$, is reachable. Assume that the current forwarder (the sensor that currently holds the sensed data) is $s_i$. If the sensor $s_i$ has a Delaunay edge $\overline{(s_i,s_p)}$ in $LDT(s_0)$, then $s_i$ could either directly forward the data to $s_p$ (long-range) or forward the data to $s_p$ through a series of forwarders (short-range) depending on the length of the Delaunay edge $\overline{(s_i,s_p)}$ compared to the length of other Delaunay edges adjacent to $s_i$. The WLDT protocol uses a $1-$*lookahead* scheme, which helps the sensors choose appropriate forwarders. Therefore, $s_i$ selects its forwarder $s_j$ only if there is at least one path from $s_j$ to $s_p$ along the Delaunay edges, where the $x$-coordinate of any forwarder along this path is less than the $x$-coordinate of $s_p$. This means that the checkpoint $s_p$ will be reached using only positive progress from any forwarder between the source $s_0$ and the sink $s_m$. Using the same argument as above, we can prove that every checkpoint between $s_0$ and $s_m$ is reachable. ■

The sink $s_m$ is also a checkpoint, and hence should be reachable. Corollary 9.2 below follows directly from Lemma 9.3.

*Corollary 9.2:* Any sensed data originated from the source $s_0$ is guaranteed to reach the sink $s_m$, assuming no packet loss. ■

The total energy consumption in data forwarding depends on the distance between the sending and receiving nodes. Lemma 9.4 approximates the minimum length of any edge between any pair of sensors in the transmission graph $G=(S,T)$.

*Lemma 9.4:* The *minimum edge length* in the transmission graph $G=(S,T)$ can be approximated by

$$d_{\min}=\left(\frac{E_{elec}}{\varepsilon}\right)^{1/\alpha} \tag{9.4}$$

*Proof:* We need to compute the *minimum transmission distance* used by a sensor in data transmission. According to the energy model [102] discussed earlier in Chap. 2, $E_{rx}=\kappa E_{elec}$ and $E_{tx}(d)=\kappa E_{elec}+\kappa\varepsilon d^{\alpha}$, where $d$ is the transmission distance used by a sender. By Assumption 4, $E_{rx}$ represents the minimum energy which could be spent by a given sensor. It is always the case that when a sensor transmits, it always consumes more than when it receives. However, we can approximate the second term in the formula of $E_{tx}(d)$, which corresponds to the energy consumed by the amplifier transmitter component, with the energy consumed by the receiver in data reception. Furthermore, if we consider the values of $E_{elec}$ and ε given later in Table 1, we observe that $E_{elec}$ is much higher than ε. Thus, there must be certain value of the distance $d$ such that $\varepsilon d = E_{elec}$. Hence, the minimum value of $\kappa\varepsilon d^{\alpha}$ in $E_{tx}$ can be approximated by $E_{rx}$. That is, $\kappa\varepsilon d^{\alpha}=\kappa E_{elec}$. Thus, the minimum transmission distance that can be used by a sensor in data transmission is equal to

$$d_{\min}=(E_{elec}/\varepsilon)^{1/\alpha}$$ ■

It is worth noting that the *minimum transmission distance* $d_{\min}$ is totally different from the *minimum distance between neighbouring sensors*. While the former is quantified by using energy parameters, the latter is a geometric parameter that can be quantified based on sensor density as one possible parameter. In fact, the minimum distance between any pair of neighbouring sensors can be very small if the sensor density if very high.

Lemma 9.5 computes the length of any Delaunay edge in a localized DT, $LDT(s_i)$, based on the result of Lemma 9.4.

*Lemma 9.5:* The length of any Delaunay edge between two sensors $s_j$ and $s_k$, denoted by $\overline{(s_j, s_k)}$, in a localized DT, $LDT(s_i)$, satisfies

$$d_{\min} \leq \overline{(s_j, s_k)} \leq r$$

*Proof:* Let $G(s_i)$ be the sub-graph of the transmission graph $G = (S, T)$ induced by $NS(s_i) \cup \{s_m\}$, and $LDT(s_i)$ the localized DT of $s_i$. First, both $G(s_i)$ and $LDT(s_i)$ have the same vertex set $NS(s_i) \cup \{s_m\}$. By definition of $G(s_i)$, an edge $t_{j,k} = (s_j, s_k)$ if and only if $\delta(s_j, s_k) \leq r$. On the one hand, if $\delta(s_j, s_k) = d_{\min}$, then the corresponding Voronoi diagram should have two Voronoi cells $VC(s_j)$ and $VC(s_k)$ with a common Voronoi edge. By definition of the DT, there must be a Delaunay edge connecting $s_j$ and $s_k$ whose length is $\overline{(s_j, s_k)} = \delta(s_j, s_k)$. Thus, $\overline{(s_j, s_k)} = d_{\min}$. On the other hand, no edge between two sensors exists unless they are within the transmission range of each other. Therefore, the maximum length of any Delaunay edge cannot exceed $r$. Hence, $\overline{(s_j, s_k)} \leq r$. Both results yield

$$d_{\min} \leq \overline{(s_j, s_k)} \leq r$$ ■

Theorem 9.1 computes the minimum energy consumption and the corresponding optimum number of forwarders required for forwarding one data packet from the source $s_0$ to the sink $s_m$.

*Theorem 9.1:* The lower bound on the energy consumption required to forward one sensed data packet from the source $s_0$ to the sink $s_m$ along $[s_0, s_m]$ is given by

$$E_{\min}(s_0, s_m) = 2^{\alpha-1} \kappa \frac{\delta(s_0, s_m)}{d_{\min}} \times [(\alpha - 1) E_{elec} + 2^{1/\alpha} (\alpha - 1)^{1/\alpha - 1} \varepsilon \, {d_{\min}}^{\alpha}] \tag{9.5}$$

and the corresponding optimum number of forwarders is

$$m_{opt}(s_0, s_m) = \left(\frac{\alpha - 1}{2}\right)^{1/\alpha} \frac{\delta(s_0, s_m)}{d_{\min}} \tag{9.6}$$

*Proof:* Assume that there are $m$ forwarders including $s_0$ $(s_0, s_1, \ldots, s_{m-1})$ between the source $s_0$ and the sink $s_m$. The total energy $E_{tot}(s_0, s_m)$ spent in forwarding one sensed data packet from the source $s_0$ to the sink $s_m$ is computed as

$$E_{tot}(s_0, s_m) = E_{tot}(s_0, s_1) + E_{tot}(s_m) + \sum\nolimits_{i=1..m-1} E_{tot}(s_i)$$

where $E_{tot}(s_0, s_1) = \kappa\, E_{elec} + \kappa\, \varepsilon\, \delta^{\alpha}(s_0, s_1)$, $E_{tot}(s_m) = \kappa\, E_{elec}$, and $E_{tot}(s_i, s_{i+1}) = 2\,\kappa\, E_{elec} + \kappa\, \varepsilon\, \delta^{\alpha}(s_i, s_{i+1})$ (Chap. 2, Sect. 2.5). Thus,

$$E_{tot}(s_0, s_m) = \sum\nolimits_{i=0..m-1} 2\,\kappa\, E_{elec} + \kappa\, \varepsilon\, \delta^{\alpha}(s_i, s_{i+1})$$

In general, the distances between all pairs of consecutive forwarders are not equal. Indeed, $E_{tot}(s_0, s_m)$ reaches its minimum when all distances $\delta(s_i, s_{i+1})$ are the same. In other words, $\delta(s_i, s_{i+1}) = \frac{\delta(s_0, s_m)}{m}$. Thus, $E_{tot}(s_0, s_m) \geq E(m)$, where

$$E(m) = \sum\nolimits_{i=0..m-1} 2\,\kappa\, E_{elec} + \kappa\, \varepsilon \left(\frac{\delta(s_0, s_m)}{m}\right)^{\alpha}$$

Thus,

$$E(m) = 2\, m\, \kappa\, E_{elec} + \kappa\, \varepsilon \frac{\delta^{\alpha}(s_0, s_m)}{m^{\alpha - 1}} \tag{9.7}$$

The function $E(m)$ reaches its minimum when $\frac{\partial\, E(m^*)}{\partial\, m^*} = 0$. It is easy to check that $\frac{\partial^2\, E(m^*)}{\partial^2\, m^*} > 0$, for all $\alpha \geq 2$. That is, $E(m)$ is strictly convex as shown in Figs. 9.5 and 9.6. Hence, $m^*$ in $\frac{\partial E(m^*)}{\partial m^*} = 0$ corresponds to the minimum of $E(m)$. Thus, the optimum number of forwarders and checkpoints is given by

$$m_{opt}(s_0, s_m) = m^* = \left(\frac{\alpha - 1}{2}\right)^{1/\alpha} \frac{\delta(s_0, s_m)}{d_{\min}}$$

and hence the minimum energy consumption required in data forwarding between the source $s_0$ and the sink $s_m$ is

$$E_{\min}(s_0, s_m) = 2^{\alpha - 1}\, \kappa \frac{\delta(s_0, s_m)}{d_{\min}} \times [(\alpha - 1)\, E_{elec} + 2^{1/\alpha}\, (\alpha - 1)^{1/\alpha - 1}\, \varepsilon\, {d_{\min}}^{\alpha}]$$

■

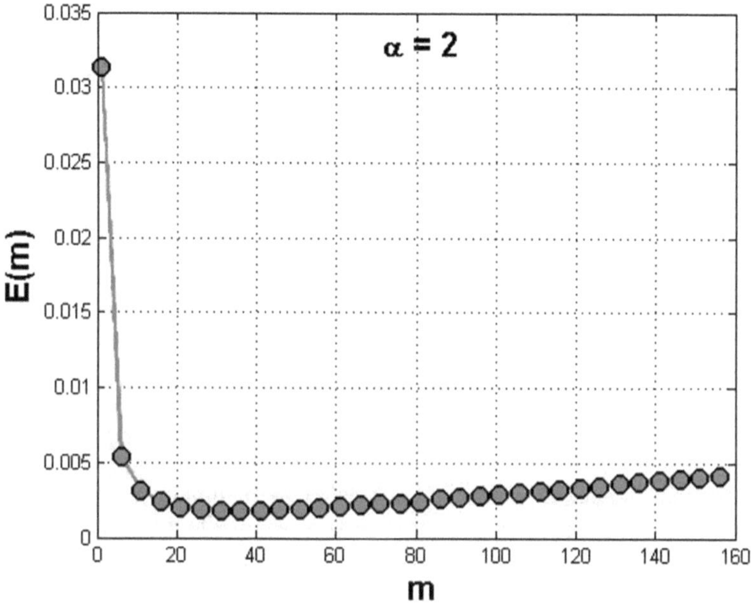


**Fig. 9.5** Energy function $E(m)$ for $\alpha = 2$

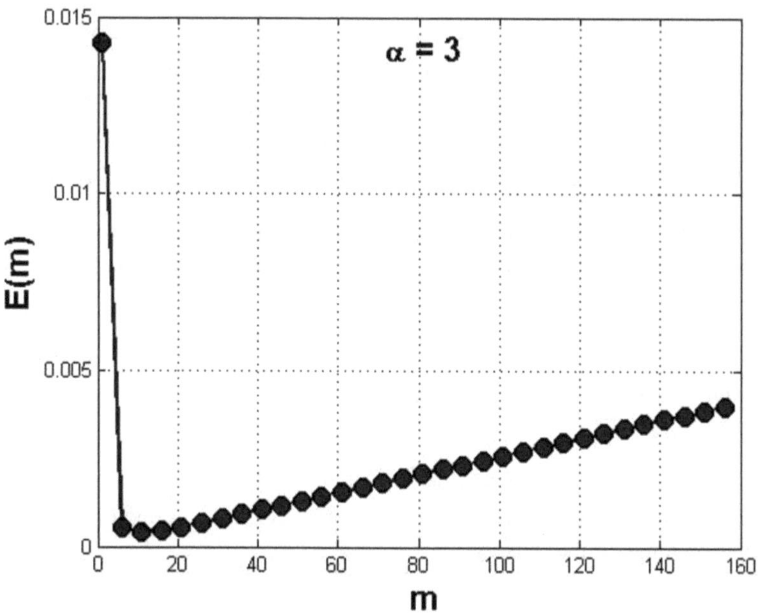


**Fig. 9.6** Energy function $E(m)$ for $\alpha = 3$

Theorem 9.1 shows that we made a good approximation of the transmission distance $d_{\min}$ between any pair of sensors in the transmission graph $G = (S,T)$. Given that $2 \le \alpha \le 4$, we have

$$0.707 \le \left(\frac{\alpha - 1}{2}\right)^{1/\alpha} \le 1.107$$

Thus, using the above result and Eq. 9.6, we obtain

$$m_{opt}(s_0, s_m) \approx \frac{\delta(s_0, s_m)}{d_{\min}}$$

Thus, we get

$$E_{\min}(s_0, s_m) \approx \kappa \frac{\delta(s_0, s_m)}{d_{\min}} (2\, E_{elec} + \varepsilon\, d_{\min}{}^{\alpha})$$

This means that the energy consumption reaches its minimum when the distance between any pair of consecutive forwarders is $d_{\min}$. The numerical values of all the constants are as follows: $\kappa = 216$, $E_{elec} = 50\,\text{nJ/bit}$, $\varepsilon_{fs} = 10\,\text{pJ/bit/m}^2$, and $\varepsilon_{mp} = 0.0013\,\text{pJ/bit/m}^2$. Figures 9.5 and 9.6 show the plot of $E(m)$ (see Eq. 9.7), where $d_{\min} = 70.71\,\text{m}$ for $\alpha = 2$ and $d_{\min} = 156.68\,\text{m}$ for $\alpha = 3$ (see Eq. 9.4). According to Theorem 9.1, the optimum number of forwarders (see Eq. 9.6) is $m_{opt}(s_0, s_m) \approx 50$ for $\alpha = 2$ and $m_{opt}(s_0, s_m) \approx 23$ for $\alpha = 3$ given that $\delta(s_0, s_m) = 3500\,\text{m}$. We find that the numerical results are consistent with our theoretical results since $50 \times 70.71$ m $\approx$ 3500 m and $23 \times 156.68$ m $\approx$ 3500 m.

Theorem 9.2 computes the expected minimum and maximum energy required by the WLDT protocol in data forwarding from the source $s_0$ to the sink $s_m$.

*Theorem 9.2:* The expected minimum and maximum energy spent in forwarding one data packet from the source $s_0$ to the sink $s_m$ are respectively given by

$$E_{\exp}^{\min}(s_0, s_m) = (2\,\kappa\, E_{elec} + \kappa\, \varepsilon\, d_{\min}{}^{\alpha}) \times (N_{\exp}^{\min}(s_0, s_m) + 1) \tag{9.8}$$

$$E_{\exp}^{\max}(s_0, s_m) = (2\,\kappa\, E_{elec} + \kappa\, \varepsilon\, d_{\min}{}^{\alpha}) \times (N_{\exp}^{\max}(s_0, s_m) + 1) \tag{9.9}$$

where:

- $d_{\min}$ : the expected minimum length of a Delaunay edge.
- $r$ : the expected distance between any pair of consecutive checkpoints ($r$ is the transmission range of the sensors).
- $N_{\exp}^{\min}(s_0, s_m) = \left\lceil \frac{\delta(s_0, s_m)}{d_{\min}} \right\rceil - 1$: the expected minimum number of forwarders

and checkpoints that lie on $[s_0, s_m]$.

- $N_{\exp}^{\max}(s_0,s_m)=\left\lceil\frac{r}{P(d_{\min},\theta_0)}\right\rceil\times\left\lceil\frac{\delta(s_0,s_m)}{P(r,\theta)}\right\rceil-1$ : the expected maximum number of forwarders and checkpoints that lie on the data forwarding path between $s_0$ and $s_m$.

- $P(r,\theta)=\delta(s_0,s_m)-\sqrt{r^2+\delta^2(s_0,s_m)-2\,r\,\delta(s_0,s_m)\cos\theta}$ : the progress made towards the sink $s_m$.

- $P(d_{\min},\theta_0)=r-\sqrt{{d_{\min}}^2+r^2-2\,d_{\min}\,r\cos\theta_0}$ : the progress made towards a checkpoint.

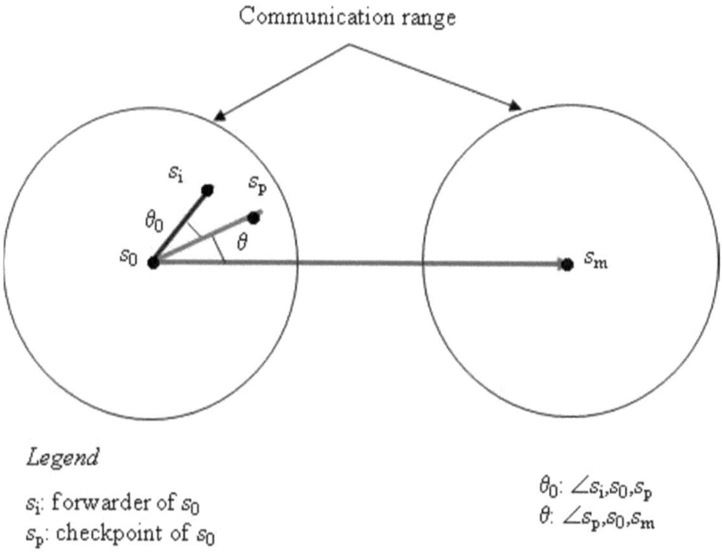


**Fig. 9.7** Progress made towards $s_p$ and $s_m$

*Proof:* The expected minimum energy consumption in data forwarding occurs when all forwarders and checkpoints between the source $s_0$ and the sink $s_m$ lie on $[s_0,s_m]$. Because the checkpoints are located at the perimeter of the transmission range of the sensors, the expected distance between any pair of consecutive checkpoints is $r$. Thus, the expected minimum number of checkpoints between the source $s_0$ and the sink $s_m$ is $\left\lceil\frac{\delta(s_0,s_m)}{r}\right\rceil-1$. Similarly, the expected minimum number of forwarders between any pair of consecutive checkpoints is $\left\lceil\frac{r}{d_{\min}}\right\rceil-1$, where the expected distance between any pair of consecutive forwarders is $d_{\min}$. Recall that our protocol prefers forwarding the sensed data via a series of forwarders linked by short Delaunay edges in $LDT(s_i)$. Thus, the expected minimum total number of forwarders and checkpoints is given by

$$N_{\exp}^{\min}(s_0,s_m) = \left\lceil \frac{r}{d_{\min}} \right\rceil \times \left\lceil \frac{\delta(s_0,s_m)}{r} \right\rceil - 1 = \left\lceil \frac{\delta(s_0,s_m)}{d_{\min}} \right\rceil - 1$$

Because $E_{tot}(s_0) = \kappa E_{elec} + \kappa \varepsilon d_{\min}{}^{\alpha}$, $E_{tot}(s_j) = 2\kappa E_{elec} + \kappa\varepsilon d_{\min}{}^{\alpha}$ ($s_j$ is a forwarder or a checkpoint) and $E_{tot}(s_m) = \kappa E_{elec}$, the expected minimum energy consumption is

$$E_{\exp}^{\min}(s_0,s_m) = [E_{tot}(s_0) + (2\kappa E_{elec} + \kappa \varepsilon d_{\min}{}^{\alpha}) \times N_{\exp}^{\min}(s_0,s_m)] + E_{tot}(s_m)$$

Thus,

$$E_{\exp}^{\min}(s_0,s_m) = (2\kappa E_{elec} + \kappa\varepsilon d_{\min}{}^{\alpha}) \times (N_{\exp}^{\min}(s_0,s_m) + 1)$$

Now, we consider data forwarding paths that do not coincide with the shortest path $[s_0,s_m]$. Let $P(r,\theta)$ be the progress made towards the sink at each forwarding action, where $\theta = \angle s_p, s_0, s_m$ is the expected maximum angle between segments $[s_p,s_0]$ and $[s_p,s_m]$ (Fig. 9.7). By definition, $P(r,\theta) = \delta(s_0,s_m) - \delta(s_p,s_m)$, where $\delta(s_p,s_m) = \sqrt{r^2 + \delta^2(s_0,s_m) - 2r\,\delta(s_0,s_m)\cos\theta}$ and $r$ is the expected distance between two consecutive checkpoints. Thus, the expected maximum number of checkpoints is $\left\lceil \frac{\delta(s_0,s_m)}{P(r,\theta)} \right\rceil - 1$. Likewise, the expected maximum number of forwarders between any pair of consecutive checkpoints is $\left\lceil \frac{r}{P(d_{\min},\theta_0)} \right\rceil - 1$, where the expected distance between any pair of consecutive forwarders is $d_{\min}$. Therefore, the expected maximum number of forwarders and checkpoints between the source $s_0$ and the sink $s_m$ is

$$N_{\exp}^{\max}(s_0,s_m) = \left\lceil \frac{r}{P(d_{\min},\theta_0)} \right\rceil \times \left\lceil \frac{\delta(s_0,s_m)}{P(r,\theta)} \right\rceil - 1$$

Thus, the expected maximum energy consumption for forwarding one sensed data packet from the source $s_0$ to the sink $s_m$ is given by

$$E_{\exp}^{\max}(s_0,s_m) = (2\kappa E_{elec} + \kappa \varepsilon d_{\min}{}^{\alpha}) \times (N_{\exp}^{\max}(s_0,s_m) + 1) \qquad \blacksquare$$

Figure 9.8 shows the plot of the function $E_{\exp}^{\min}(s_0,s_m)$ (see Eq. 9.8) for different values of the distance $\delta(s_0,s_m)$ between the source $s_0$ and the sink $s_m$. As can be seen, the minimum energy consumption grows proportionally to the

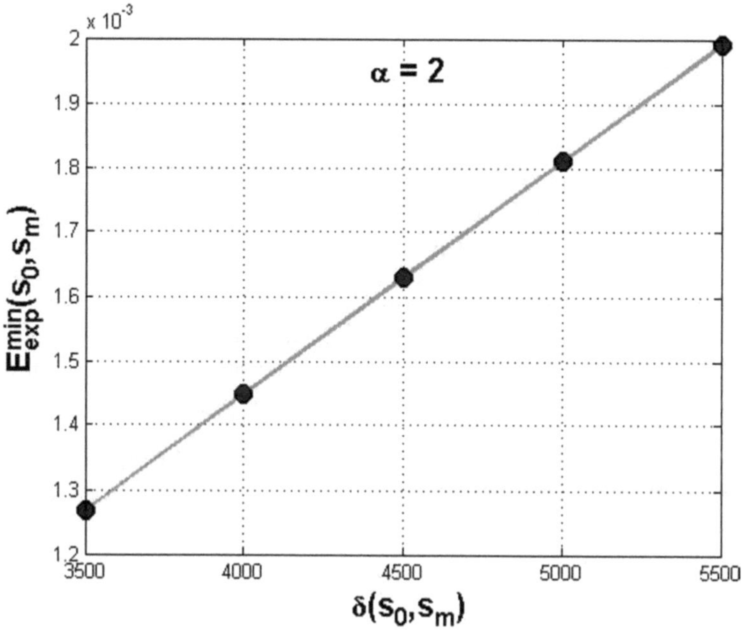


**Fig. 9.8** $E_{\mathrm{exp}}^{\mathrm{min}}(s_0, s_m)$ for $\alpha = 2$

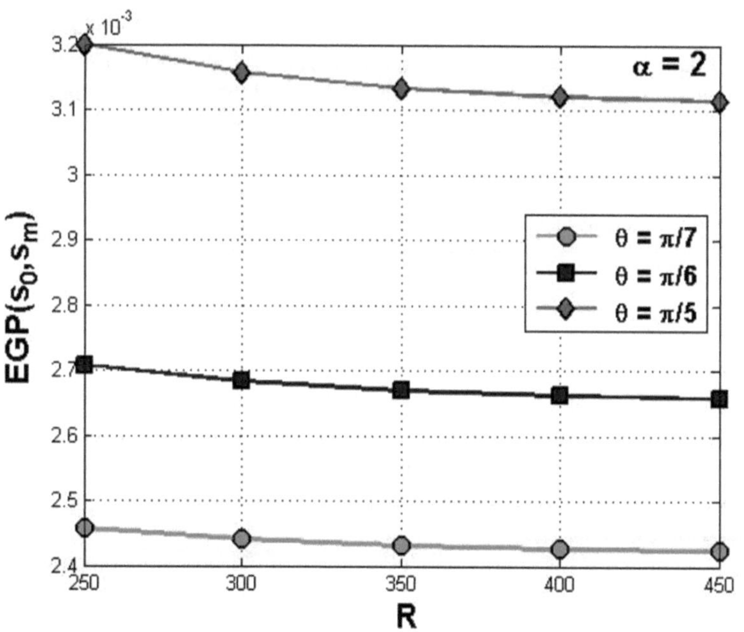


**Fig. 9.9** $E_{\mathrm{exp}}^{\mathrm{max}}(s_0, s_m)$ for $\alpha = 2$

distance separating the source and the sink. Figure 9.9 shows the plot of the function $E_{\exp}^{\max}(s_0, s_m)$ (see Eq. 9.9) for $\alpha = 2$ and different values of $r$ and $\theta$ while keeping $\delta(s_0, s_m)$ constant. We observe that when the angle $\theta$ increases, the length of the data forwarding path between the source and the sink increases. As a result, more energy consumption will be introduced. In other words, any deviation from the shortest path between the source and the sink will increase the number of forwarders and hence yields additional cost in terms of energy consumption.

We also observe that small and large values of $r$ yield more energy consumption. Because the distance between any pair of checkpoints is $r$, the number of checkpoints is determined by the value of $r$. Thus, there are an optimum number of checkpoints that need to be used to optimize the deviation from the shortest path and hence leads to a minimum value of $E_{\exp}^{\max}(s_0, s_m)$. According to Fig. 9.9, for $\theta = \pi / 7$, the optimum value or $r$ is $r_{opt}$ = 400 m, implying that the optimum number of checkpoints is $\lceil 3500/400 \rceil = 9$.

## 9.4 Short-Range Versus Long-Range

In this section, we compare the WLDT protocol with the BVGF [204] and GPSR [118] protocols. We show that WLDT outperforms BVGF and GPSR even when BVGF considers its optimal path in terms of network dilation or number of hops between the source $s_0$ and the sink $s_m$. Precisely, we compute the energy gain percentage in data forwarding from a source to the sink in comparison with the BVGF [204] and GPSR [118] protocols, which we slightly updated so they account for energy in the selection of a next forwarder. Note that the BVGF and GPSR protocols forward data via long range. Moreover, we prove that the presence of checkpoints in WLDT introduces an energy gain in comparison with a similar protocol, called WLDT w/c (or WLDT *without checkpoints*), which forwards sensed data via short range but does not use checkpoints.

### *9.4.1 Energy Gain*

The BVGF and GPSR protocols follow a greedy forwarding approach, where the data are forwarded to the sensor with the closest distance to the sink, thus enabling long-range data forwarding. The WLDT protocol, however, enables short-range transmission of the data between a source and its checkpoint or between any pair of consecutive checkpoints.

Theorem 9.3 computes the energy gain percentage of our protocol compared to the BVGF protocol.

*Theorem 9.3:* The energy gain percentage of our protocol compared to the BVGF protocol along the shortest path $[s_0, s_m]$ between the source $s_0$ and the sink $s_m$ is given by

$$EGP(s_0,s_m) = 1 - \frac{3\,\varepsilon^{1/\alpha}\,E_{elec}^{\;1-1/\alpha}\,r}{2\,E_{elec} + \varepsilon\,r^{\alpha}} \tag{9.10}$$

where $r$ is the radius of the transmission range of the sensors.

*Proof:* The total energy consumption required by BVGF [204] to forward a data packet from the source $s_0$ to the sink $s_m$ is

$$E_{BVGF}(s_0,s_m) = (2\,\kappa\,E_{elec} + \kappa\,\varepsilon\,r^{\alpha})\,\frac{\delta(s_0,s_m)}{r}$$

In fact, the BVGF protocol performs $\frac{\delta(s_0,s_m)}{r}$ forwardings of the data packet along $[s_0,s_m]$, where two consecutive forwarders are separated by a distance equal to $r$. Likewise, the total energy consumption required by the WLDT protocol is

$$E_{WLDT}(s_0,s_m) = (2\,\kappa\,E_{elec} + \kappa\,\varepsilon\,d_{\min}{}^{\alpha})\,\frac{\delta(s_0,s_m)}{d_{\min}}$$

Indeed, the WLDT protocol requires $\frac{\delta(s_0,s_m)}{d_{\min}}$ forwardings of the sensed data along $[s_0,s_m]$. The energy gain percentage of the WLDT protocol compared to the BVGF protocol is

$$EGP(s_0,s_m) = 1 - \frac{E_{WLDT}(s_0,s_m)}{E_{BVGF}(s_0,s_m)} = 1 - \frac{3\,\varepsilon^{1/\alpha}\,E_{elec}^{\;1-1/\alpha}\,r}{2\,E_{elec} + \varepsilon\,r^{\alpha}} \qquad \blacksquare$$

Theorem 9.4 computes the energy gain percentage of WLDT compared to the GPSR using the non-direct data forwarding path between $s_0$ and $s_m$.

*Theorem 9.4:* The energy gain percentage of our protocol compared to the GPSR protocol along the non-shortest path between the source $s_0$ and the sink $s_m$ is given by

$$EGP(s_0,s_m) = 1 - \frac{3\,E_{elec}}{2\,E_{elec} + \varepsilon\,r^{\alpha}} \times \frac{r}{P(d_{\min},\theta_0)} \tag{9.11}$$

where $P(d_{\min},\theta_0) = r - \sqrt{d_{\min}{}^2 + r^2 - 2\,d_{\min}\,r\cos\theta_0}$ is the progress made towards the sink along any segment between any pair of consecutive forwarders (checkpoint), $\theta_0 = \angle s_{i+1}, s_i, s_m$ is the angle between the segments $[s_i,s_{i+1}]$ and $[s_i,s_m]$, $s_i$ and $s_{i+1}$ are two consecutive forwarders, and $r$ is the radius of the transmission range of sensors.

*Proof:* We assume that the distance between any pair of consecutive forwarders in GPSR and checkpoints in WLDT is equal to $r$. Thus, the total energy consumption of GPSR is given by

$$E_{GPSR}(s_0,s_m) = (2\,\kappa\,E_{elec} + \kappa\,\varepsilon\,r^{\alpha})\frac{\delta(s_0,s_m)}{P(r,\theta)}$$

since the sensed data will be forwarded $\frac{\delta(s_0,s_m)}{P(r,\theta)}$ times. On the other hand, the total energy consumption of WLDT is

$$E_{WLDT}(s_0,s_m) = (2\,\kappa\,E_{elec} + \kappa\,\varepsilon\,d_{\min}{}^{\alpha})\frac{r}{P(d_{\min},\theta_0)}\times\frac{\delta(s_0,s_m)}{P(r,\theta)}$$

where

$$P(r,\theta) = \delta(s_0,s_m) - \sqrt{r^2 + \delta^2(s_0,s_m) - 2r\delta(s_0,s_m)\cos\theta}$$

and

$$P(d_{\min},\theta_0) = r - \sqrt{d_{\min}{}^2 + r^2 - 2d_{\min}r\cos\theta_0}$$

In fact, there are $\delta(s_0,s_m)/P(r,\theta)$ checkpoints between $s_0$ and $s_m$ and $r/P(d_{\min},\theta_0)$ forwarders between any pair of consecutive checkpoints. Thus,

$$\begin{aligned}EGP(s_0,s_m) &= 1 - \frac{E_{WLDT}(s_0,s_m)}{E_{GPSR}(s_0,s_m)}\\ &= 1 - \frac{3\,r\,E_{elec}}{(2\,E_{elec} + \varepsilon\,r^{\alpha})\,P(d_{\min},\theta_0)}\end{aligned}$$

■

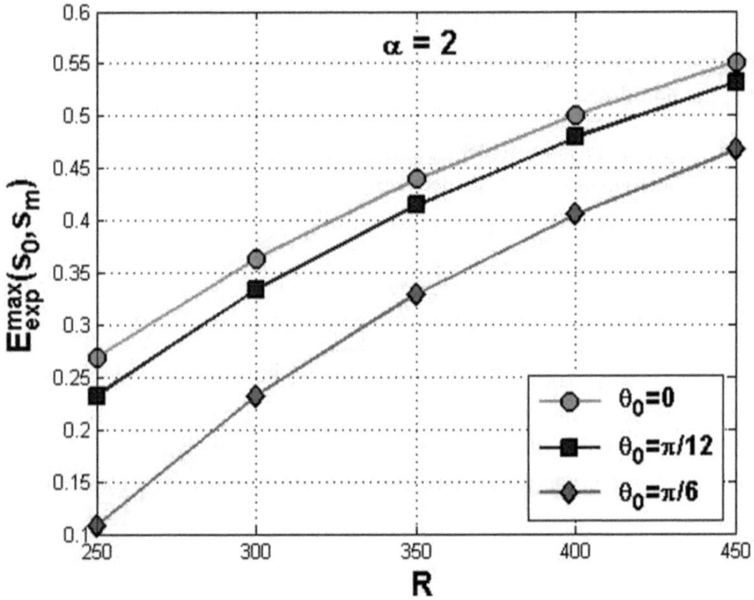


**Fig. 9.10** Impact of $r$ and $\theta_0$ on $EGP(s_0,s_m)$

Figure 9.10 shows the impact of the radius of the transmission range of sensors, $r$, and the angle $\theta_0$ on $EGP(s_0,s_m)$ (see Eq. 9.11), assuming $\alpha = 2$. We observe that the maximum energy gain percentage is obtained when every series of forwarders lie on the segment between their corresponding checkpoints, i.e., $\theta_0 = 0$. When $\theta_0 = 0$, the BVGF and GPSR protocols are similar in the sense that both of them forward the data to the closest sensors to the sink on the shortest path $[s_0,s_m]$. We find that the energy gain percentage of WLDT compared to BVGF and GPSR is about $55\%$. As $\theta_0$ increases, the length of the path between two consecutive checkpoints increases and hence more energy will be spent to forward the sensed data towards the next checkpoint. Notice that when $r$ increases, GPSR would consume more energy as $E_{GPSR}(s_0,s_m) \propto r$, while WLDT transmits data over short Delaunay edges. Thus, $EGP(s_0,s_m)$ increases with $r$ as shown in Fig. 9.10. Thus, WLDT achieves significant energy savings for higher values of $r$ compared to the GPSR protocol.

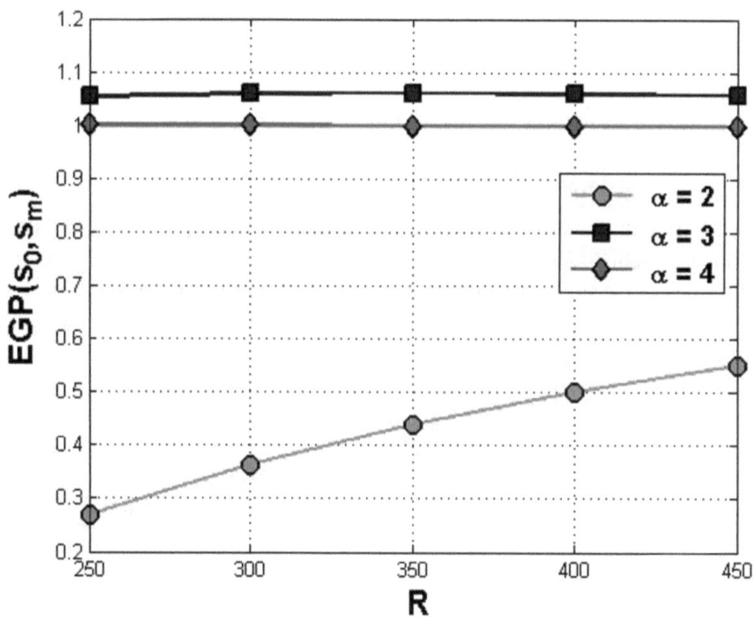


**Fig. 9.11** Impact of $\alpha$ on $EGP(s_0,s_m)$

Figure 9.11 shows the impact of $r$ and $\alpha$ on $EGP(s_0,s_m)$, assuming $\theta_0 = 0$. As $r$ and $\alpha$ increase, $EGP(s_0,s_m)$ increases and reaches 100% for $\alpha = 3$. Moreover, for $\alpha = 3$, the energy gain percentage is more than 100%.

### 9.4.2 Controlled Short-Range Data Forwarding

As mentioned earlier, the presence of checkpoints helps build short data forwarding paths between the source $s_0$ and the sink $s_m$ by reducing their deviation from the shortest path $[s_0, s_m]$. Theorem 9.5 states that the use of checkpoints yields short data forwarding paths and hence significant energy savings.

*Theorem 9.5:* The energy gain percentage of our protocol due to the presence of checkpoints in forwarding one sensed data packet from the source $s_0$ to the sink $s_m$ is given by

$$EGP(s_0, s_m) = 1 - \frac{rP(d_{\min}, \theta + \theta_0)}{P(r, \theta)P(d_{\min}, \theta_0)} \tag{9.12}$$

where

$$P(d_{\min}, \theta_0) = R - \sqrt{d_{\min}{}^2 + R^2 - 2\,R\,d_{\min}\,\cos\theta_0}$$

$$P(d_{\min}, \theta + \theta_0) = \delta(s_0, s_m) - \sqrt{\delta^2(s_0, s_m) + d_{\min}{}^2 - 2\,d_{\min}\,\delta(s_0, s_m)\cos(\theta + \theta_0)}$$

$$P(r, \theta) = \delta(s_0, s_m) - \sqrt{\delta^2(s_0, s_m) + r^2 - 2\,r\,\delta(s_0, s_m)\cos\theta}$$

and $\phi$ is an infinitesimal angle in radian.

*Proof:* In order to measure the energy gain introduced by checkpoints, let us consider a similar protocol to WLDT, which forwards the sensed data through short distances but does not use checkpoints. For the sake of clarity of the notation, let us call this protocol *NoCP* (No Checkpoint). The progress made towards the sink using the *NoCP* scheme is given by

$$P(d_{\min}, \theta + \theta_0) = \delta(s_0, s_m) - \sqrt{\delta^2(s_0, s_m) + d_{\min}{}^2 - 2\,d_{\min}\,\delta(s_0, s_m)\cos(\theta + \theta_0)}$$

where $\theta = \angle s_{i+1}, s_i, s_m$, $s_i$ and $s_{i+1}$ are two consecutive forwarders, and $d_{\min}$ is the distance between any pair of consecutive forwarders. Figure 9.7 illustrates this scenario. Therefore, the number of forwarders between the source $s_0$ and the sink $s_m$ is given by $NF(NoCP) = \dfrac{\delta(s_0, s_m)}{P(d_{\min}, \theta + \theta_0)}$. Thus, the total energy consumption required by the *NoCP* protocol for forwarding one sensed data packet from the source $s_0$ to the sink $s_m$ is

$$E_{NoCP}(s_0, s_m) = (2\,\kappa\,E_{elec} + \kappa\,\varepsilon\,d_{\min}{}^{\alpha}) \times NF(NoCP)$$

On the other hand, our protocol requires that the sensed data be forwarded through checkpoints. These checkpoints lie on or closely to the shortest path $[s_0, s_m]$. Thus, the progress made towards the sink using the WLDT protocol is given by

$$P(r,\theta) = \delta(s_0,s_m) - \sqrt{\delta^2(s_0,s_m) + r^2 - 2\,r\,\delta(s_0,s_m)\cos\theta}$$

where $\phi < \theta$ and $r$ is the distance between any pair of consecutive checkpoints. In order to compare the WLDT protocol to the $NoCP$ protocol, the progress made towards any checkpoint is $P(d_{\min},\theta_0)$, which is given by

$$P(d_{\min},\theta_0) = R - \sqrt{d_{\min}{}^2 + R^2 - 2\,R\,d_{\min}\cos\theta_0}$$

Thus, the total number of forwarders and checkpoints is

$$N = \frac{\delta(s_0,s_m)}{P(r,\theta)} \times \frac{r}{P(d_{\min},\theta_0)}$$

In fact, there are $\frac{\delta(s_0,s_m)}{P(r,\theta)}$ checkpoints between the source $s_0$ and the sink $s_m$ and $\frac{r}{P(d_{\min},\theta_0)}$ forwarders between any pair of consecutive checkpoints. Therefore, the total energy consumption for forwarding one sensed data packet from the source $s_0$ to the sink $s_m$ is given by

$$E_{WLDT}(s_0,s_m) = (2\,\kappa\,E_{elec} + \kappa\,\varepsilon\,d_{\min}{}^{\alpha}) \times N$$

Thus, the energy gain percentage of WLDT compared to $NoCP$ is given by

$$EGP(s_0,s_m) = 1 - \frac{E_{WLDT}(s_0,s_m)}{E_{NoCP}(s_0,s_m)} = 1 - \frac{rP(d_{\min},\theta+\theta_0)}{P(r,\theta)P(d_{\min},\theta_0)} \qquad \blacksquare$$

Figures 9.12 and 9.13 show the impact of $r$ and $\delta(s_0,s_m)$ on $EGP(s_0,s_m)$ (see Eq. 9.12), respectively, where $\theta = \pi/3$ and $\theta_0 = \pi/7$. As can be seen, regardless of the value of the radius *R* of the communication range of the sensors, there is always an energy gain percentage between 0.842 and 0.837 when we use checkpoints. This result shows the benefits of using checkpoints, which shorten the data forwarding paths between sources and the sink and hence yields significant energy savings. This will extend the network lifetime. However, this gain is inversely proportional to *R*. We observe that the energy gain percentage using checkpoints reaches its maximum, which is about 0.842, for $r = 250$ m and its minimum, which is about 0.837, for $r = 450$ m. We should mention that the performance of the *NoCP* protocol does not depend on *R* at all while that of WLDT does. Indeed, any increase in the value of *R* will reduce the number of checkpoints between the source sensor and the sink. But it will lengthen the data forwarding path between any pair of consecutive checkpoints. The presence of more checkpoints guarantees short forwarding paths between consecutive checkpoints, and hence less energy consumptions of the sensors.

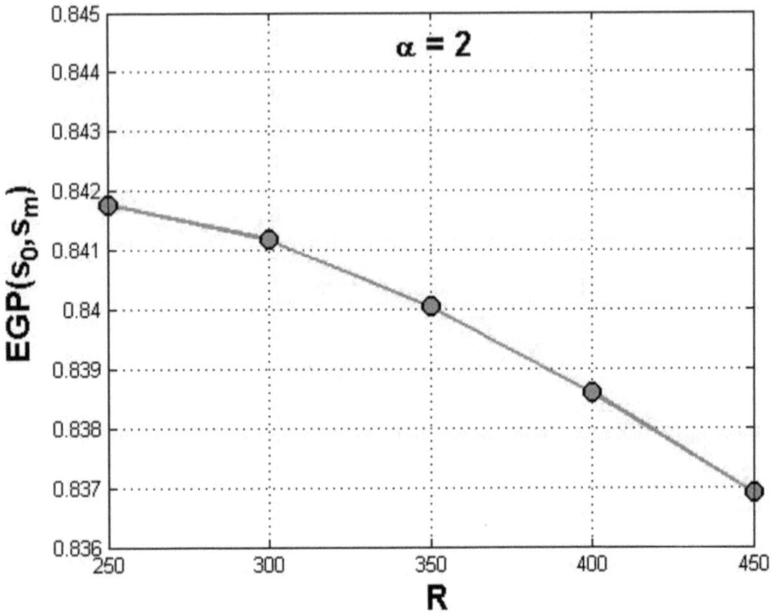


**Fig. 9.12** Impact of $r$ on $EGP(s_0, s_m)$

## 9.5 Discussion

In this section, we briefly discuss sensor reliability issues, which are left for future work. When the sensors fail, WLDT may also fail to forward the data to the sink.

1. *What would WLDT do when the checkpoints fail?* We need to consider the following two cases:

*Case 1. A checkpoint goes down before forwarding the sensed data to its next forwarder:* Any checkpoint $s_p$ is required to send back an acknowledgment message to its previous forwarder $s_k$ from which it has received the sensed data packet. In other words, as soon as the checkpoint $s_p$ forwards the sensed data towards the sink, it replies back with a message saying that the data have been successfully forwarded to its next forwarder. Therefore, when the forwarder $s_k$ does not hear from its checkpoint $s_p$ before certain time-out, it will understand that $s_p$ went down. In this case, the forwarder $s_k$ will have to act like a checkpoint. Thus, the sensor $s_k$ will run phase 1 of the WLDT protocol to identify a new checkpoint and forward the sensed data to it through a series of forwarders.

*Case 2. A checkpoint goes down before having the sensed data forwarded to it:* If a forwarder, say $s_k$, learns that its checkpoint $s_p$ has disappeared, it will ignore it and find another checkpoint. Then, it forwards the sensed data towards this new checkpoint through a series of forwarders.

2. *What if a forwarder fails?* If a forwarder disappears before forwarding the sensed data to its next forwarder or checkpoint, the sensed data will be lost. To solve this problem, an acknowledgment with a time-out could be used to check whether a forwarder has successfully forwarded the sensed data towards the sink. Therefore, any forwarder has to send back an acknowledgment message to inform the previous forwarder (or checkpoint) that the sensed data packet has been sent out to the next forwarder. Otherwise, the previous forwarder will have to select another sensor as forwarder so the sink can receive the sensed data.

## 9.6 Related Work

In this section, we discuss a sample of energy-aware data forwarding protocols for wireless sensor networks as well as those protocols using the notion of proxies in data forwarding in wireless networks.

Xing et al. [204] proposed a greedy geographic routing protocol, called Bounded Voronoi Greedy Forwarding (BVGF), which allows sensing-covered networks to achieve a lower routing path length compared to other existing protocols. The nodes eligible to act as the next hops are the ones whose Voronoi regions are traversed by the segment line joining the source and the destination. The BVGF protocol chooses as the next hop the neighbour that has the shortest Euclidean distance to the destination among all eligible neighbours. This protocol does not help the sensors deplete their battery power uniformly. Each sensor has, indeed, only one next hop to forward its data to the sink. Thus, any data forwarding path between a source sensor and the sink will always have the same chain of next hops, thus suffering from battery power depletion. Bose and Morin [46] proposed a Voronoi routing for Delaunay triangulations that moves data along the nodes whose Voronoi regions intersect the straight line between the sender and the receiver. The major problem of this algorithm is that it requires the construction of the Voronoi diagram and the Delaunay triangulation of all the wireless nodes. This strategy is very expensive in distributed environments, such as sensor networks. Also, this protocol considers the same path between the source and destination, and hence would deplete the battery power of the sensors quickly. Karp and Kung [118] proposed a Greedy Perimeter Stateless Routing (GPSR) protocol for mobile wireless ad hoc networks. GPSR forwards data packets through long distances and hence consumes much energy. Our protocol, however, forwards sensed data through short Delaunay edges and hence achieves significant energy savings. Li et al. [135] studied compass routing [124], random compass routing [124], greedy routing [47], and most forwarding routing [183] on different graphs. Lindsey et al. [141] presented a scheme, called PEGASIS (Power-Efficient Gathering in Sensor Information Systems), where each node can receive from and send to close neighbours. Choi and Das [63] proposed an applicative indirect routing

(AIR) protocol for ad hoc wireless networks using proxy candidates, which are defined as the neighbours that are shared by the sender and the receiver.

## 9.7 Summary

In this chapter, we proposed a data forwarding protocol for wireless sensor networks, which helps sensors save their energy by forwarding the sensed data towards the sink over short distances [20]. In addition, preference is always given to the sensors with high remaining energy and whose locations lie on or closely to the shortest path between the source and the sink. The objective of our protocol is to prolong the network lifetime by controlling sensed data transmission. Specifically, the proposed protocol builds a data forwarding path between the source and the sink as a sequence of sub-paths, each of which is composed of a series of forwarders between two endpoints, called checkpoints. These checkpoints are selected based on their remaining energy and closeness to the sink based on the geometric properties of Delaunay triangulation. The data should be forwarded to the sink through these checkpoints.

To demonstrate the effectiveness of the WLDT protocol, we presented theoretical results supported by extensive numerical results. We computed lower bound on the energy cost of forwarding one sensed data packet from a source to the sink as well as the corresponding optimum number of forwarders between them. We proved that the proposed protocol yields significant energy savings. Also, we compared WLDT with BVGF and GPSR, which enable data forwarding through long distances, and have found that WLDT achieves an energy gain percentage in the order of 55% for the free space model and close to 100% for the multi-path model. Moreover, we proved that the checkpoints help shorten data forwarding paths between the sources and the sink. We found that WLDT yields an energy gain percentage compared to a protocol similar to WLDT, which favours short-range data forwarding but does not use checkpoints. These energy gain percentages are in the order of 84.2% for $R = 250$ m and 83.7% for $R = 450$ m. These significant energy savings will definitely increase the network lifetime. There is an ongoing debate on short-range versus long-range data forwarding in multi-hop wireless sensor networks. We believe that the primary concern of wireless sensor networks is energy savings to prolong the network lifetime. The WLDT protocol supports the short-range strategy to achieve this goal.

# Chapter 10
# Trade-Off between Energy and Delay in Geographic Forwarding on Always-On Sensors

This chapter proposes a data forwarding protocol for wireless sensor networks that trades off between energy and delay. Specifically, this protocol helps achieve minimum energy consumption while ensuring uniform battery power depletion of the sensors and meeting the required delay constraints in the sense that data gathering points must receive the sensed data within a specified time bound. Given that these are conflicting goals, this trade-off is formulated as a multi-objective optimization problem whose solution is an input to our data forwarding scheme.

## 10.1 Introduction

Data forwarding is an essential component and critical determinant of the effectiveness of wireless sensor networks, where source sensors (or simply *sources*) send their sensed data possibly through multi-hop wireless links to the sink. The longevity of multi-hop wireless sensor networks requires that the load of data forwarding be balanced amongst all the sensors so they deplete their battery power uniformly. Indeed, battery power (or *energy*) is the most crucial resource in wireless sensor networks, especially when battery recharging or replenishing is impossible. Thus, sensors should use energy-efficient data forwarding protocols that guarantee *uniform energy depletion* of the sensors. This will keep the sensors operating for longer periods of time, thus extending the network lifetime. However, some sensing applications should satisfy strict source-to-sink delay (or simply *delay*) constraints in the sense that data gathering points, also known as *sinks*, must receive the sensed data originated from *source sensors* within a specified time bound. Therefore, ensuring the longevity of wireless sensor networks becomes a challenging issue, especially for sensing applications with strict delay constraints [152, 153] that must be satisfied at the sink so it can make decisions in a timely fashion regarding the collected data. Hence, appropriate data forwarding protocols should be designed to achieve minimum energy consumption while ensuring uniform battery power depletion of the sensors and meeting the required delay constraints, thus leading to a multi-objective optimization problem.

Given that minimum energy consumption, minimum delay, and uniform energy depletion are conflicting goals, which have to be dealt with simultaneously, finding a trade-off between them is necessary. Indeed, minimizing energy consumption requires transmitting the sensed data over short distances; recall that the

H.M. Ammari: Challenges and Opportunities of Connected $k$-Covered WSNs, SCI 215, pp. 201–240.
springerlink.com

energy ($E_{mt}$) spent in data transmission over a physical distance $d$ between a pair of transmitting and receiving points, is proportional to $d$, i.e., $E_{mt} \propto d^{\alpha}$, with $2 \leq \alpha \leq 4$ being the path-loss exponent. Minimizing delay, however, requires minimizing the number of intermediate forwarders between a source and the sink. This goal could be achieved by maximizing the distance between any pair of consecutive forwarders. Furthermore, a reduced search space of candidate forwarders yields an unbalanced distribution of the data forwarding load amongst the sensors, thus causing a non-uniform depletion of their available energy. Indeed, the candidate forwarders located in a small search space would suffer heavy depletion of their energy as they will be frequently selected as forwarders. In contrast, a large search space ensures a more balanced data forwarding load amongst the sensors and hence helps achieve uniform energy depletion of the sensors. Haenggi [98] and Haenggi and Puccinelli [99], however, took an extreme position by arguing that long-hop routing is a very competitive strategy compared to short-hop routing, thus sacrificing the very scarce energy resource of the sensors. Haenggi [98] provided 12 reasons explaining the advantages of long-range over short-range forwarding. We believe that a more balanced approach should be used to account for delay and energy uniformity [19].

The remainder of this chapter is organized as follows. Section 10.2 introduces the concepts of slicing the communication range of the sensors and proxy forwarders. It also characterizes the uniform energy depletion of the sensors. Section 10.3 presents an approach to trade-off between energy consumption, delay, and energy depletion in data forwarding based on the needs of the sensing applications. Then, it discusses our proposed data forwarding protocol, which trades off energy with delay, and shows how to relax the assumptions stated in Sect. 10.4. Section 10.5 evaluates the performance of our protocol. Section 10.6 reviews sample data forwarding protocols for wireless sensor networks. Section 10.7 summarizes the chapter.

## 10.2 A Slicing Approach

This section shows how to slice the communication range of the sensors into *concentric circular bands* (CCBs). It also characterizes the *uniform battery power depletion* of the sensors.

### *10.2.1 Slicing of Communication Range*

The idea of slicing the communication range of the sensors stems from the simple fact that any sensor has higher preference to some of its neighbours than to others. This notion of preference becomes apparent in the next-forwarder selection process when a sensor has to decide to which sensor it wishes to forward its data so it

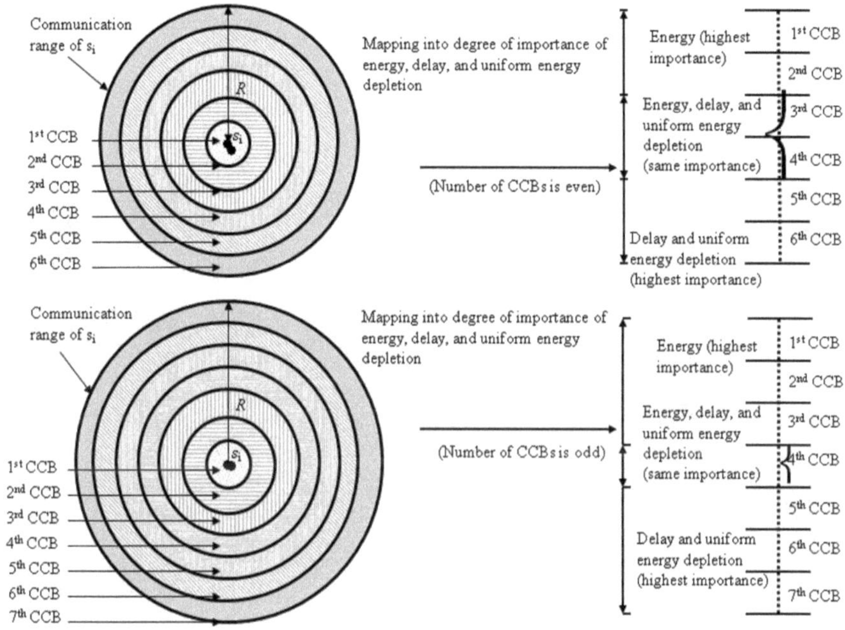


**Fig. 10.1** Slicing of the communication range of sensors

reaches the sink while meeting some energy and/or delay constraints. The slicing approach is based on an approximation of a *minimum transmission distance* $d_{\min}$ in data transmission. As will be seen, the communication range slicing approach helps a sensor classify its neighbours depending on which of the above-mentioned metrics it wishes to optimize.

From Lemma 9.4, the minimum *transmission distance* that can be used in data transmission can be approximated by

$$d_{\min} = (E_{elec}/\varepsilon)^{1/\alpha} \tag{10.1}$$

To achieve a better balance between minimum energy consumption, minimum delay, and uniform energy depletion, we propose to slice the communication range $CD(\zeta_i, R)$ of a sensor $s_i$ into $n_{ccb} = \lceil R/d_{\min} \rceil$ CCBs, each of which is centred at $s_i$ and has a width of $d_{\min}$. As will be seen, a set of CCBs can be divided into three categories (Fig. 10.1). The inner CCBs favour minimizing energy consumption over minimizing delay and uniform energy depletion; the middle CCBs give the same degree of interest to the three performance metrics; and the outer CCBs favour minimizing delay and uniform energy depletion over minimizing energy consumption of the sensors.

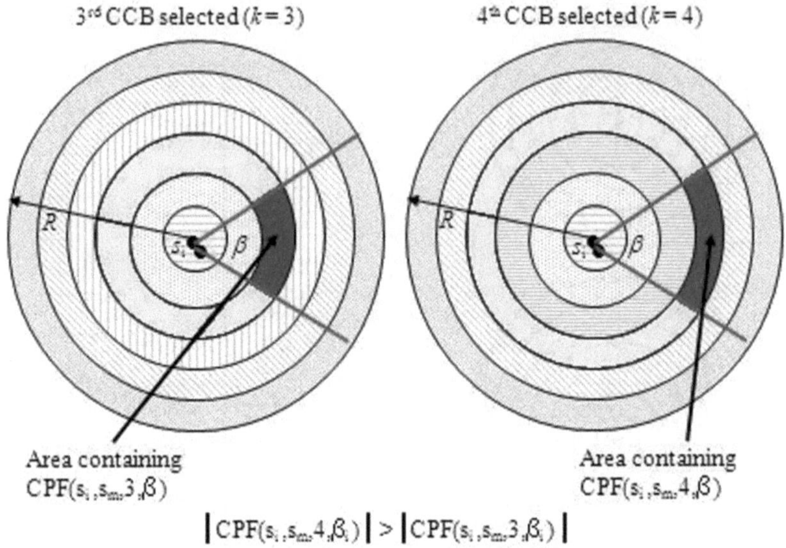


**Fig. 10.2** Impact of $k$ on the size of the subset $CPF(s_i,s_m,k,\beta)$

## 10.2.2 Selection of Candidate Proxy Forwarders

We consider a set of sensors and a single sink $s_m$, all in a planar field. Next, we define the notion of *candidate proxy forwarder.*

*Definition 10.1:* From a neighbour set $NS(s_i)$ of a sensor $s_i$, we define $CPF(s_i,s_m,k,\beta)$ as a subset of sensors, called *candidate proxy forwarder set* of $s_i$, which belong to the $k^{th}$ CCB and located within a zone determined by a wedge with an angle $\beta$ centred at $s_i$ (Fig. 10.2). ■

The size of $CPF(s_i,s_m,k,\beta)$ depends on the values of $\beta$ and $k$, where $1 \le k \le n_{ccb}$ and $0 < \beta < \pi$. We design our trade-off protocol between energy and delay in such a way that when a source selects a specific concentric circular band, say $k$, from which it will designate a sensor as a forwarder, any future proxy forwarder will have to use the same value of $k$. We use this design decision for the sake of ease of the analysis of our proposed protocol based on this trade-off. Also, this decision would help us find some theoretical results in terms of upper bounds on some specific metrics. We will further relax this decision by allowing the sensors to select their own values of $k$ before forwarding data originated from a source to the sink.

### 10.2.3 Uniform Energy Depletion Characterization

Now, we propose to characterize *uniform energy depletion*. First, we define this notion with respect to our slicing approach.

*Definition 10.2:* We say that *uniform energy depletion* is achieved if each sensor guarantees that a large number of its neighbours located in a given CCB are equally likely to act as candidate proxy forwarders, and hence participate in forwarding the sensed data to the sink. ■

As can be seen from Fig. 10.2, the size of the subset $CPF(s_i, s_m, k, \beta)$ increases with $k$. On one hand, high values of $k$ yield a large subset of sensors to participate in the selection process of a proxy forwarder. This leads to a uniform depletion of the battery power of sensors. On the other hand, when the size of $CPF(s_i, s_m, k, \beta)$ is small, the same sensors will be frequently selected as proxy forwarders and hence their remaining energy drains faster compared to the other sensors in the network, leading to a non-uniform energy depletion. Thus, uniform energy depletion of sensors can be achieved by maximizing the size of $CPF(s_i, s_m, k, \beta)$, i.e., maximizing $k$. In other words, the uniform energy depletion can be characterized by a large size of subset $CPF(s_i, s_m, k, \beta)$, implying high values of $k$.

It is worth noting that higher values of $k$ mean that the proxy forwarders would be selected from CCBs that are closer to the boundary of the communication range of the sensors. This will introduce an additional computational cost due to the selection of a proxy forwarder from the subset $CPF(s_i, s_m, k, \beta)$, which is large, compared to choosing a proxy forwarder from a CCB that is far away from the boundary of the communication range of the sensor, and hence a small set of candidate proxy forwarder is considered. Thus, the origin of this "additional computational cost" is due to the size of the set of candidate proxy forwarder. Indeed, any sensor would need to select a candidate proxy forwarder that has the highest remaining energy as will be stated in Sect. 5.3.3. This step would need the sensor to *sort* its neighbouring sensors based on their remaining energy. However, the cost of this *sorting* step depends on the size of the set of candidate proxy forwarders. Clearly, choosing proxy forwarders from CCBs closer to the boundary of the communication ranges of the sensors would require higher sorting cost compared to choosing from CCBs farther from the boundary of the communication ranges of the sensors.

## 10.3 Trading-Off Energy with Delay

In this section, we discuss our approach that trades off between minimum energy consumption, minimum delay, and uniform energy depletion in data forwarding in wireless sensor networks. Precisely, we propose a data forwarding protocol that trades off between these goals by slicing the communication range of sensors into *concentric circular bands*. In particular, we benefit from an approach that we call *weighted scale-uniform-unit sum* and which will be used by source sensors to solve this multi-objective optimization problem. Our proposed data forwarding protocol, called TED (short for *Trade-off Energy with Delay*), makes use of a solution to a multi-objective optimization problem to find a "best" trade-off between minimum energy consumption, minimum delay, and uniform sensors' battery power depletion. We present several numerical results to show the effectiveness of TED. Then, we relax several assumptions to enhance the practicality of TED. We evaluate TED performance through simulations and find that it is *near optimal* with respect to the $energy \times delay$ metric. This simulation study seems to be an essential step to gain more insight into TED before implementing it on a sensor testbed. We also compute upper bounds on these three metrics and the optimal values of $k$ corresponding to their optimum trade-offs.

## 10.3.1 Simple Analytical Bounds

### 10.3.1.1 Data Forwarding along Shortest Paths

Lemma 10.1 computes upper bounds on the expected number of candidate proxy forwarders, energy consumption, and delay. We omit its proof since it is verbatim.

*Lemma 10.1:* Let $\lambda$ be the sensor spatial density (i.e., the number of sensors per unit area) and $c = qd + td$. The expected total number of candidate proxy forwarders, energy consumption, and delay associated with the $k^{th}$ CCB in forwarding a data packet from a source $s_0$ to the sink $s_m$ along the shortest path $[s_0, s_m]$ are computed as

$$|CPF_{\exp}(s_0, s_m, k, \beta)| = \frac{\lambda\,\beta\,(2k-1)\,\delta(s_0, s_m)\,d_{\min}}{2k} \tag{10.2}$$

$$E_{\exp}(s_0, s_m, k) = a\left(\frac{2\,E_{elec}}{k\,d_{\min}} + \varepsilon\,k^{\alpha-1}\,d_{\min}{}^{\alpha-1}\right)\delta(s_0, s_m) \tag{10.3}$$

$$D_{\exp}(s_0, s_m, k) = \frac{c\,\delta(s_0, s_m)}{k\,d_{\min}} \tag{10.4}$$

respectively, and their respective upper bounds are given by

$$|CPF_{\exp}(s_0, s_m, \beta)| \le \frac{\lambda\,\beta\,(2R - d_{\min})\,\delta(s_0, s_m)\,d_{\min}}{2R}$$

$$E_{\exp}(s_0, s_m) \le a\left(\frac{2\,E_{elec}}{R} + \varepsilon\,R^{\alpha-1}\right)\delta(s_0, s_m)$$

$$D_{\exp}(s_0, s_m) \le \frac{c\,\delta(s_0, s_m)}{d_{\min}}$$

*Proof:* The area of the $k^{th}$ CCB is $\pi(2k-1)\,d_{\min}{}^2$, where $1 \le k \le n_{ccb}$. Thus, the size of the subset $CPF(s_i, s_m, k, \beta)$ is $\dfrac{\lambda\,\beta\,(2k-1)\,d_{\min}{}^2}{2}$. Given that the expected total number of forwarding of the sensed data is $N_f = \dfrac{\delta(s_0, s_m)}{k\,d_{\min}}$, the expected total number of candidate proxy forwarders between a source $s_0$ and the sink $s_m$ is given by

$$|\,CPF_{\exp}(s_0, s_m, k, \beta)\,| = \sum_{i=0}^{N_f - 1} |\,CPF_{\exp}(s_i, s_m, k, \beta)\,|$$
$$= \frac{\lambda\,\beta\,(2k-1)\,\delta(s_0, s_m)\,d_{\min}}{2k}$$

Because $|\,CPF_{\exp}(s_0, s_m, k, \beta)\,|$ is concave, its upper bound corresponds to the largest value of $k$, i.e., $k = n_{ccb}$. That is,

$$|\,CPF_{\exp}(s_0, s_m, \beta)\,| \le \frac{\lambda\,\beta\,(2R - d_{\min})\,\delta(s_0, s_m)\,d_{\min}}{2R}$$

Also, the expected delay is given by

$$D_{\exp}(s_0, s_m, k) = \frac{c\,\delta(s_0, s_m)}{k\,d_{\min}}$$

It is easy to check that an upper bound on delay corresponds to $k = 1$ since $D_{\exp}(s_0, s_m, k)$ is inversely proportional to $k$. Thus,

$$D_{\exp}(s_0, s_m) \le \frac{c\,\delta(s_0, s_m)}{d_{\min}}$$

The expected total energy consumption along the shortest path between the source $s_0$ and the sink $s_m$ is

$$E_{\exp}(s_0, s_m, k) = \frac{a\,(2E_{elec} + \varepsilon\,(k\,d_{\min})^{\alpha})\,\delta(s_0, s_m)}{k\,d_{\min}}$$
$$= a\left(\frac{2E_{elec}}{k\,d_{\min}} + \varepsilon\,k^{\alpha-1}\,d_{\min}{}^{\alpha-1}\right)\delta(s_0, s_m)$$

To find an upper bound on the energy consumption, we solve

$$\frac{\partial E_{\exp}(s_0, s_m, k)}{\partial k} = 0$$

As $2 \le \alpha \le 4$, the unique solution to this equation is given by

$$k^* = \frac{2}{(\alpha - 1)^{1/\alpha}}$$

Since $\frac{\partial^2 E_{\exp}(s_0, s_m, k^*)}{\partial^2 k^*} \geq 0$, the function $E_{\exp}(s_0, s_m, k)$ reaches its lower bound at $k = k^*$. Thus, we have

$$E_{\exp}(s_0, s_m) \geq a\left(\frac{E_{elec}}{d_{\min}} + \varepsilon\,(2\,d_{\min})^{\alpha-1}\right)\delta(s_0, s_m)$$

Afterwards, $E_{\exp}(s_0, s_m, k)$ starts growing proportionally to $k$. Hence, an upper bound on the energy consumption corresponds to the last CCB, i.e., $k = n_{ccb}$. Thus, the distance between $s_0$ and its proxy forwarder is equal to $d_{\max} = R$ and hence the minimum number of data forwarding is

$$N_f(s_i, s_m) = \frac{\delta(s_i, s_m)}{R}$$

Thus, we obtain

$$E_{\exp}(s_0, s_m) \leq a\left(\frac{2\,E_{elec}}{R} + \varepsilon\, R^{\alpha-1}\right)\delta(s_0, s_m). \qquad ■$$

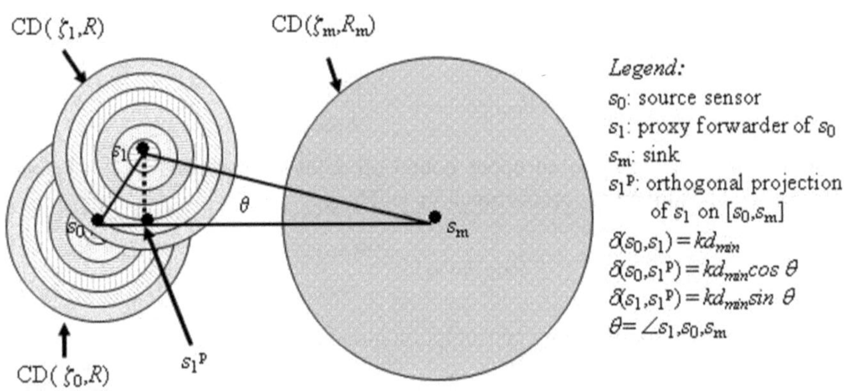


**Fig. 10.3** Non-shortest path between $s_0$ and $s_m$

#### 10.3.1.2 Data Forwarding along Non-direct Paths

In general, shortest paths between the senders and the sink are not always available given that our data forwarding protocol is based on the remaining energy of sensors. Precisely, the sensors that lie on the shortest path between a sender and the sink could be used as proxy forwarders a few times before their remaining energy get smaller than that of other sensors. The following analysis focuses on non-direct paths between sources and the sink based on the angle $\theta = \angle s_{i+1}, s_i, s_m$ (Fig. 10.3). In Lemma 10.2, we assume that the progress that is made towards the sink, which is denoted by $\psi(k,\theta)$, is constant. This assumption is to simplify the analysis so the formulas that we obtain for the expected total energy consumption, delay, and number of candidate proxy forwarders, would look like the ones given in Eqs. (10.5), (10.6), and (10.7). Otherwise, we would obtain three summations that depend on the value of the angle, $\theta$. Note that the maximum value of $\theta$ is $\beta/2$, i.e., $\theta$ could be any value in the interval $[0,\beta/2]$, where the value of $\beta$ determines the size of the candidate proxy forwarder set as shown in Fig. 10.2. Thus, in practice, the value of $\theta$ could be chosen between 0 and $\beta/2$. Lemma 10.2 is a generalization of Lemma 10.1.

*Lemma 10.2:* Let $\psi(k,\theta)$ be a constant progress that is made towards the sink $s_m$ at each forwarding action, i.e., $\theta$ is constant (Fig. 10.3). The expected total energy consumption, delay, and number of candidate proxy forwarders considered when a data packet is disseminated along a non-direct path from a source $s_0$ to the sink $s_m$ with respect to the $k^{th}$ CCB are given by

$$E_{\exp}(s_0,s_m,k,\theta) = \frac{a\,(2\,E_{elec} + \varepsilon\, k^{\alpha}\, d_{\min}{}^{\alpha})\,\delta(s_0,s_m)}{\psi(k,\theta)} \tag{10.5}$$

$$D_{\exp}(s_0,s_m,k,\theta) = \frac{c\,\delta(s_0,s_m)}{\psi(k,\theta)} \tag{10.6}$$

$$|\,CPF_{\exp}(s_0,s_m,k,\theta)\,| = \frac{\lambda\,\beta\,(2\,k-1)\,d_{\min}{}^{2}\,\delta(s_0,s_m)}{2\,\psi(k,\theta)} \tag{10.7}$$

respectively, where $1 \le k \le n_{ccb}$ and $\psi(k,\theta) = \delta(s_0,s_1^p) = k\,d_{\min}\cos\theta$.

*Proof:* By Pythagorean theorem, we have $\psi(k,\theta) = \delta(s_0,s_1^p) = k\,d_{\min}\cos\theta$, where $s_1^p$ is the orthogonal projection of $s_1$ on $[s_0,s_m]$ and $\theta_{\max} = \beta/2$. Also, we have $N_f = \delta(s_0,s_m)/\psi(k,\theta)$. Thus, we obtain

$$E_{\exp}(s_0,s_m,k,\theta) = \frac{a\,(2\,E_{elec} + \varepsilon\, k^{\alpha}\, d_{\min}{}^{\alpha})\,\delta(s_0,s_m)}{\psi(k,\theta)}$$

Similarly, the expected delay is given by

$$D_{\exp}(s_0,s_m,k,\theta) = \frac{c\,\delta(s_0,s_m)}{\psi(k,\theta)}$$

Using the same reasoning as in Lemma 10.1, we obtain

$$|\,CPF_{\exp}(s_0,s_m,k,\theta)\,| = \frac{\lambda\,\beta\,(2\,k-1)\,d_{\min}{}^{2}\,\delta(s_0,s_m)}{2\,\psi(k,\theta)}$$

■

**Table 10.1** Parameters setting

| $\delta(s_0, s_m)$ | $R$ | $c$ | $\lambda$ |
|---|---|---|---|
| 3500 m | 350 m | 0.001 | 0.001 |
| $\alpha = 2$ | $\alpha = 3$ | $\alpha = 4$ | |
| $d_{min}$=70.71m<br>$n_{ccb}$=5 | $d_{min}$=156.68 m<br>$n_{ccb}$=3 | $d_{min}$=44.29 m<br>$n_{ccb}$=8 | |

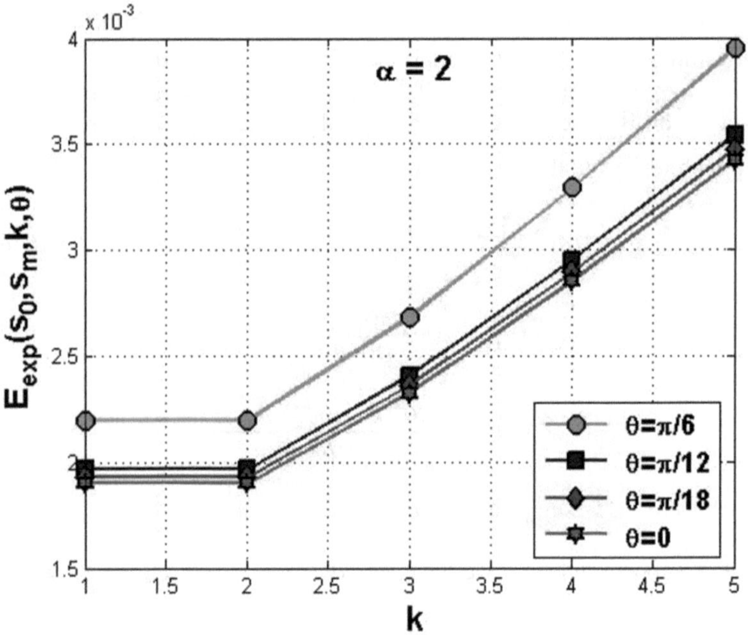


**Fig. 10.4** Impact of CCB id $(k)$ and angle $\theta$ on the energy consumption for $\alpha = 2$

#### 10.3.1.3 Numerical Results

Figure 10.4 shows the impact of $k$, $\theta$, and $\alpha$ on $E_{\exp}(s_0, s_m, k, \theta)$ (Eq. 10.5). Notice that the energy consumption increases with $k$ and reaches its minimum at $k^* = 2$ as computed earlier. This result can be proved as follows. Let $E_{\exp}(s_0, s_m, k) = a[A_1(k) + A_2(k)]\,\delta(s_0, s_m)$, where $A_1(k) = \dfrac{2\,E_{elec}}{k\,d_{\min}}$ and $A_2(k) = \varepsilon\, k^{\alpha-1}\, d_{\min}{}^{\alpha-1}$. Thus, $A_2(k)$ becomes a dominant factor (i.e., $A_2(k) \geq A_1(k)$) for some $k$ that makes $E_{\exp}(s_0, s_m, k)$ grow proportionally to $k$.

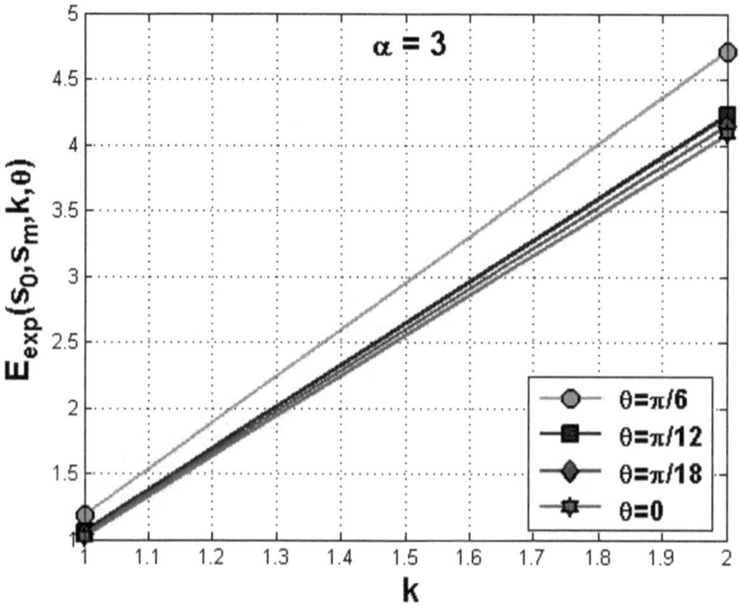


**Fig. 10.5** Impact of CCB id $(k)$ and angle $\theta$ on the energy consumption for $\alpha = 3$

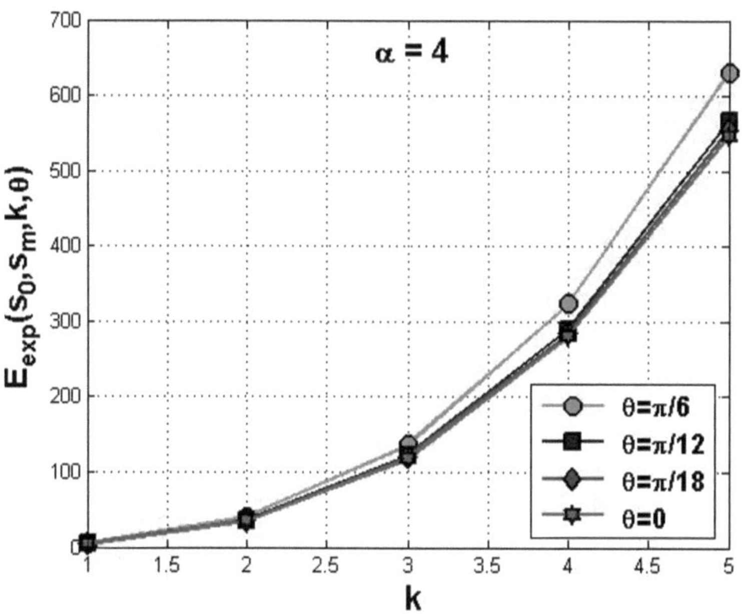


**Fig. 10.6** Impact of CCB id $(k)$ and angle $\theta$ on the energy consumption for $\alpha = 4$

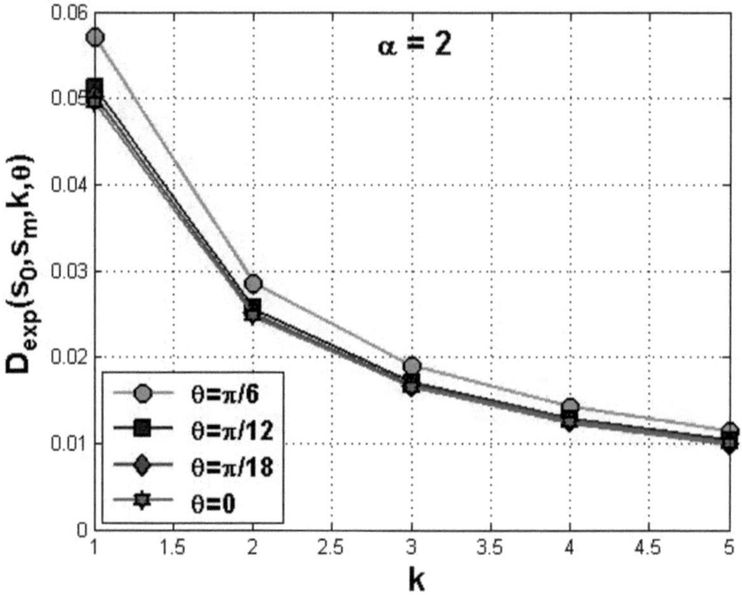


**Fig. 10.7** Impact of CCB id $(k)$ and angle $\theta$ on the delay for α = 2

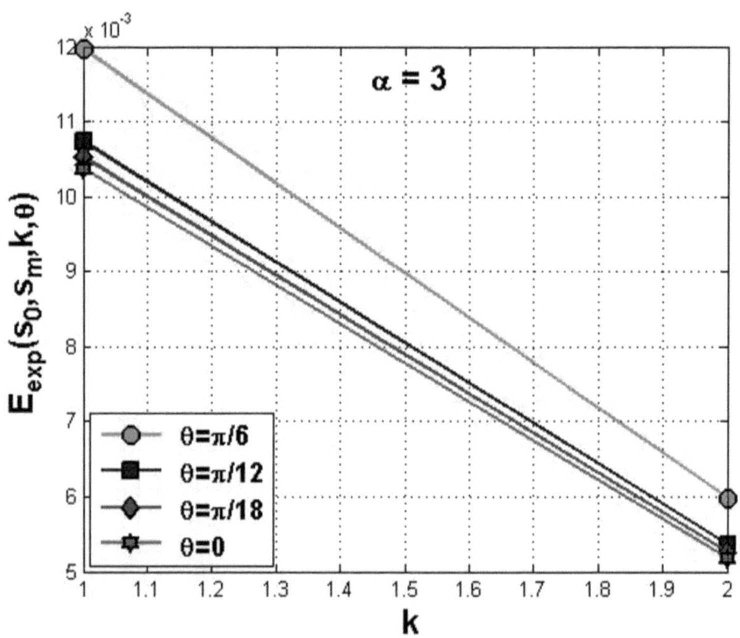


**Fig. 10.8** Impact of CCB id $(k)$ and angle $\theta$ on the delay for α = 3

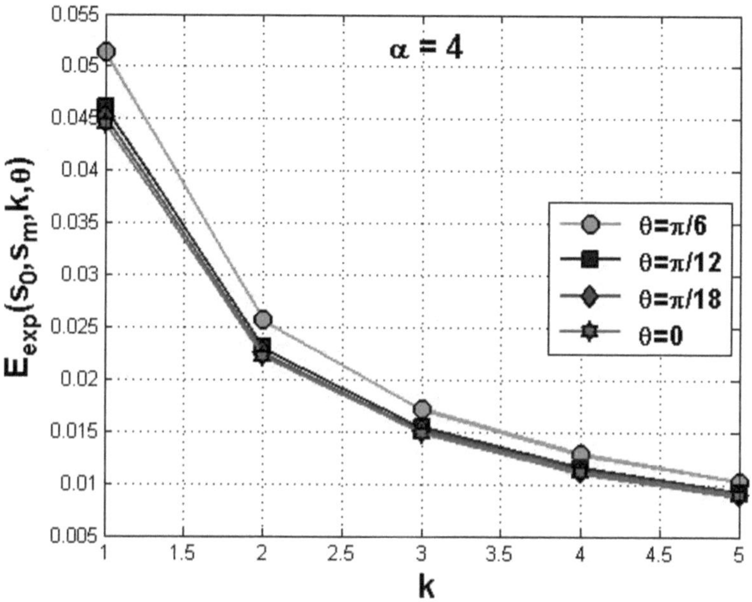


**Fig. 10.9** Impact of CCB id $(k)$ and angle $\theta$ on the delay for α = 4

In fact, $A_2(k) \geq A_1(k) \Rightarrow k^{\alpha} \geq 2$, implying $k \geq 2$ as $2 \leq \alpha \leq 4$. Moreover, more energy consumption would be incurred as we deviate from the shortest path $[s_0, s_m]$ between a source $s_0$ to the sink $s_m$, i.e., as $\theta$ increases.

Figure 10.5 shows that $D_{\exp}(s_0, s_m, k, \theta)$ (Eq. 10.6) decreases with $k$ since fewer proxy forwarders would be needed. Also, $\alpha$ has an impact on the delay. Indeed, $D_{\exp}(s_0, s_m, k, \theta)$ is inversely proportional to $d_{\min}$ which in turn depends on $\alpha$. Similarly, more delay would be incurred as we deviate from $[s_0, s_m]$.

### *10.3.2 Multi-objective Optimization Approach*

In this section, we discuss a *weighted scale-uniform-unit sum* (WES) approach [119] to solve our multi-objective optimization problem for trading off between the above-mentioned three metrics, namely minimum energy consumption, minimum delay, and uniform energy depletion.

#### 10.3.2.1 Overview of the WES Approach

Assume we want to minimize a multi-objective function

$$F(x) = (F_1(x), \ldots, F_n(x))^{\mathrm{T}}$$

where $F_i(x)$ is an objective function, for $1 \leq i \leq n$. WES is a simple approach that introduces a weighting coefficient $w_i\, c_i$ for each $F_i(x)$, where $w_i$ is a weight

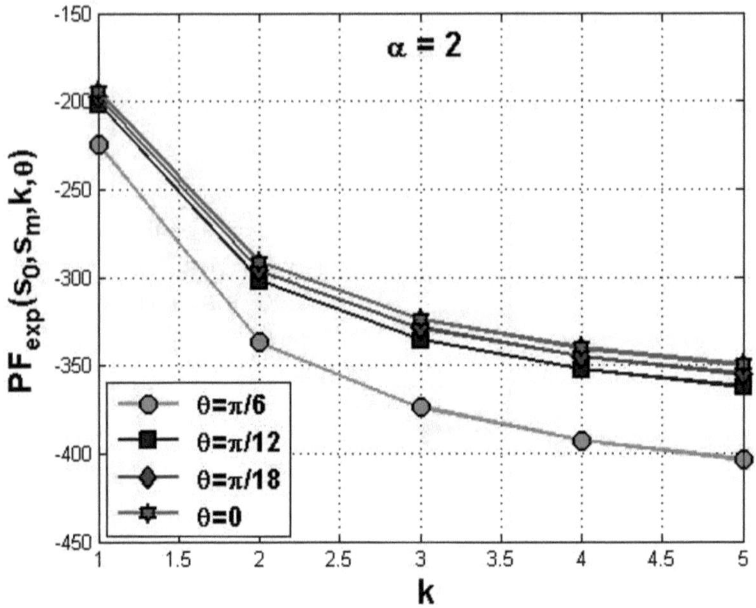


**Fig. 10.10** Impact of CCB id, $k$ , and angle $\theta$ on $PF_{\text{exp}}(s_0, s_m, k)$ for α = 2

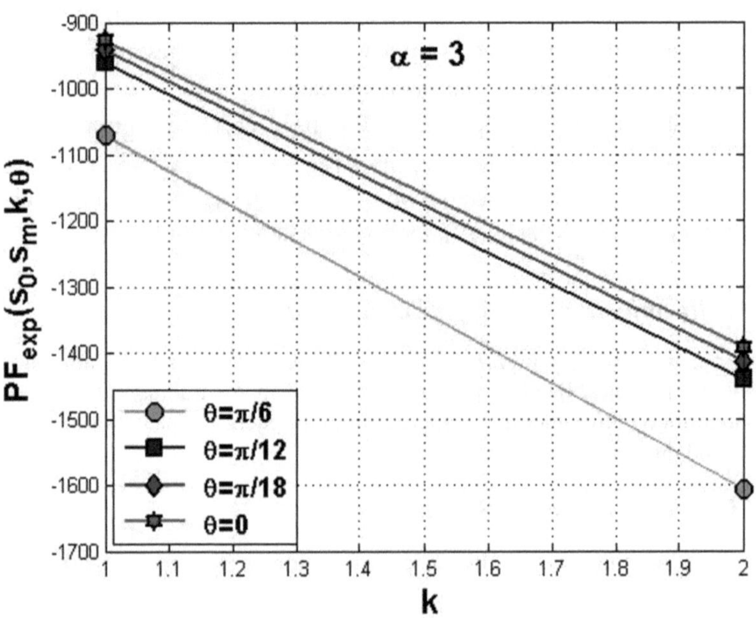


**Fig. 10.11** Impact of CCB id, $k$ , and angle $\theta$ on $PF_{\text{exp}}(s_0, s_m, k)$ for α = 3

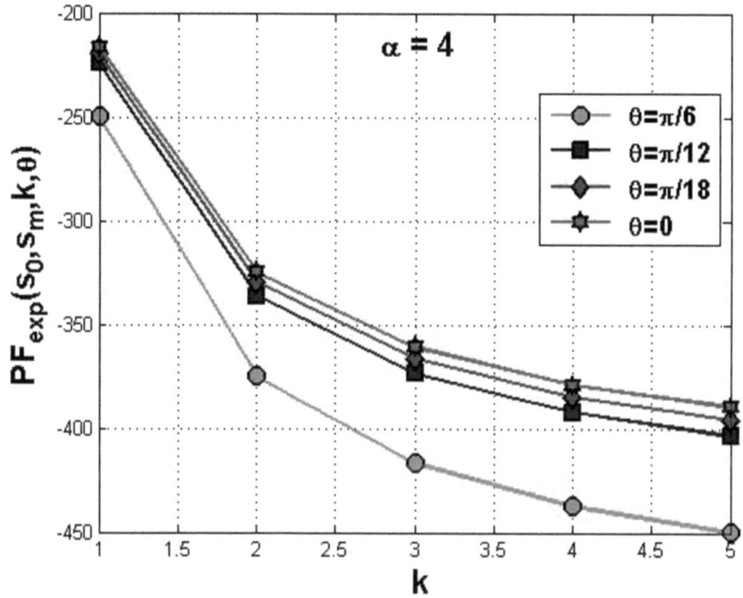


**Fig. 10.12** Impact of CCB id, $k$ , and angle $\theta$ on $PF_{\exp}(s_0, s_m, k)$ for $\alpha = 4$

selected by a network designer to reflect the relative importance of $F_i(x)$ and $c_i$ is a coefficient that not only scales $F_i(x)$ but also helps produce a one-dimensional function $F(x)$. A survey on similar approaches for solving multi-objective optimization problems can be found in [154]. Using WES, a multi-objective optimization problem can be formulated as follows:

$$\textit{Minimize } F(x) = \sum_{i=1}^{n} w_i \, c_i \, F_i(x)$$

$$\textit{Subject to } x \in X$$

where

$$c_i = \frac{\mu}{F_i^{\max}},$$

$$F_i^{\max} = \max\{F_i(x) : \forall x \in X\},$$

$$w_i \geq 0, \ \sum_{i=1}^{n} w_i = 1,$$

$$\mu = \max\{F_i^{\max} : 1 \leq i \leq n\},$$

and $X$ is a set of admissible solutions.

It is assumed that $F_i^{\max} \neq 0 : 1 \leq i \leq n$.

#### 10.3.2.2 Solving the Trade-Off Problem Using WES

Notice that the function $CPF_{\exp}(s_0,s_m,k)$ (Eq. 10.2) is *concave* for any $\theta$ and in particular for $\theta = 0$. In fact,

$$\frac{\partial^2 \mid CPF_{\exp}(s_0,s_m,k) \mid}{\partial^2 k} < 0\,, \text{ for } 1 \le k \le n_{ccb}.$$

Thus, we consider its opposite function given by

$$PF_{\exp}(s_0,s_m,k) = - \mid CPF_{\exp}(s_0,s_m,k) \mid \tag{10.8}$$

which is a *convex* function (Fig. 10.6), so we can formulate our unconstrained multi-objective optimization problem using the WES approach. Now, uniform energy depletion requires minimizing $PF_{\exp}(s_0,s_m,k)$. Notice that $E_{\exp}(s_0,s_m,k)$, $D_{\exp}(s_0,s_m,k)$, and $PF_{\exp}(s_0,s_m,k)$ (Eqs. 10.3, 10.4, and 10.8, respectively) are conflicting objective functions. Minimizing $E_{\exp}(s_0,s_m,k)$ requires minimizing $k$, whereas minimizing $PF_{\exp}(s_0,s_m,k)$ and $D_{\exp}(s_0,s_m,k)$ requires maximizing $k$. Moreover, $E_{\exp}(s_0,s_m,k)$, $D_{\exp}(s_0,s_m,k)$, and $PF_{\exp}(s_0,s_m,k)$ are strictly convex [93] on the interval $[1..n_{ccb}]$ given that

$$\frac{\partial^2 E_{\exp}(s_0,s_m,k)}{\partial^2 k} > 0,$$

$$\frac{\partial^2 D_{\exp}(s_0,s_m,k)}{\partial^2 k} > 0,$$

$$\text{and } \frac{\partial^2 PF_{\exp}(s_0,s_m,k)}{\partial^2 k} > 0, \text{ for } 1 \le k \le n_{ccb}.$$

In addition, because the feasible set $\{1,2,...,n_{ccb}\}$ is a convex set, the WES approach yields correct solutions [58].

Let $\mathrm{M}(k) = (E_{\exp}(s_0,s_m,k), D_{\exp}(s_0,s_m,k), PF_{\exp}(s_0,s_m,k))^{\mathrm{T}}$ be our multi-objective function, which we wish to minimize. $E_{\exp}(s_0,s_m,k)$ reaches its maximum value

$$E_{\exp}^{\max} = a\left(\frac{2\,E_{elec}}{R} + \varepsilon\, R^{\alpha-1}\right)\delta(s_0,s_m) \text{ at } k = n_{cbb} \text{ for } \alpha \in \{2,4\}$$

and

$$E_{\exp}^{\max} = a\left(\frac{2\,E_{elec}}{d_{\min}} + \varepsilon\, d_{\min}{}^{\alpha-1}\right)\delta(s_0,s_m) \text{ at } k = 1 \text{ for } \alpha = 3.$$

Also, $D_{\exp}(s_0,s_m,k)$ and $PF_{\exp}(s_0,s_m,k)$ attain their maximum

$$D_{\exp}^{\max} = \frac{c\,\delta(s_0,s_m)}{d_{\min}}$$

and

$$PF_{\exp}^{\max} = \frac{\lambda\,\pi\,d_{\min}\,\delta(s_0,s_m)}{6},$$

respectively, at $k=1$. Using the WES approach, where the weights $w_1$, $w_2$, and $w_3$ indicate the relative importance of $E_{\exp}(s_0,s_m,k)$, $D_{\exp}(s_0,s_m,k)$, and $PF_{\exp}(s_0,s_m,k)$, respectively, our unconstrained multi-objective optimization problem can be written as follows:

$$\begin{aligned}&\text{(MO)}\ \ \textit{Minimize}\ \mathrm{M}(k)\\ &\textit{Subject to}\ 1\le k\le n_{cbb}\end{aligned}$$

where

$$\mathrm{M}(k)=\begin{cases} w_1 E_{\exp}(s_0,s_m,k)+w_2\dfrac{E_{\exp}^{\max}D_{\exp}(s_0,s_m,k)}{D_{\exp}^{\max}}+w_3\dfrac{E_{\exp}^{\max}PF_{\exp}(s_0,s_m,k)}{PF_{\exp}^{\max}} \\ \quad \textit{if}\ E_{\exp}^{\max}=\max\{E_{\exp}^{\max},D_{\exp}^{\max},PF_{\exp}^{\max}\} \\ w_1\dfrac{D_{\exp}^{\max}E_{\exp}(s_0,s_m,k)}{E_{\exp}^{\max}}+w_2 D_{\exp}(s_0,s_m,k)+w_3\dfrac{D_{\exp}^{\max}PF_{\exp}(s_0,s_m,k)}{PF_{\exp}^{\max}} \\ \quad \textit{if}\ D_{\exp}^{\max}=\max\{E_{\exp}^{\max},D_{\exp}^{\max},PF_{\exp}^{\max}\} \\ w_1\dfrac{PF_{\exp}^{\max}E_{\exp}(s_0,s_m,k)}{E_{\exp}^{\max}}+w_2\dfrac{PF_{\exp}^{\max}D_{\exp}(s_0,s_m,k)}{D_{\exp}^{\max}}+w_3 PF_{\exp}(s_0,s_m,k) \\ \quad \textit{if}\ PF_{\exp}^{\max}=\max\{E_{\exp}^{\max},D_{\exp}^{\max},PF_{\exp}^{\max}\} \end{cases}$$

$0\le w_1,w_2,w_3\le 1$ with $w_1+w_2+w_3=1$,

$E_{\exp}^{\max}=\max\{E_{\exp}(s_0,s_m,k):1\le k\le n_{ccb}\}$, $D_{\exp}^{\max}=\max\{D_{\exp}(s_0,s_m,k):1\le k\le n_{ccb}\}$, and $PF_{\exp}^{\max}=\max\{PF_{\exp}(s_0,s_m,k):1\le k\le n_{ccb}\}$

Let us now study the non-linear multi-objective function $\mathrm{M}(k)$, which depends on the maximum values $E_{\exp}^{\max}$, $D_{\exp}^{\max}$, and $PF_{\exp}^{\max}$ of their corresponding objective functions. Hence, we consider the following three cases depending on the values $E_{\exp}^{\max}$, $D_{\exp}^{\max}$, and $PF_{\exp}^{\max}$.

**Case 1:** $E_{\exp}^{\max}=\max\{E_{\exp}^{\max},D_{\exp}^{\max},PF_{\exp}^{\max}\}$

Let $k_1^*$ be a solution to (MO).

$$\frac{\partial\,\mathrm{M}(k_1^*)}{\partial\,k_1^*}=0\Rightarrow\qquad k_1^*=\sqrt[\alpha]{\frac{2\,E_{elec}}{(\alpha-1)\,\varepsilon\,d_{\min}{}^{\alpha}}+\frac{w_2\,c\,E_{\exp}^{\max}}{w_1(\alpha-1)\,a\,\varepsilon\,d_{\min}^{\alpha}D_{\exp}^{\max}}+\frac{w_3\,\pi\,\lambda\,E_{\exp}^{\max}}{6(\alpha-1)\,w_1\,a\,\varepsilon\,d_{\min}^{\alpha-2}\,PF_{\exp}^{\max}}}$$

Notice that $\dfrac{\partial^2\,\mathrm{M}(k_1^*)}{\partial^2\,k_1^*}>0$. Thus, $k_1^*$ corresponds to the minimum of $\mathrm{M}(k_1^*)$.

Furthermore, varying the weights $w_1$, $w_2$, and $w_3$ from 0 to 1, where $w_1 + w_2 + w_3 = 1$, generates the corresponding minimum solutions of $\mathrm{M}(k_1^*)$.

**Case 2:** $D_{\exp}^{\max} = \max\{E_{\exp}^{\max}, D_{\exp}^{\max}, PF_{\exp}^{\max}\}$

Let $k_2^*$ be a solution to (MO).

$$\frac{\partial \mathrm{M}(k_2^*)}{\partial k_2^*} = 0 \Rightarrow \quad k_2^* = \sqrt[\alpha]{\frac{2\,E_{elec}}{(\alpha-1)\,\varepsilon\, d_{\min}{}^{\alpha}} + \frac{w_2\, c\, E_{\exp}^{\max}}{w_1(\alpha-1)\,a\,\varepsilon\, d_{\min}^{\alpha} D_{\exp}^{\max}} + \frac{w_3\,\pi\,\lambda\, E_{\exp}^{\max}}{6(\alpha-1)\,w_1\,a\,\varepsilon\, d_{\min}^{\alpha-2}\, PF_{\exp}^{\max}}}$$

**Case 3:** $PF_{\exp}^{\max} = \max\{E_{\exp}^{\max}, D_{\exp}^{\max}, PF_{\exp}^{\max}\}$

Let $k_3^*$ be a solution to (MO).

$$\frac{\partial \mathrm{M}(k_3^*)}{\partial k_3^*} = 0 \Rightarrow \quad k_3^* = \sqrt[\alpha]{\frac{2\,E_{elec}}{(\alpha-1)\,\varepsilon\, d_{\min}{}^{\alpha}} + \frac{w_2\, c\, E_{\exp}^{\max}}{w_1(\alpha-1)\,a\,\varepsilon\, d_{\min}^{\alpha} D_{\exp}^{\max}} + \frac{w_3\,\pi\,\lambda\, E_{\exp}^{\max}}{6(\alpha-1)\,w_1\,a\,\varepsilon\, d_{\min}^{\alpha-2}\, PF_{\exp}^{\max}}}$$

Notice that WES generates a unique, optimum solution ($k_1^* = k_2^* = k_3^*$) to the multi-objective optimization problem (MO) regardless of the outcome of the comparison between the maximum values $E_{\exp}^{\max}$, $D_{\exp}^{\max}$, and $PF_{\exp}^{\max}$. Although there are other methods for solving multi-objective optimization problems, such as multi-objective optimization genetic algorithm (MOGA) [84], WES fits well our purpose. Indeed, the TED protocol is designed for sensing applications with different requirements in terms of energy and delay. That is, real-world applications could give priority to one or more metrics rather than others and vice-versa. The flexibility of WES enables a network designer to emphasize the importance of a specific objective function over the others by choosing a suitable weight vector. Particularly, the different objective functions could be given the same preference by using equal weights. Moreover, we obtain a *Pareto-optimum* solution [58].

#### 10.3.2.3 Numerical Results

In Fig. 10.7, we plot $\mathrm{M}(k)$ with respect to different weight vectors $(w_1, w_2, w_3)$, where $w_1 + w_2 + w_3 = 1$, while varying the path-loss exponent $2 \le \alpha \le 4$. As expected, the minimum of $\mathrm{M}(k)$ depends on the weights assigned by a network designer to the individual objective functions. When $\alpha = 2$, the minimum of $\mathrm{M}(k)$ is obtained at $k = 2$ for high values of $w_1$ ($w_1 \in \{0.8, 0.7, 0.6\}$), meaning that the network designer wishes to minimize the energy consumption of sensors, which could be achieved for $k = 2$ as was proved theoretically. However, when $w_1$ has comparable values to those of $w_2$ and $w_3$ ($(w_1, w_2, w_3) \in \{(0.5, 0.3, 0.2), (0.4, 0.4, 0.2)\}$), $\mathrm{M}(k)$ reaches its minimum at $k = 3$ as the network designer wants to achieve some balance between the three objective

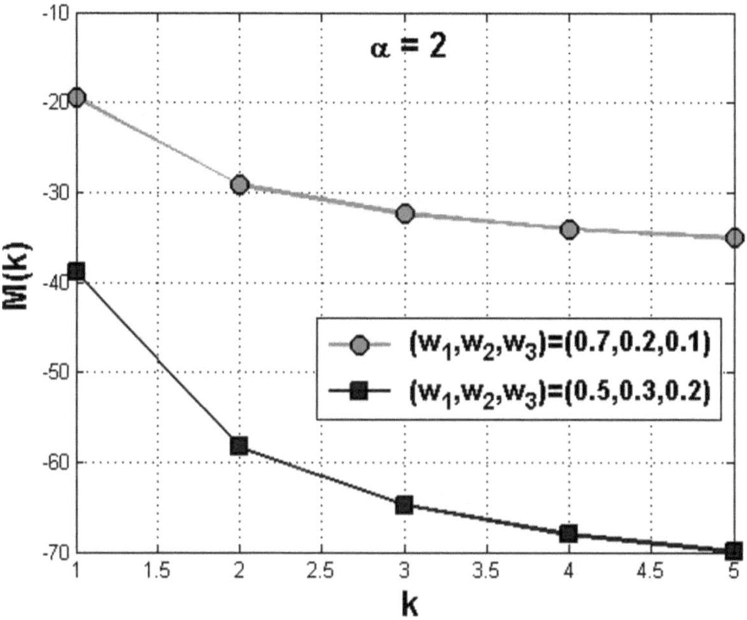


**Fig. 10.13** Trade-off between the three metrics with $(w_1 > w_2, w_3)$ for $\alpha = 2$

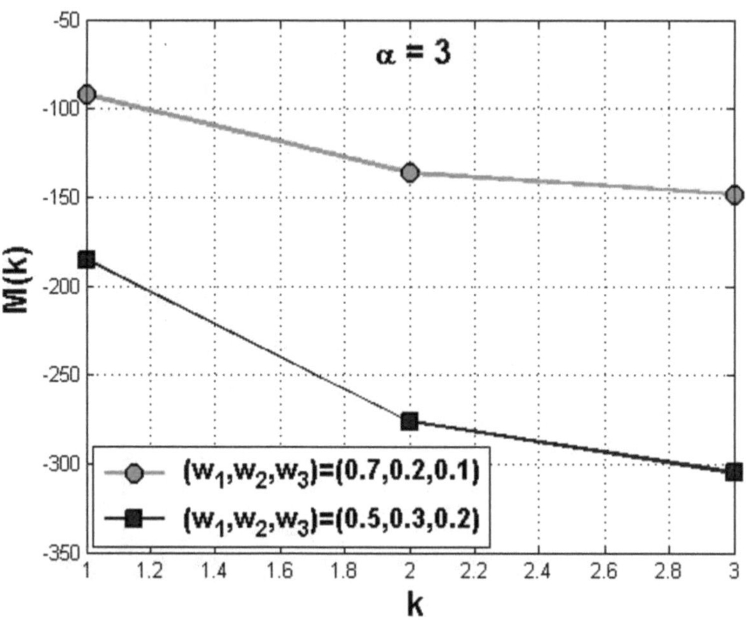


**Fig. 10.14** Trade-off between the three metrics with $(w_1 > w_2, w_3)$ for $\alpha = 3$

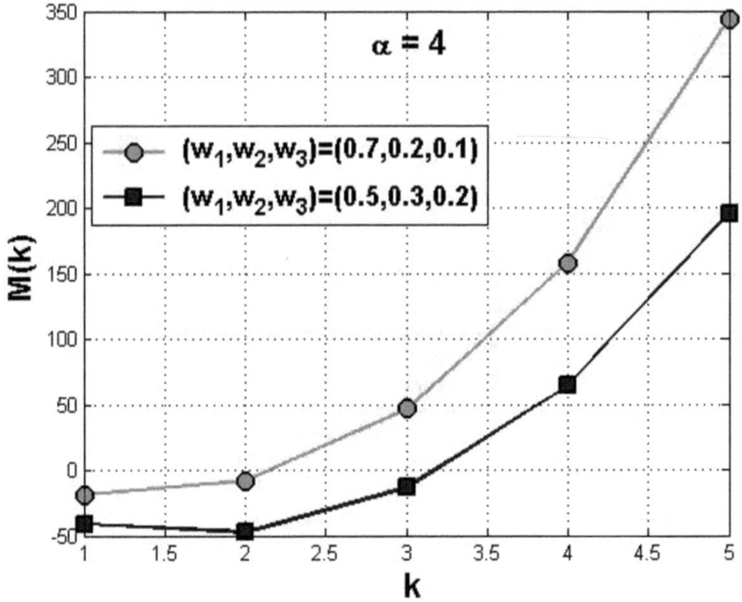


**Fig. 10.15** Trade-off between the three metrics with $(w_1 > w_2, w_3)$ for $\alpha = 4$

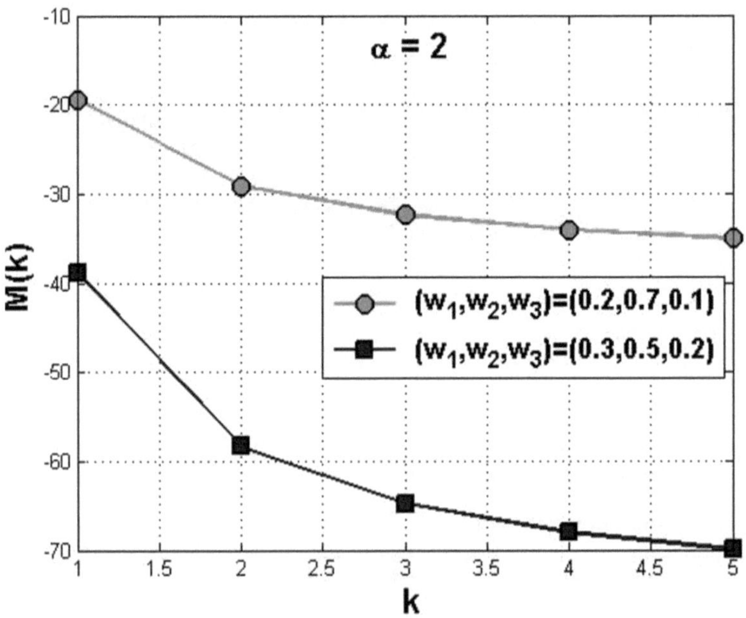


**Fig. 10.16** Trade-off between the three metrics with $(w_2 > w_1, w_3)$ for $\alpha = 2$

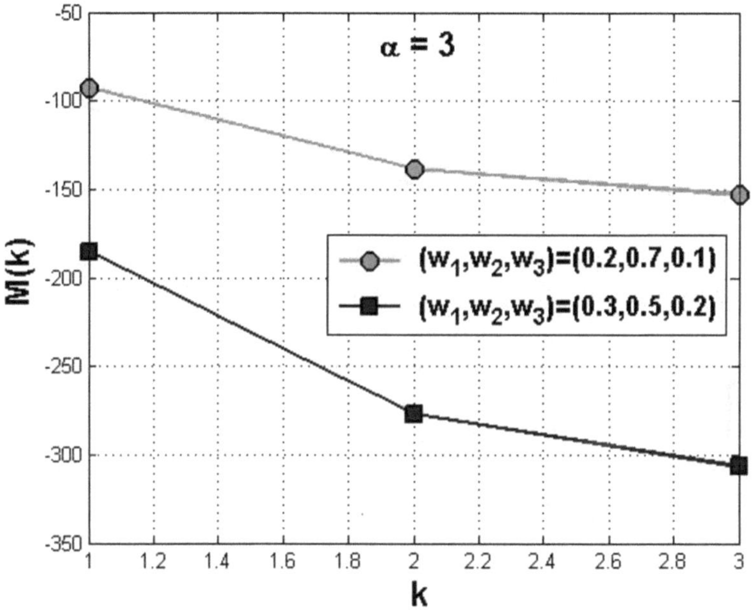


**Fig. 10.17** Trade-off between the three metrics with $(w_2 > w_1, w_3)$ for α = 3

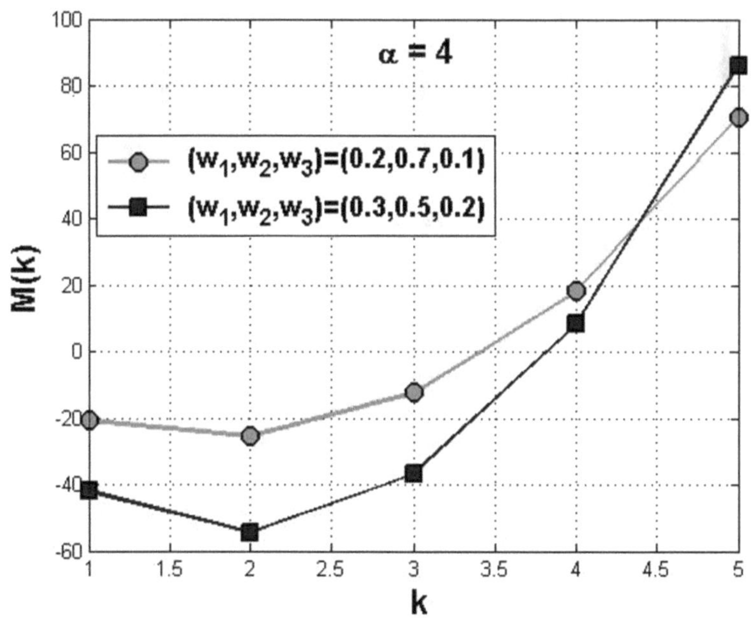


**Fig. 10.18** Trade-off between the three metrics with $(w_2 > w_1, w_3)$ for α = 4

functions. When $\alpha = 3$, the best trade-off corresponds to $k = 2$ for $((w_1, w_2, w_3) \in \{(0.8,0.1,0.1),(0.7,0.2,0.1)\})$, which favour minimizing energy consumption over minimizing delay and guaranteeing uniform energy depletion, and $k = 3$ for lower values of $w_1$. However, when $\alpha = 3$, $\mathrm{M}(k)$ attains its minimum at $k = 3$ and $k = 4$, depending on the weights $w_1$, $w_2$, and $w_3$. It is clear that as $w_1$ decreases, the value of $k$ corresponding to the minimum of $\mathrm{M}(k)$ increases, meaning that outer CCBs are more preferred.

Figures 10.16–10.18 plot $\mathrm{M}(k)$ favouring minimizing delay over the other two metrics using appropriate values of $w_1$, $w_2$, and $w_3$. When $\alpha = 2$, the minimum of $\mathrm{M}(k)$ is obtained at the outmost CCB, i.e., $k = 5$. Similarly, when $\alpha = 3$, the best trade-off occurs at $k = 3$. We observe from Figs. 10.16 and 10.17 that the value of $k$ increases as we increase the value of $w_2$. Recall that the minimum delay is reached for the outer CCBs. For $\alpha = 4$, the minimum of $\mathrm{M}(k)$ is reached at $k = 2$.

The plots of $\mathrm{M}(k)$ in Figs. 10.19–10.21 use the weights $w_3 > w_1 \geq w_2$, which favour uniform battery power depletion over the other two metrics. As can be observed from Figs. 10.19–10.21, the minimum of $\mathrm{M}(k)$ occurs at $k = 5$ for $\alpha = 2$. Also, when $\alpha = 3$, $\mathrm{M}(k)$ attains its minimum at $k = 3$ (outmost CCB). However, when $\alpha = 4$, the minimum of $\mathrm{M}(k)$ is reached at $2 \leq k \leq 3$.

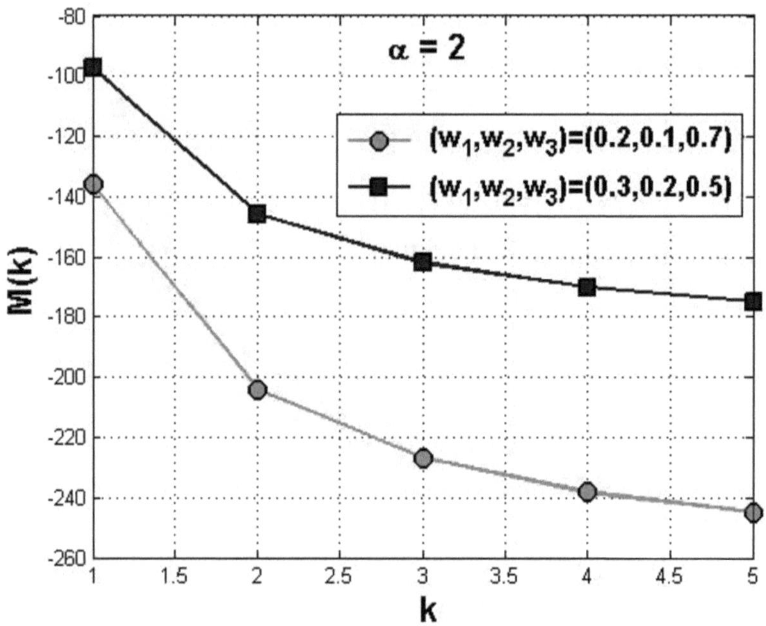


**Fig. 10.19** Trade-off between the three metrics with $(w_3 > w_1, w_2)$ for $\alpha = 2$

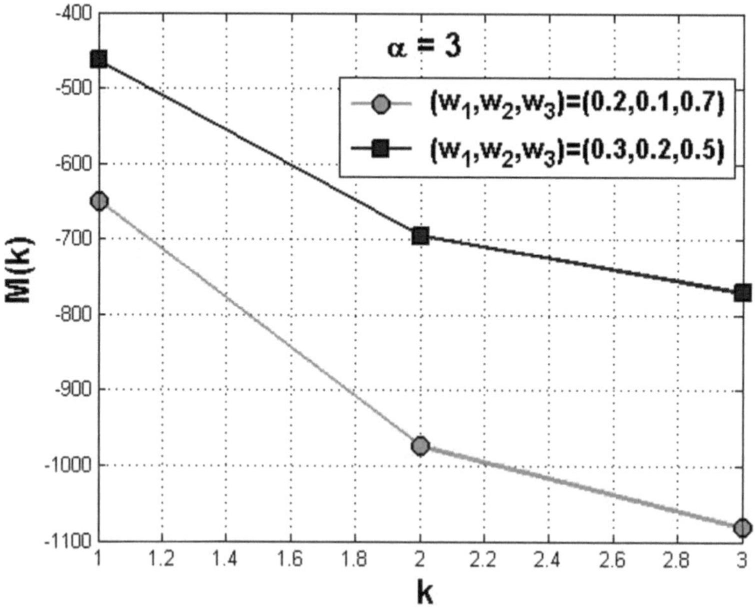


**Fig. 10.20** Trade-off between the three metrics with $(w_3 > w_1, w_2)$ for α = 3

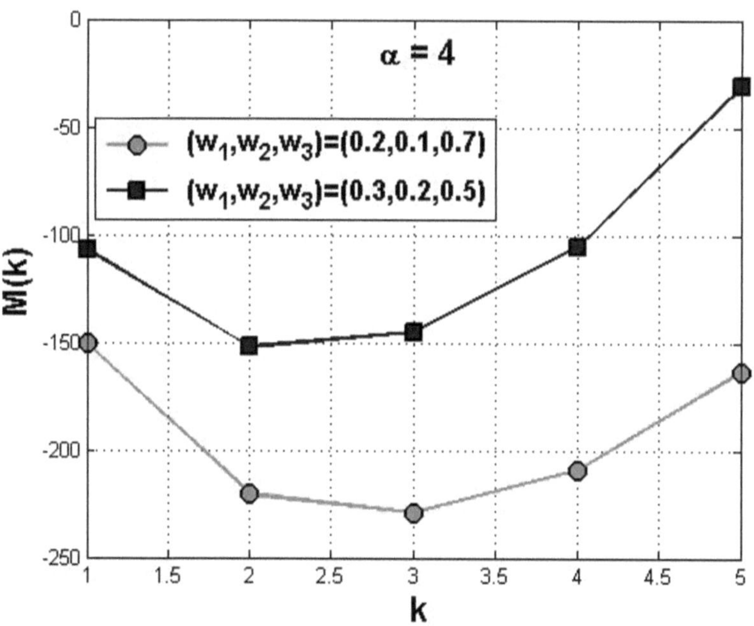


**Fig. 10.21** Trade-off between the three metrics with $(w_3 > w_1, w_2)$ for α = 4

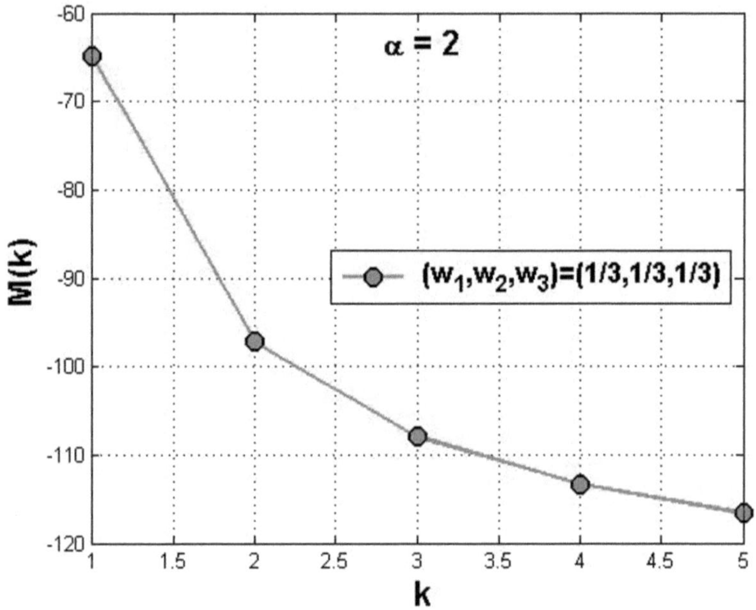


**Fig. 10.22** Trade-off between the three metrics with $(w_1 = w_2 = w_3 = 1/3)$ for $\alpha = 2$

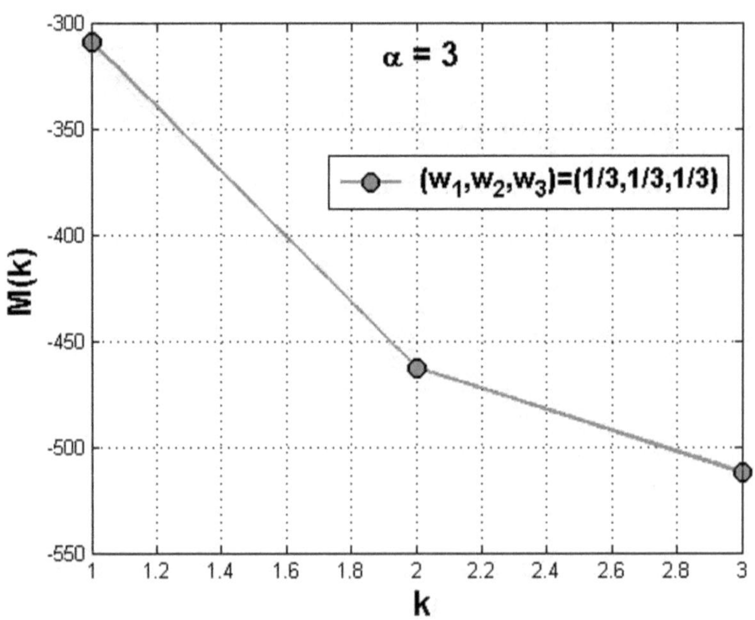


**Fig. 10.23** Trade-off between the three metrics with $(w_1 = w_2 = w_3 = 1/3)$ for $\alpha = 3$

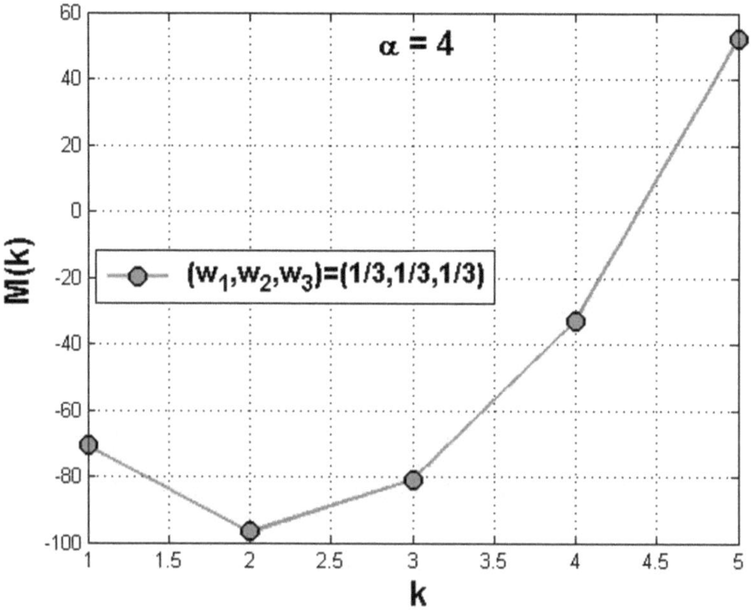


**Fig. 10.24** Trade-off between the three metrics with $(w_1 = w_2 = w_3 = 1/3)$ for $\alpha = 4$

Figures 10.22–10.24 show that the best trade-off between the three objective functions with the same weight corresponds to $k = 5$ for $\alpha = 2$, $k = 3$ for $\alpha = 3$, and $k = 2$ for $\alpha = 4$. It is worth noting that minimizing delay and guaranteeing uniform energy depletion are not conflicting metrics since both of them require maximizing $k$. Thus, the weights $w_2$ and $w_3$ can be viewed as a combined weight against the weight $w_1$.

### *10.3.3 TED Detailed Description*

The TED protocol given in Fig. 10.25 has three phases: *communication range slicing*, *concentric circular band selection*, and *proxy forwarder selection.*

#### 10.3.3.1 Communication Range Slicing

This phase is run by each sensor only once at the beginning of the sensing task. Each sensor creates a table with $n_{ccb}$ entries, each including a subset of neighbours located in the corresponding CCB. Sensors located at the boundaries of two consecutive CCBs are assigned to the inner one.

#### 10.3.3.2 Concentric Circular Band Selection

When a source $s_0$ wishes to disseminate its data to the sink $s_m$, it computes the *id* of the CCB to be used. This *id* is denoted by $k$, where $1 \le k \le n_{ccb}$. The value of $k$ corresponds to the optimum solution of the multi-objective optimization problem defined earlier. It is the responsibility of a source to choose the appropriate weights associated with each of the three metrics. The selection of the values of these weights is guided by the requirements of the underlying sensing application in terms of energy savings and delay. In other words, a source may wish to favour one of the metrics over the others or find a "best" trade-off between them. Once a source $s_0$ has selected the corresponding weights, it solves the multi-objective optimization problem to find the value of $k$ that meets the application needs.

#### 10.3.3.3 Proxy Forwarder Selection

A source $s_0$ selects its proxy forwarder $s_{PF1}$ amongst the subset $CPF(s_0, s_m, k)$, such that

$$E_{rem}(s_{PF1}) = \max\{E_{rem}(s_j) : s_j \in CPF(s_0, s_m, k, \theta)\}$$

where $E_{rem}(s_j)$ is the remaining energy of sensor $s_j$. In other words, the protocol takes into consideration the remaining energy of sensors and prefers the ones with high available energy, thus avoiding sensors with little remaining energy whose selection would possibly lead to network disconnections.

When a proxy forwarder receives sensed data, it runs the same protocol to select its next proxy forwarder. This process continues until the sink receives data originated from a source.

#### 10.3.3.4 Is *k* Fixed for All Proxy Forwarders or Not?

Recall that WES generates a unique, optimum solution $(k = k_1^* = k_2^* = k_3^*)$ to the multi-objective optimization problem (MO). Furthermore, the value of $k$ depends only on the weighting coefficients $w_1$, $w_2$, and $w_3$, which are associated with each of the three metrics, i.e., minimum energy consumption, minimum delay, and uniform energy depletion. The question that we should address is whether the same value of $k$ computed by a source is used by all proxy forwarders. To answer this question, we consider the following two scenarios for computing the value of $k$. In the first one, the TED protocol requires all proxy forwarders on the data forwarding path between $s_0$ and $s_m$ to use the same value of $k$ to identify their proxy forwarders. In other words, the formulation of the multi-objective optimization problem and its solving is done only once by a source $s_0$ and any proxy forwarder should use the value of $k$ found by $s_0$. In the second scenario, however, each of the senders (i.e., sources and proxy forwarders) computes its own value of $k$ based on its own values of $w_1$, $w_2$, and $w_3$. The second scenario seems more practical for the following reason. As we get closer to the sink, sensors could get more active in data forwarding towards the sink. In particular, sensors located around the sink act as relay of all data originated from all sources.

**Algorithm**: **TED**

**Begin**

**// Actions executed by a source** $s_0$

**Phase 1: Slice the communication range** $CD(\zeta_0, R)$ of $s_0$

**1.** Slice $CD(\zeta_0, R)$ into $n_{ccb}$ CCBs

**Phase 2: Select an appropriate CCB using** $k$

**2.** Select the appropriate weights $0 \le w_1, w_2, w_3 \le 1$ such that $w_1 + w_2 + w_3 = 1$ to solve the multi-objective optimization problem: (MO) *Minimize* $M(k)$

*Subject to* $1 \le k \le n_{ccb}$

**3.** Choose a CCB id, $k$, which is a solution to (MO)

**4.** **If** sink $s_m \in NNS(s_0)$ and $s_m \in k'^{th}$ with $k' \le k$ **Then**

**Begin**

**5.** Forward the sensed data directly to the sink $s_m$

**6.** Break;

**End**

**7.** **Else**

**Begin**

**Phase 3: Select a proxy forwarder from** $k^{th}$ **CCB**

**8.** Identify a subset candidate proxy forwarders $CPF(s_0, s_m, k)$ from the $k^{th}$ CCB

**8.** **If** this subset is empty **Then**

**Begin**

**9.** Randomly pick the closest $q^{th}$ non-empty lower/higher CCB

**10.** $k = q$;

**End**

**11.** Determine the first proxy forwarder $s_{PF1}$ such that $E_{rem}(s_{PF1}) = \max\{E_{rem}(s_j) : s_j \in CPF(s_0, s_m, k, \theta)\}$

**12.** Forward the sensed data packet to $s_{PF1}$

**// Actions executed by any proxy forwarder**

**13.** **While** (sensed data has not reached $s_m$) **Do**

**Begin**

**14.** **If** $s_m \in NNS(s_{PFi})$ and $s_m \in k'^{th}$ CCB with $k' \le k$ **Then**

**Begin**

**15.** Forward the sensed data directly to $s_m$

**16.** Break;

**End**

**17.** **Else** Replace $s_0$ with $s_{PFi}$ and run steps 5-15

**End**

**End**

**End**

**Fig. 10.25** The TED data forwarding protocol

Specifically, the data traffic model would have an impact on the energy consumption. In fact, if each sensor is required to transmit data to the sink, sensors nearer the sink would consume more energy than all other sensors in the network. Thus, for these sensors, it would be more important for them to minimize their energy consumption as much as possible by sending/forwarding data to the sink through short distances. This implies that the value of $w_1$ is higher than those of $w_2$ and $w_3$. Therefore, in our design and simulation of the TED protocol, we consider the second scenario in which the value of $k$ is not the same for a source and all subsequent proxy forwarders when forwarding sensed data towards the sink.

## 10.4 Relaxation of Several Key Assumptions

The assumptions that we have made about wireless sensor networks, namely sensor homogeneity (Assumption 1), communication disk model (Assumption 2), highly dense wireless sensor networks (Assumption 3), energy consumption dominated by energy spent in data transmission and reception (Assumption 4), and always-on sensor model (Assumption 5), form the basis for most of the existing data forwarding and topology control protocols for wireless sensor networks. In this section, we discuss how to relax these assumptions and assess their impact on the TED protocol to promote its applicability to real-world sensing applications.

### *10.4.1 Relaxing the Sensor Homogeneity Model*

In real-world scenario, our assumption (Assumption 1) may not be true. Indeed, sensors may have different capabilities, particularly in terms of communication ranges. Several studies [75, 209] showed that wireless sensor networks with heterogeneous sensors, possessing unequal energy levels and different sensing, processing, and communication capabilities, have increased reliability and lifetime. In the second scenario (i.e., each sensor computes its own value of $k$), there is no change to TED. For the first scenario (i.e., same value of $k$ for all source and proxy forwarders), however, it may happen that the number of CCBs of some proxy forwarder $s_i$, say $n_{ccb}(s_i)$, is less than $k$ (i.e., solution to the multi-objective optimization problem $(MO)$). In this case, the sensor $s_i$ selects its proxy forwarder from its $k'^{th}$ CCB, where $k' = n_{ccb}(s_i) < k$. Given that $k' < k$, the data forwarding process will consume less energy than expected. Also, given that the proxy forwarder $s_i$ selects its last CCB, a minimum delay and a uniform energy depletion will be ensured. Thus, we obtain the best trade-off with respect to the three goals.

### 10.4.2 Relaxing the Communication Disk Model

The circular radio model (Assumption 2) is used to simplify the analysis and may be unrealistic. Indeed, empirical studies have found that the communication range of radios is highly probabilistic and irregular [198, 221, 224]. For tractability, we consider convex communication ranges that are not necessarily disks. Using the notion of largest enclosed disk (Definition 3, see Chap. 2), each sensor would be able to run the TED protocol by slicing the largest enclosed disk of its communication range. As in Sect. 10.4.1, there is no change to TED protocol in the second scenario. For the first scenario, however, the number $n_{ccb}(s_i)$ of CCBs of a sensor $s_i$ may be less than the value of $k$ computed by a source. Similarly, sensor $s_i$ would consider its last CCB from which it would select its proxy forwarder. Using the same argument as in Sect. 10.4.1, we can see that we obtain the best trade-off between all the three individual objective functions.

### 10.4.3 Relaxing the Dense Network Model

One of the main characteristics of wireless sensor networks is their high density compared to ad hoc wireless networks [6, 7]. Thus, our assumption (Assumption 3) is realistic. Now, what would happen if TED is used by a sparse wireless sensor network? In this case, a sensor may not find any sensor located in the selected $k^{th}$ CCB. Although some proxy forwarder can be located in another CCC, say the $i^{th}$ CCB, where $i \neq k$, and hence selected to participate in the forwarding process of the sensed data, it may not be always possible to find the best trade-off required by the underlying sensing application. Thus, the TED protocol would be able to trade-off energy with delay with respect to the specific needs of a sensing application only when the network has some density such that each CCB has some sensors that would be selected to act as proxy forwarders. This is, indeed, a valid assumption since wireless sensor networks are densely deployed in general to provide high-quality monitoring.

### 10.4.4 Relaxing the Energy Consumption Model

The slicing approach of the communication range of sensors depends on the minimum transmission distance $d_{\min}$. Recall that $d_{\min}$ is computed based on our assumption (Assumption 4) which considers only the energy spent in data transmission and reception. Nevertheless, sensors spend energy in computation and sensing. However, both types of energy are negligible compared to the one spent in communication. Also, neither the energy spent in computation nor the one spent in sensing is quantifiable. Hence, if either type of energy has some known formula, the value of $d_{\min}$ would be modified accordingly. Precisely, this value would be smaller than the one that we have computed in Lemma 9.4. Hence, the number of CCBs would be higher, which would have an impact on TED performance.

### *10.4.5 Relaxing the Always-On Sensors Model*

Most of the existing geographic routing and data forwarding protocols assume that the radios of all sensors are turned on all the time (Assumption 5). In particular, all sensors in the network are awake during the forwarding activity. However, in real-world scenarios, sensors switch between on and off states to save their limited energy. It is not even practical to keep a sensor awake all the time while it is in reality active for only some short periods of time. Indeed, keeping sensors always on may cause failures that have severe impact on the performance of the wireless sensor network. Precisely, the network could be partitioned into at least two non-communicating sub-networks and hence the existence of the whole network may become meaningless. Therefore, it is important to duty-cycle sensors so that they deplete their energy resources uniformly and slowly [205, 218]. Unfortunately, duty-cycling may create a problem for routing the current message to the next hop while it is asleep. The challenge is how to duty-cycle sensors running TED while guaranteeing good routing performance [161, 226, 227]. The handling of highly dynamic wireless sensor networks that experience time-varying connectivity due to sensor duty-cycling is left as one of our future work.

## 10.5 Simulation Results

Our approach to trading-off minimum energy consumption, minimum delay, and uniform energy depletion is unique in several ways as described below, and hence it is impossible to make a fair quantitative comparison between TED and other existing approaches, such as the ones given in [142, 143, 160, 208, 226, 227] and reviewed in Sect. 10.6. First, TED allows a network designer to optimize the above-mentioned three metrics according to the specific needs of the underlying sensing application by using weights that specify the interest in each of these three metrics. Second, TED works at the network layer and does not assume any sleep–wakeup scheduling protocol, where sensors are duty-cycled (i.e., turned *on* or *off*) to save energy. TED applies to always-on and many-to-one WSNs, where sensors are always on to collect data about a specific phenomenon in a target field and send them to a single sink for further analysis and processing. Third, TED does not assume any aggregation of sensed data originated from sources towards the sink. In other words, all data should be received by the sink without undergoing any fusion or aggregation at any intermediate sensor. On the other hand, PEGASIS (Power-Efficient GAthering in Sensor Information Systems) [142, 143] is a simple and elegant data aggregation protocol in that it forms a chain amongst sensors so that each sensor receives from and transmits to a close neighbour and only one designated sensor sends the combined sensed data to the sink in each round. Precisely, all sensors take turns to directly transmit the combined data to the sink. PBBF (*Probability-Based Broadcast Forwarding*) [160] works at the MAC layer and assume a sleep–wakeup mechanism for all sensors. Indeed, PBBF benefits from the redundancy in broadcast communication and forwards packets using a probability-based approach with two parameters that need to be used by a sleep–wakeup scheduling protocol. The first parameter is a probability $p$ that a sensor rebroadcasts a packet in the current active time while not all its neighbours are guaranteed to be awake to receive the broadcast. The second parameter is a probability $q$ that a sensor would remain awake after the active time when it would be asleep. Furthermore, the goal of PBBF applies to many-to-many WSNs and ensures that a sensor receives at least one copy of each broadcast packet with high probability, while reducing the latency due to sleeping. Also, PTW (*Pipelined Tone Wakeup*) [208] helps sensors forward their data to their final destinations based on a wakeup scheme that helps to achieve a balance between energy saving and end-to-end delay. Also, GeRaF assumes a sleep–wakeup scheduling protocol, where each sensor keeps forwarding data until at least one of its neighbouring sensor is awake and able to receive them.

## 10.5.1 Simulation Settings

In this section, we present the performance results of the TED protocol for the free-space model $(\alpha = 2)$ based on simulation programs written in the C programming language. We consider a square deployment field of side length is equal to $1000\,\text{m}$ where sensors are randomly and uniformly distributed. Furthermore, we assume that every sensor continuously generates constant bit rate (CBR) data of 1024 bits/s (i.e., 4 data packets of size 256 bits per second). Moreover, the radius of the communication range of the sensors is equal to 250 m. Also, we assume that the total number of deployed sensors in the default case is 1000, which corresponds to a sensor spatial density equal to $\lambda = 0.001$ sensor per $\text{m}^2$. In addition, we assume that all sensors have the same amount of initial energy that is equal to $1\,\text{J}$.

## 10.5.2 Impact of Selection Space Size

In this experiment, we consider one metric, namely *number of communication rounds*, and show the impact of the size of the selection space of proxy forwarders on TED. We assume that all sensors use the same value of the angle $\beta$, which together with a value of $k$ define the size of the space of the $k^{th}$ CCB from which proxy forwarders are selected. As can be seen from Fig. 10.26, as the value of $\beta$ increases, the number of communication rounds increases too. The existence of

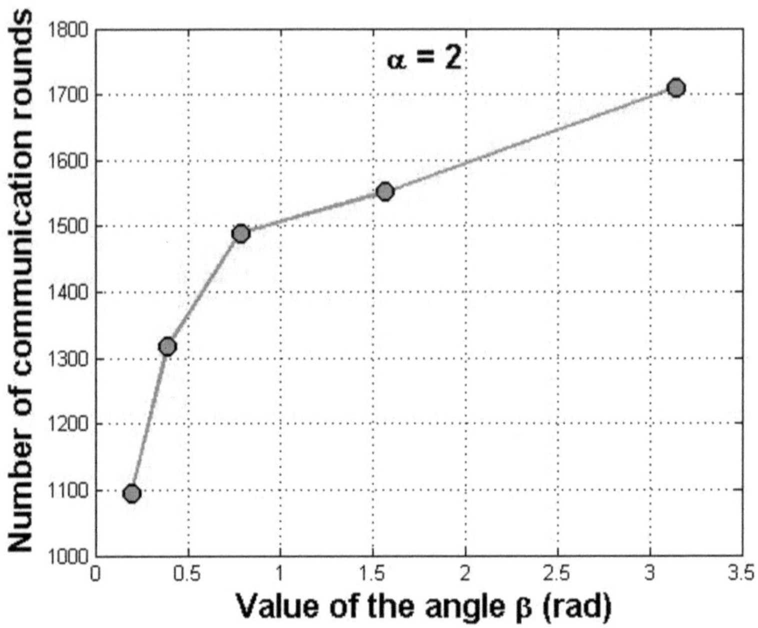


**Fig. 10.26** Uniform energy depletion

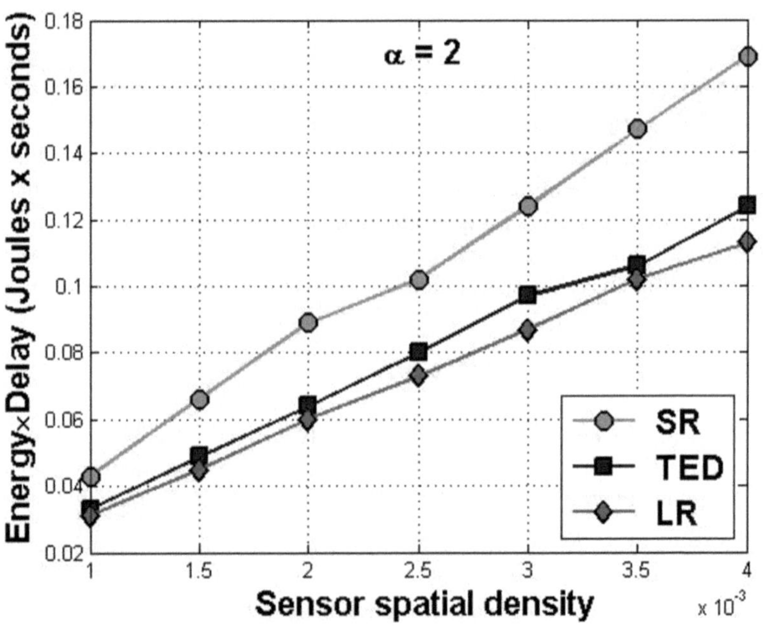


**Fig. 10.27** Impact of sensor density

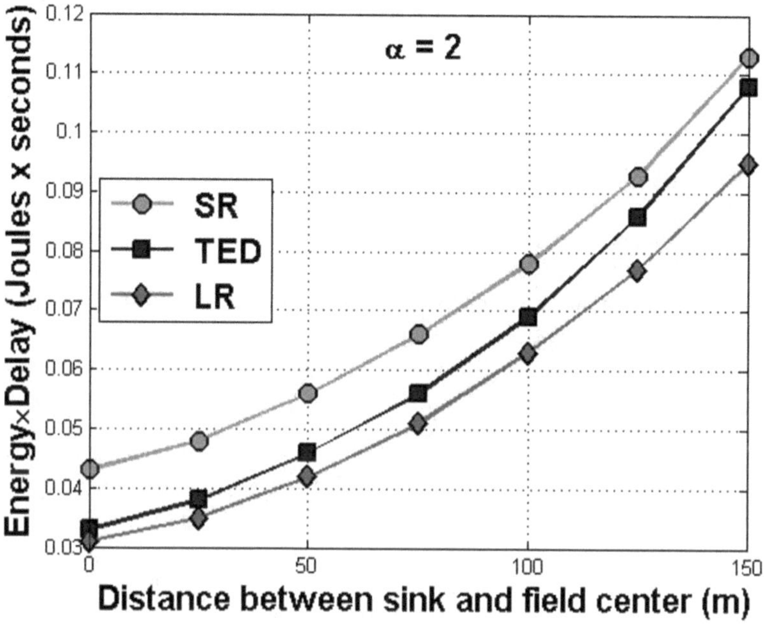


**Fig. 10.28** Impact of location of the sink

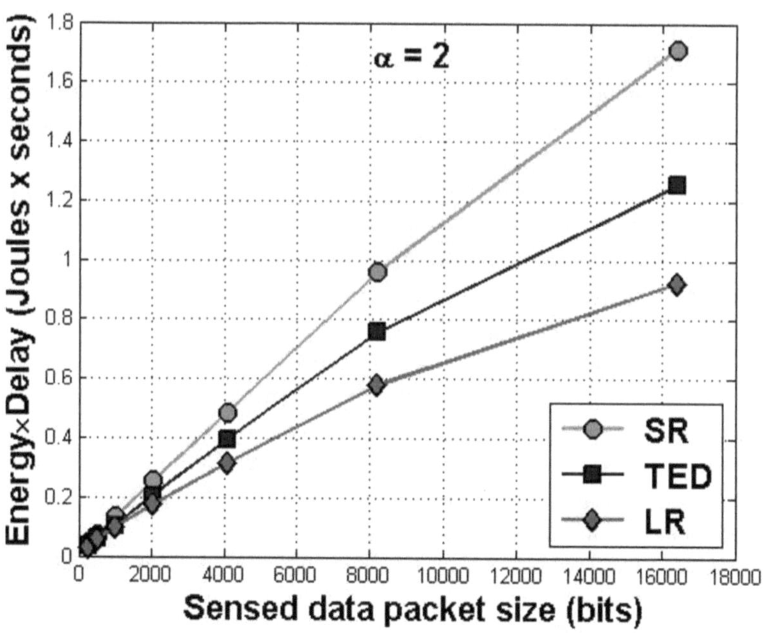


**Fig. 10.29** Impact of data packet size

more communication rounds means that there are more communication paths between sensors and the sink so that data originated from the former would reach the latter. In the absence of a balance of the load of data forwarding amongst all sensors, some sensors would be more heavily used as proxy forwarders than others, and hence die quickly. Thus, holes (i.e., void regions) in the network would appear and the network would consequently disconnect, thus prohibiting sensed data from reaching the sink. This situation appears if the space from which sensors select their proxy forwarders is small. Hence, the same neighbours would be selected frequently to forward data towards the sink, thus depleting their energy faster than others. Thus, uniform energy would guarantee that most of sensors would evenly participate in data forwarding to the sink and deplete their energy as slowly and uniformly as possible. However, we should mention that the energy sink-hole problem [137, 140, 160, 164, 199, 200] cannot be avoided while using a static sink.

### 10.5.3 Using the Energy × Delay Metric

In this experiment, we consider another metric, namely $energy \times delay$, which was introduced by Lindsey et al. [142] to evaluate the performance of their PEGASIS protocol. Since it is impossible to compare our TED protocol with any existing approach, we consider two instances of TED, namely *short-range forwarding* (SRF) and *long-range forwarding* (LRF). Using SRF, sensors forward data over short distances. With LRF, sensors forward data over long distances. SRF performs the best in terms of energy consumption but yields a highest delay, and hence provides us with lower bound on energy. LRF performs the best in terms of delay but consumes the maximum energy, and hence provides us with lower bound on delay. On the other hand, TED helps find a balance between energy with delay. Thus, we expect that TED performs like SRF when sensors care more about energy consumption ($w_1 >> w_2, w_3$), and performs like LRF when sensors care more about delay and uniform battery depletion ($w_2, w_3 >> w_1$). Given that our goal is to find a trade-off between energy and delay, we consider the $energy \times delay$ metric to compare TED against SRF and LRF.

Figure 10.27 shows that LRF has the lowest $energy \times delay$, whereas SRF yields the highest $energy \times delay$. As expected, Fig. 10.27 shows that $energy \times delay$ for the three schemes increases with the sensor spatial density. Figure 10.28 shows that $energy \times delay$ depends on the location of the sink. It is worth mentioning that the optimum location of the sink in terms of energy-efficient data gathering corresponds to the centre of the deployment field [148]. Figure 10.28 shows that the best performance of the three protocols is obtained only when the sink is located at the centre of the deployment field. Thus, the obtained results are conforming to the claim given in [148]. Figure 10.29 shows the impact of sensed data packet size

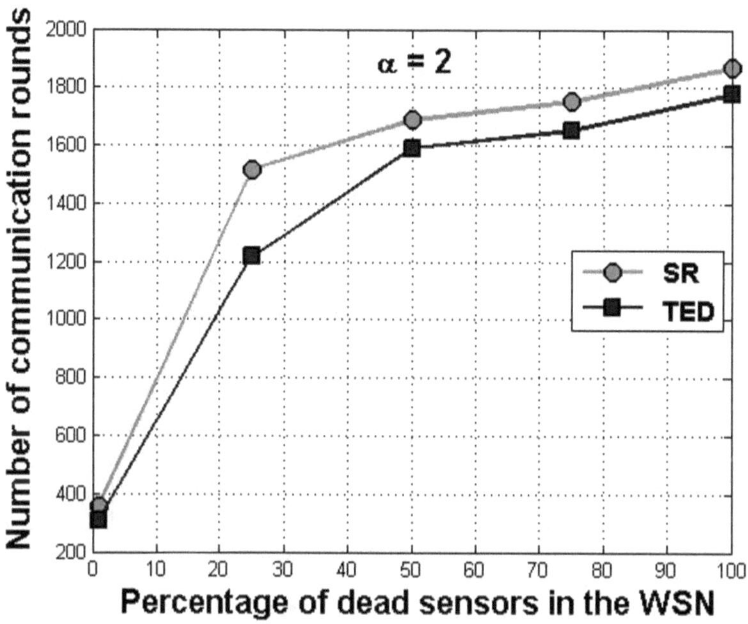


**Fig. 10.30** TED compared to SRF

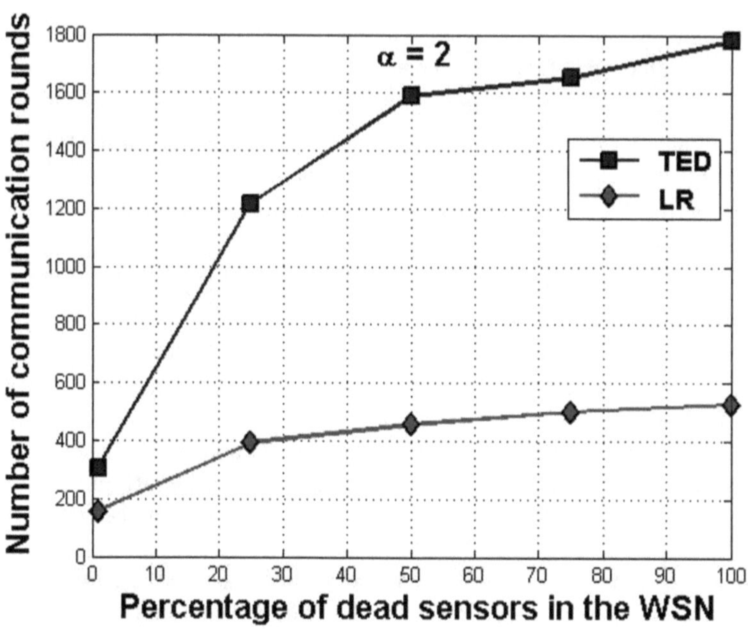


**Fig. 10.31** TED compared to LRF

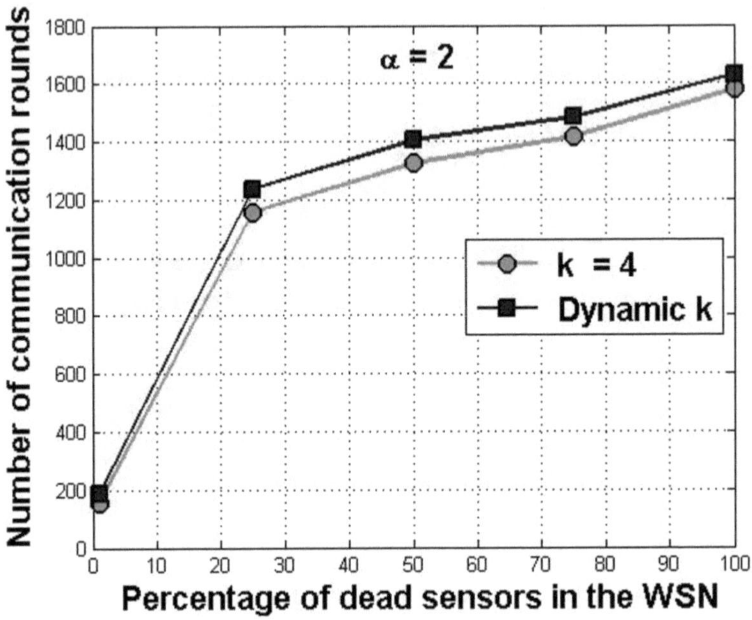


**Fig. 10.32** Impact of variability of $k$

on $energy \times delay$. Recall that the energy a sensor spends in data transmission $(E_{mt})$ depends on the size of the sensed data forwarded to the sink. As can be observed, TED has $energy \times delay$ close to the one produced by LRF. Given the orders of magnitude of $energy$ and $delay$, we can claim that $energy \times delay$ reaches its minimum for the smallest value of $delay$. Our intuition matches the results of the analysis of the function $E_{\exp}(s_0, s_m, k) \times D_{\exp}(s_0, s_m, k)$. Indeed, it is easy to check that this function reaches its minimum at $k = 5$, which corresponds to LRF as it always selects proxy forwarders from the last CCB. Thus, LRF is *optimal* with respect to the $energy \times delay$ metric, and hence TED is *near optimal*. Although LRF has the best $energy \times delay$, it is not suitable for energy-constrained WSNs. SRF, on the other hand, is not appropriate for time-critical sensing applications.

As can be seen in Fig. 10.30, SRF yields better network lifetime than TED. However, as shown in Fig. 10.31, we find that TED outperforms LRF in terms of network lifetime. This is due to the nature of forwarding schemes of SRF and LRF.

Compared to SRF and LRF, the TED protocol is more balanced in terms of energy and delay, and hence is a best candidate for WSNs where both energy savings and delay are (equally) important. Also, TED offers more flexibility to WSN designers to meet the specific needs of sensing applications in terms of energy and delay.

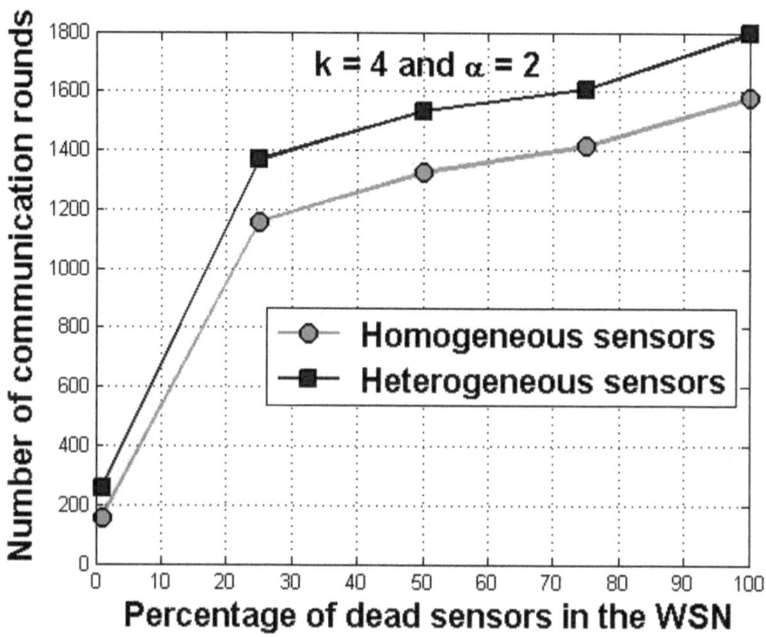


**Fig. 10.33** Impact of sensor heterogeneity

## *10.5.4 Impact of Variability of k*

In this experiment, we focus on the variability of $k$ whose value of $k$ may not be the same for all the proxy forwarders. We have considered two cases: In the first case, the source sensor computes the value of $k$ and each proxy forwarder involved in the forwarding data towards the sink has to use the same value of $k$. In the second case, computing the value of $k$ is done locally. In fact, each proxy forwarder computes its proper value of $k$ to identify the CCB from which it would select its next proxy forwarder. Figure 10.32 shows the simulation results for both cases. Although both results are very similar, we find that the variability of $k$ yields better performance. It is worth mentioning that the sensors nearer the static sink represent hot-spot traffic points in the sense that they are heavily used in the process of data forwarding to the sink. This situation creates a problem known as the *energy sink-hole* problem, which could possibly isolate the sink, thus disconnecting the network. These sensors care more about minimizing their energy consumption by forwarding the sensed data to the sink over short distances although it may be within their communication range. Thus, their selection scheme enables them to extend their individual lifetime and, consequently, the network lifetime.

### *10.5.5 Impact of Sensor Heterogeneity*

In this experiment, we evaluate the performance of our TED protocol in the presence of heterogeneous sensors. We assume that the sensors may not necessarily have the same amount of initial energy and the same radii of their communication ranges. The initial energy of the sensors is randomly picked from the interval [1 J, 2 J]. Similarly, the radii of their communication ranges are randomly selected from the interval [250 m, 350 m]. Given that the sensors may not have the same radii of their communication range, they would not have the same number of CCBs when they slice their communication ranges. Thus, a value of $k$ that is initially computed by the source sensor may exceed the total number of CCBs a proxy forwarder would have in its communication range. In that case, a proxy forwarder would simply choose the last CCB from which it would select its next proxy forwarder. Therefore, the value of $k$ would not be the same for all the proxy forwarders, i.e., $k$ is dynamic. It is as if each sensor would have its proper value of $k$. Figure 10.33 shows the results for homogeneous sensors and heterogeneous sensors network setup. Not surprisingly, we find that the performance of TED using heterogeneous sensors outperforms that of the network with homogeneous sensors. Here, we show that sensor heterogeneity helps extend the lifetime of the individual sensors, thus prolonging the network lifetime. Indeed, Yarvis et al. [209] presented several analytical, simulations, and real testbed results showing the potential benefit and impact of energy and link heterogeneity on sensor nets, where all the sensors report their sensed data to a single static sink.

## 10.6 Related Work

This section reviews a sample of data forwarding protocols for wireless sensor networks that trade off energy with other metrics, such as delay and robustness.

Yang and Vaidya [208] proposed a wakeup scheme, called Pipelined Tone Wakeup (PTW), which achieves a balance between energy saving and end-to-end delay. The PTW scheme is based on an asynchronous wakeup pipeline that overlaps the wakeup procedures with the packet transmissions. It uses wakeup tones which allow a large value of duty-cycle ratio without causing a large wakeup delay at each hop. Miller et al. [160] studied the trade-off between energy, latency and reliability. They presented a Probability-Based Broadcast Forwarding (PBBF) scheme which minimizes energy usage and optimizes latency and reliability. Zorzi and Rao [226] proposed a transmission technique for wireless sensor networks called geographic random forwarding (GeRaF), where relay nodes are decided only after the transmission has started. The packet duplication problem is solved using a scheme for contention amongst receivers. Zorzi and Rao [227] also gave a detailed description of a MAC protocol and an evaluation of the latency and energy performance. Bandyppadhyay and Coyle [39] proposed a transmission scheduling scheme using a collision-free protocol for gathering sensor data. They also studied many trade-offs between energy usage, sensor density, temporal, and spatial sampling rates. Sohrabi et al. [180] proposed a sequential assignment routing (SAR) protocol which is used by sensors to select a path amongst multiple ones to the sink node. The SAR protocol selects a path based on the energy resources and the priority level of a packet. Lindsey et al. [141] presented a scheme, called PEGASIS, where each node can receive from and send to close neighbours. The data gathered by nodes in each round have to be collected and transmitted to the base station by only one designated node in order to reduce energy consumption and extend the network lifetime. PEGASIS, which outperforms LEACH protocol [102], considered $energy \times delay$ as the optimization metric per round of data gathering in wireless sensor networks [142, 143]. Krishnamachari et al. [125] showed by well-selected examples that when robustness and energy efficiency are the main concern, single-path routing outperforms multipath routing under the assumption of perfectly reliable source and destination sensors. Choi and Das [62] proposed a data gathering scheme which trades off coverage and data reporting latency while enhancing energy conservation. Hynh and Hong [111] proposed an energy*delay-aware routing protocol for wireless sensor networks using cluster-based and chain-based approaches. Each communication round consists of a cluster and chain formation phase and a data transmission phase. The construction of the network configuration, which is defined by the cluster of sensors and their cluster heads, is accomplished by a base station. Soltan et al. [182] proposed a mobility-aware multi-hop routing scheme for a hierarchical wireless sensor network using mobile relays with a goal to optimize the network lifetime, delay, and local storage size. They proposed to solve an optimization problem to maximize the network lifetime under local storage, delay, and maintenance cost constraint.

## 10.7 Summary

In this chapter, we studied a communication range slicing-based approach to trade-off between conflicting objectives of sensing applications. Our approach aims to slice the communication range of the sensors into *concentric circular bands* and classify them with a goal to satisfy specific requirements of sensing application in terms of energy consumption, delay, and energy depletion. In this chapter, we formulated the trade-off between these three conflicting goals as a multi-objective optimization problem which is solved using a *weighted scale-uniform-unit sum* (WES) approach [119]. Specifically, we proposed a data forwarding protocol for wireless sensor networks, which exploits a solution to a multi-objective optimization problem to find an optimum trade-off between three conflicting goals, namely minimum energy consumption, minimum delay, and uniform energy depletion. To account for the uniform energy depletion of the sensors, we proposed an approach to characterize it based on the size of the subset of candidate forwarders of the data towards the sink. Though there are other efficient methods, such as multi-objective optimization genetic algorithm (MOGA) [84], the WES approach offers more flexibility to a network designer to find solutions that satisfy some associated priorities of the objective functions. Theoretical results show that an optimum trade-off between the three abovementioned goals exists. Moreover, it depends on the weighting coefficients, which are introduced to reflect the relative importance of the individual objective functions and to address the problem of their different units and order of magnitude. For tractability, the communication ranges of the sensors are supposed to be circular and have the same radius. Then, we discussed ways of relaxing these assumptions to enhance the practicality of TED. Also, we evaluate the performance of TED through extensive simulations and compare it with existing ones. We find that the performance of TED is *near optimal* with respect to the $energy \times delay$ metric. This simulation study seems to be an essential step to gain more insight into TED before implementing it on a sensor testbed. To the best of our knowledge, although the design of energy-efficient data forwarding protocols for wireless sensor networks has received much attention, there is no previous work that jointly considered minimum energy consumption, minimum delay, and uniform energy depletion to find the best trade-off between them.

# Chapter 11
# Energy Sink-Hole Problem with Always-On Sensors in Two-Dimensional Deployment Fields

This chapter investigates the *energy sink-hole* problem, which is inherent to static always-on wireless sensor networks, where the sensors located around a static sink act as relays to the sink on behalf of *all* other sensors, thus suffering from severe energy depletion. It presents a theoretical analysis of the problem showing that it can be solved provided that the sensors *adjust* their communication ranges. However, this solution imposes a severe restriction on the size of a deployment field. To overcome this limitation, we propose a sensor deployment strategy based on energy heterogeneity with a goal that *all* the sensors deplete their energy at the same time. Also, it proposes an energy-efficient protocol to solve the energy sink-hole problem using sensor mobility and our newly introduced concept, called *energy-aware Voronoi diagram.*

## 11.1 Introduction

One way to extend the lifetime of a wireless sensor network is through load balancing so that all sensors deplete their energy slowly and uniformly during their monitoring activity. Particularly, the behaviour of the sink has an impact on the network lifetime. Indeed, static always-on wireless sensor network (i.e., the radios of the sensors are turned on all the time) are much affected by the *energy sink-hole* problem, whereas sensors located around a sink suffer from severe battery power depletion problem. Indeed, the sensors close to the sink act as relays to the sink on behalf of *all* other sensors, and hence deplete their battery power more quickly, thus leading to possible disconnection of the network and disruption of the sensed data from reaching the sink. It was proved that it is impossible to guarantee uniform energy depletion of all the sensors in static, uniformly distributed, always-on wireless sensor network with constant data reporting to the sink when sensors use their maximum communication range to transmit sensed data to the sink [137, 138, 140, 164, 199, 200].

The deployment of static sink and sensors in real-world applications is very common, and hence efficient solutions should be provided to tackle the *energy sink-hole* problem, which is inherent to static wireless sensor networks. We believe that the network lifetime depends on three key design metrics, namely type of data forwarding (long range vs. short range), type of sensors (homogeneous vs. heterogeneous), and type of sink (static vs. mobile). This motivates us to account for these three design metrics due to the energy sink-hole problem. First, we consider the *transmission distance* that distinguishes between short-range and

H.M. Ammari: Challenges and Opportunities of Connected $k$-Covered WSNs, SCI 215, pp. 241–269.
springerlink.com 

long-range forwarding. Second, we consider *sensor heterogeneity* when deploying sensors for its ability to improve the reliability of the network and extend its lifetime [75, 209]. Third, when sensors have the same initial energy, we consider *sink mobility* for its ability to evenly distribute the data forwarding load amongst all the sensors to extend the network lifetime [195]. In this chapter, we propose our solutions to the energy sink-hole problem based on the above-mentioned metrics [17].

The remainder of this chapter is organized as follows: Section 11.2 analyzes the energy sink-hole problem and proposes a restricted solution [17]. Section 11.3 exploits energy heterogeneity to solve the energy sink-hole problem [17]. Section 11.4 makes use of the sink mobility and our new proposed concept of energy

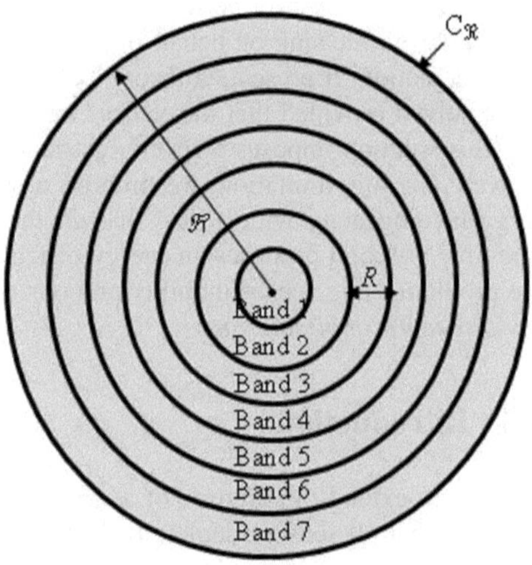


**Fig. 11.1** Slicing field into circular bands

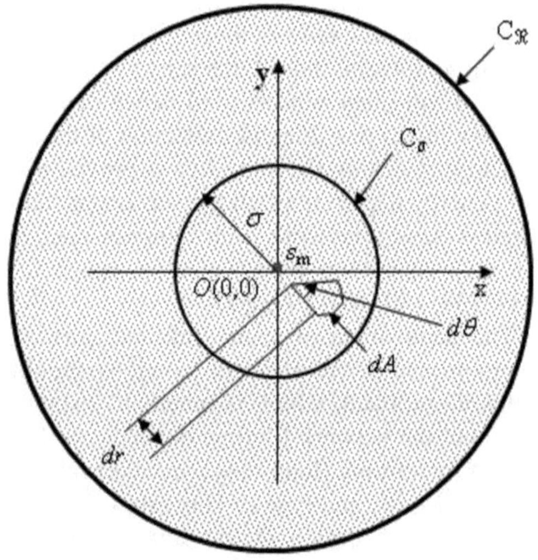


**Fig. 11.2** Circular field with a centred static sink

aware Voronoi diagram [17] to solve the energy sink-hole problem for homogeneous wireless sensor networks. Section 11.5 reviews related work while Sect. 11.6 concludes the chapter.

## 11.2 Energy Sink-Hole Problem Analysis

We consider always-on wireless sensor networks, where the sensors constantly report their sensed data to a single static sink. Hence, the sensor cannot be turned off while monitoring a physical phenomenon. We assume that the sensors are static and uniformly distributed in a circular field of radius $\Re$ with sensor spatial density $\lambda$ (Fig. 11.1).

First, we discuss a *base protocol*, where the network has a static sink and uses a short-path routing protocol [89]. The static sink is supposed to be located at its optimum position in terms of energy-efficient data gathering, which corresponds to the centre of the field [148]. We will show that the sensors around the sink have higher energy consumption than all other sensors.

### *11.2.1 Base Protocol Average Energy Consumption*

The model that we use to compute the maximum average energy consumption of sensors is similar to the model in [88]. The average energy consumption of a node located in an area of size $A_2$ that forwards traffic for other nodes located in another area of size $A_1$ is proportional to $A_1 + A_2 / A_2$. Our model focuses on the nodes within a distance $\ell \leq \sigma \leq R$ from the sink, where $R$ is the radius of the nominal communication range of sensors and $\ell << R$. Specifically, we consider a circle $C_\sigma$ of radius $\sigma$ around the sink $s_m$ which includes the most active forwarders to the sink, where $\ell \leq \sigma \leq R$ and $\ell$ is an infinitesimal value, as shown in Fig. 11.2. The number of sensors inside and outside $C_\sigma$ is $\lambda \pi \sigma^2$ and $\lambda \pi (\Re^2 - \sigma^2)$, respectively, where $\Re$ is the radius of the circular field $C_\Re$ and $\lambda$ is the sensor spatial density. Let $r$ be the radius of an infinitesimal circular region whose area is $dA = r\, dr\, d\theta$, where $0 \leq r \leq \sigma$ and $0 \leq \theta \leq 2\pi$. The average distance between a sensor in $C_\sigma$ and the sink $s_m$ is computed as

$$d_{\text{avg}} = \frac{1}{\pi \sigma^2} \int_{C_\sigma} r\, dA = \frac{2\sigma}{3}$$

The energy consumption rate per sensor in $C_\sigma$ is given by

$$ER(C_\sigma) = ER(C_\sigma, s_m) + ER(\overline{C_\sigma}, s_m)$$

where $ER(C_\sigma, s_m)$ is the average energy consumption rate per sensor in $C_\sigma$ to directly send its data to the sink, and $ER(\overline{C_\sigma}, s_m)$ is the average energy consumption

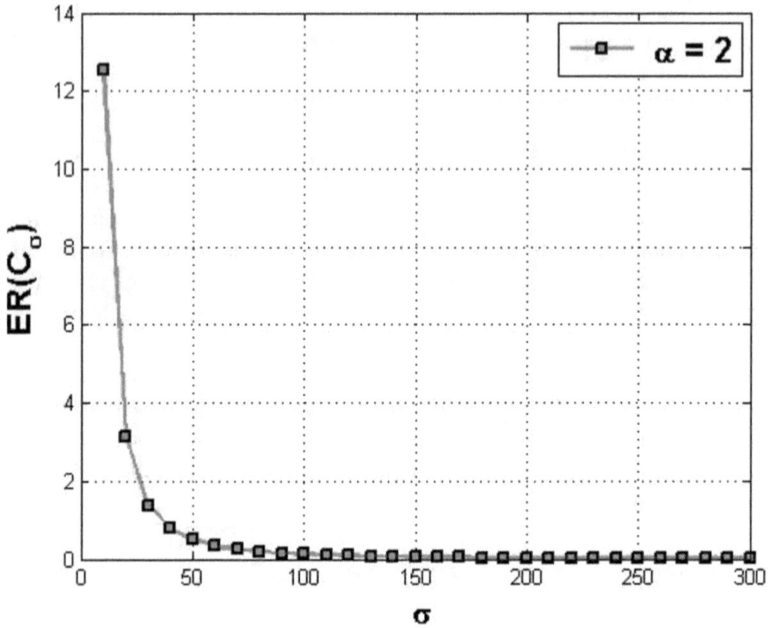


**Fig. 11.3** Plot of $ER(C_\sigma)$

rate per sensor in $C_\sigma$ to forward a subset of sensed data packets originated from sensors in $\overline{C_\sigma} = C_\Re - C_\sigma$ to the sink. Using the energy model in Chap. 2 (see Sect. 2.4), we obtain

$$ER(C_\sigma, s_m) = \frac{1}{\lambda \pi \sigma^2} \times \int_{C_\sigma} ER_{tx}(r)\, \lambda\, dA = (E_{elec} + \frac{2\varepsilon}{\alpha+2} \sigma^\alpha)\, b$$

$$ER(\overline{C_\sigma}, s_m) = \frac{\pi(\Re^2 - \sigma^2)}{\pi \sigma^2} \left( 2\, E_{elec} + \varepsilon \left( \frac{2\sigma}{3} \right)^\alpha \right) b$$

Thus, the energy consumption rate per sensor using the base protocol is given by

$$ER(C_\sigma) = b\, \varepsilon \left( \frac{2}{\alpha+2} + \left( \frac{2}{3} \right)^\alpha \right) \sigma^\alpha + \left( \frac{2\Re^2}{\sigma^2} - 1 \right) b\, E_{elec}$$

Figure 11.3 shows that $ER(C_\sigma)$ increases significantly as we approach the centre of the field (i.e., location of the sink). It is clear that all sensed data will be forwarded by sensors whose distance from the sink is at most equal to $R = 300\,\text{m}$. Hence, we slice the field $C_\Re$ into *concentric circular bands* (or simply *bands*) of width $R$ (Fig. 11.1) and assume that the field has $k > 1$ bands such that $\Re = k\, R$.

### *11.2.2 Nominal Communication Range-Based Data Forwarding*

Under the above-mentioned field slicing method, the energy consumption rate of sensors depends on which band they belong to as well as the communication range used in reporting their data to the sink. We assume that sensors are homogeneous and use their nominal communication range.

Lemma 11.1 proves that sensors located in the $k^{th}$ (i.e., outmost) band consume less energy than all other sensors.

*Lemma 11.1:* Assume a uniform sensor distribution with spatial density $\lambda$ and sensors constantly transmit their data to the sink using their nominal communication range of radius $R$. Sensors located in the $k^{th}$ band have longer lifetime than all sensors in the network.

*Proof:* Any sensor located in the $k^{th}$ band has only to report its own data to the sink. Indeed, given that the communication range of sensors coincides with the width of the bands, no one of the sensors in the $k^{th}$ band can participate in forwarding data to the sink on behalf of others. Using the energy model presented in Chap. 2, the energy consumption rate per sensor in the $k^{th}$ band is computed as

$$ER(k) = ER_{tx}(R) = (\varepsilon\, R^{\alpha} + E_{elec})\, b$$

Let $E_{init}$ be the initial energy of a given sensor. The average lifetime of a given sensor in the $k^{th}$ band is equal to $\frac{E_{init}}{ER(k)}$. To the contrary, all sensors located in any other band forward data on behalf of others. Precisely, a sensor in the $i^{th}$ band forwards data originated from sensors in the $j^{th}$ band, where $i < j \le k$. The number of sensors in the $i^{th}$ band is equal to

$$N(i) = \lambda\,\pi\,(i^2 - (i-1)^2)\,R^2 = \lambda\,\pi\,(2i-1)\,R^2$$

Let $s_i$ be an arbitrary sensor located in the $i^{th}$ band. Thus, under uniform sensor distribution and constant data reporting, the average number of messages forwarded by $s_i$ per unit of time is given by

$$M(i) = \frac{\sum_{l=i}^{k} N(l)}{N(i)} = \frac{\lambda\,\pi\,R^2(k^2 - (i-1)^2)}{\lambda\,\pi\,R^2(2i-1)} = \frac{k^2 - (i-1)^2}{2i-1}$$

Hence, the average energy consumption rate of a sensor in the $i^{th}$ band is given by

$$ER(i) = (M(i) - 1)\,(ER_{tx}(R) + ER_{rx}) + ER_{tx}(R)$$

Using the energy model in Chap. 2 and the value of $M(i)$, the above equation leads to

$$ER(i) = \frac{k^2-(i-1)^2}{2i-1} b\,\varepsilon\, R^{\alpha} + \left(\frac{2k^2-2i^2}{2i-1}+1\right) b\, E_{elec} \quad (1)$$

Hence, the average lifetime of the sensors in the $i^{th}$ band is $\frac{E_{init}}{ER(i)}$, where $i<k$. It is easy to check that $\frac{E_{init}}{ER(i)} < \frac{E_{init}}{ER(k)}$, meaning that the lifetime of the sensors in the $k^{th}$ band is longer than that of the sensors in all other bands. ■

Lemma 11.2, which follows from Lemma 11.1, states that uniform energy depletion cannot be guaranteed under the assumption of constant data reporting by sensors using their nominal transmission range. Thus, all the sensors do not have same lifetime.

*Lemma 11.2:* Assume a uniform sensor distribution with sensor spatial density $\lambda$. Also, suppose that sensors are always on and constantly report their sensed data to the sink using their nominal communication range of radius $R$. It is impossible for a given pair of sensors in two different bands to have the same energy consumption rate in their lifetime. ■

Next, we investigate the case where sensors may use their *adjustable communication range* to transmit or forward data to the sink.

## *11.2.3 Adjustable Communication Range-Based Data Forwarding*

Given that the sensor distribution is uniform and that sensors constantly report their data to the sink, the transmission distance remains the key parameter to check whether it is possible to guarantee uniform energy depletion of sensors. We consider the following two problems.

### 11.2.3.1 Perfect Uniform Energy Depletion

In the case of *perfect uniform energy depletion*, sensors in *all* bands consume energy at the same rate. Precisely, we want to compute the number $k$ of bands of a field such that the sensors located in the first and $k^{th}$ bands have the same lifetime.

Let $\mathrm{N} = \lambda\,\pi\,\Re^2$ be the total number of sensors forming the network. Now, consider two arbitrary sensors $s_1$ and $s_k$ that belong to the first and $k^{th}$ bands, respectively. Given that sensors are uniformly distributed in the field, the average number of sensors in the first and the remaining $(k-1)$ bands, denoted by $N(1)$ and $N(2 \to k)$, respectively, are equal to

$$N(1) = \lambda\,\pi\, R^2$$

$$N(2 \to k) = \lambda\,\pi\,(\Re^2 - R^2) = \lambda\,\pi\,(k^2-1)\, R^2$$

Moreover, the sensors in the *first band* and, in particular, sensor $s_1$ acts as forwarder of the data coming from *all other bands*. Thus, the average number of messages forwarded by $s_1$ (including its own message) per unit of time is

$$M(1) = \frac{\mathrm{N}}{N(1)} = \frac{\lambda \pi \Re^2}{\lambda \pi R^2} = k^2$$

Out of these $k^2$ messages, $(k^2 - 1)$ was sent by sensors located in the $(k-1)$ remaining bands. To simplify the analysis, we assume that the sensor $s_1$ uses the transmission distance $d_1$. Hence, the average rate of energy consumption of $s_1$ is given by

$$\begin{aligned} ER(d_1,1) &= \left( \sum_{l=1}^{k^2-1} ER_{tx}(d_1) + ER_{rx} \right) + ER_{tx}(d_1) \\ &= (k^2\, \varepsilon\, d_1^{\alpha} + (2k^2 - 1)\, E_{elec})\, b \end{aligned} \quad (2)$$

On the other hand, the average rate of energy consumption of the sensor $s_k$ is computed as

$$ER(D_k, k) = ER_{tx}(D_k) = (\varepsilon\, D_k^{\alpha} + E_{elec})\, b$$

where $D_k$ is the transmission distance used by $s_k$. The metric of energy depletion uniformity requires that the average rate of energy consumption of all sensors is the same. Hence, equating the above two equations, i.e., $ER(d_1,1) = ER(D_k,k)$, yield

$$D_k^{\alpha} - k^2\, d_1^{\alpha} - \frac{2E_{elec}}{\varepsilon}(k^2 - 1) = 0 \quad (3)$$

where $0 < D_k \le R$, $0 < d_1 \le R$, $2 \le \alpha \le 4$, and $k > 1$. Notice that under uniform sensor distribution and constant data reporting, it is possible to guarantee uniform energy consumption of sensors located in the first and $k^{th}$ bands if the transmission distances $d$ and $D_k$ satisfy Eq. 3. Thus, given that sensors in the $k^{th}$ band do not forward data on behalf of others, their transmission distance $D_k$ should be larger than that used by sensors in the lower $(k-1)$ bands. Achieving the goal of energy depletion uniformity requires that sensors in the lower $(k-1)$ bands adjust their transmission distances according to $D_k$. Particularly, the transmission distance $d_1$ for sensors in the *first* band is given by

$$d_1 = \frac{1}{k^{2/\alpha}} \left( D_k^{\alpha} - \frac{2E_{elec}}{\varepsilon}(k^2 - 1) \right)^{1/\alpha} \quad (4)$$

Lemma 11.3 [18, 33, 34] approximates the minimum transmission distance $d_{\min}$ a given sensor can use for transmitting its own data or forwarding data on behalf of others to the sink.

*Lemma 11.3:* The minimum transmission distance used by a sensor when it sends/forwards data to the sink can be approximated by

$$d_{\min} = \left(\frac{E_{elec}}{\varepsilon}\right)^{1/\alpha} \qquad \blacksquare$$

From Lemma 11.3, it follows that a physical solution to Eq. 3 exists if and only if

$$\frac{1}{k^{2/\alpha}}\left(D_k^{\alpha} - \frac{2E_{elec}}{\varepsilon}(k^2-1)\right)^{1/\alpha} \geq \left(\frac{E_{elec}}{\varepsilon}\right)^{1/\alpha}$$

The above inequality implies that guaranteeing uniform energy depletion of all sensors is possible if and only if the number $k$ of bands of the field satisfies

$$k \leq \sqrt{\frac{\varepsilon}{3\,E_{elec}} D_k^{\alpha} + \frac{2}{3}}$$

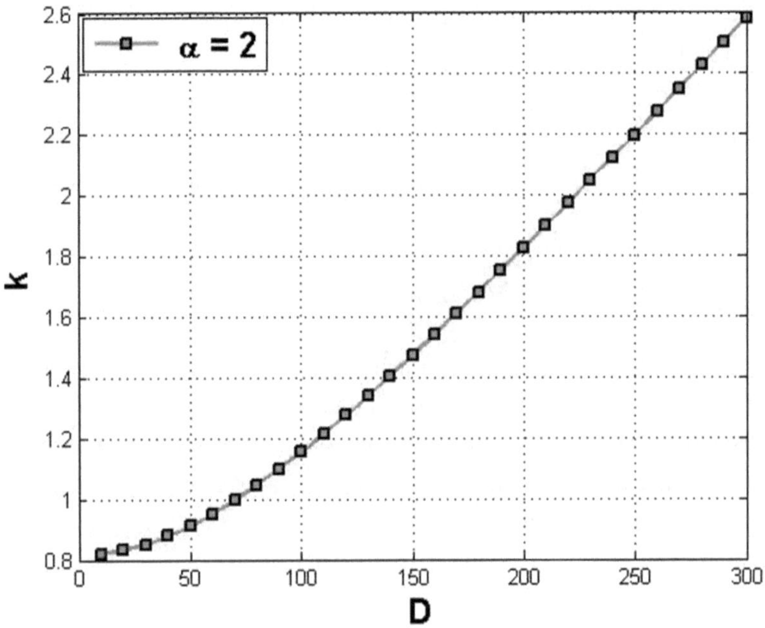


**Fig. 11.4** Plot of $k_u = \sqrt{0.000067\,D_k^{\,2} + 2/3}$ for $\alpha = 2$

Theorem 11.1 generalizes Eq. 4 and states the conditions under which uniform energy depletion of sensors is achieved.

*Theorem 11.1:* Assume that sensors can adjust their communication ranges when they transmit sensed data to the sink. Let $D_k$ be the transmission distance used by sensors in the $k^{th}$ band. Then, uniform energy depletion of sensors in the network can be guaranteed provided that

(*Condition 1*) sensors in the $i^{th}$ band with $i < k$ use a transmission distance $d_i$ given by

$$d_i = \left( \left( D_k{}^{\alpha} - \frac{2E_{elec}}{\varepsilon} \frac{(k^2 - i^2)}{2i-1} \right) \times \frac{2i-1}{k^2-(i-1)^2} \right)^{1/\alpha}$$

(*Condition 2*) the number of bands in a circular field is upper-bounded by

$$k_u = \sqrt{\frac{\varepsilon}{3\,E_{elec}} D_k{}^{\alpha} + \frac{2}{3}}$$ ■

**Table 11.1** Values of $E_{elec}$ and $\varepsilon$ depending on $\alpha$

| | $E_{elec}$ | $\varepsilon$ |
|---|---|---|
| $\alpha = 2$ | $50nJ/bit$ | $10pJ/bit/m^2$ |
| $2 < \alpha \leq 4$ | $50nJ/bit$ | $0.0013pJ/bit/m^2$ |

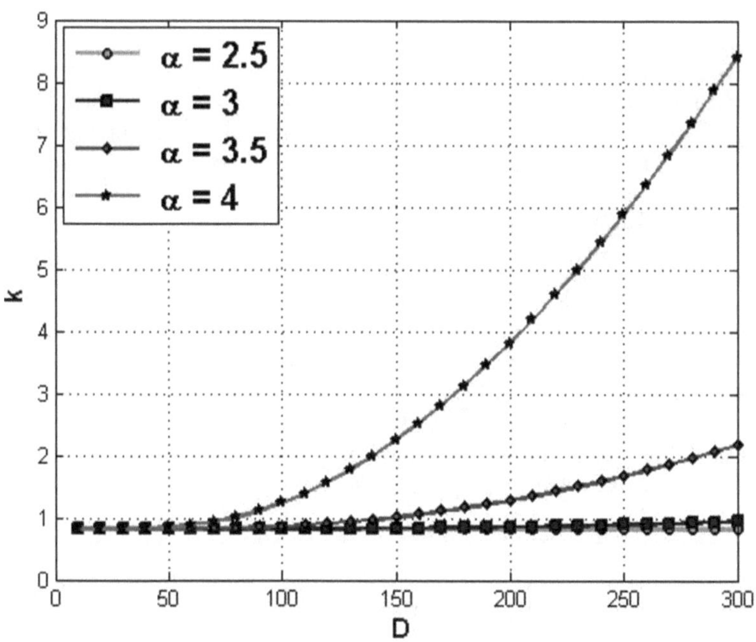


**Fig. 11.5** Plot of $k_u = \sqrt{87 \times 10^{-10} D_k{}^{\alpha} + 2/3}$ for $2 < \alpha \leq 4$

Table 11.1 summarizes the values of the constants $\varepsilon \in \{\varepsilon_{fs}, \varepsilon_{mp}\}$ and $E_{elec}$ [102], which depend on $\alpha$. The upper bound on $k$ depends on whether the free-space $(\alpha = 2)$ or multi-path $(2 < \alpha \le 4)$ propagation model is used, and is computed as

$$k_u = \begin{cases} \sqrt{0.000067\, D_k^{\;2} + 2/3} & \text{if } \alpha = 2 \\ \sqrt{87 \times 10^{-10} D_k^{\;\alpha} + 2/3} & \text{if } 2 < \alpha \le 4 \end{cases}$$

Assume that the radius of the nominal communication range of sensors is equal to $R = 300\,\text{m}$. Fig. 11.4 plots the function $k_u$ for the free-space model while Fig. 11.5 plots $k_u$ for the multi-path model for different values of $\alpha$. As can be seen from Fig. 11.4, the maximum number of bands for $\alpha = 2$ is $k_u = 2$ and can be obtained only when sensors located in the $k^{th}$ band use a transmission distance equal to at least $D_k = 225$ m. However, for $\alpha \in \{2.5, 3, 3.5\}$, as shown in Fig. 11.5, there is no solution to the problem since $k_u = 1$, i.e., uniform energy depletion cannot be guaranteed for those values of $\alpha$. For $\alpha = 4$, however, the number of bands can vary from 2 to 8. For instance, for $k_u = 8$, the transmission distance used by sensors in the $k^{th}$ band should be $D_k \ge 290\,\text{m}$. The corresponding value of the transmission distance $d$ used by sensors in the lower $(k-1)$ bands can be computed based on Eq. 4 for each pair of values $(k_u, D_k)$.

**Discussion:** In [164], it was proved that that unbalanced energy depletion is unavoidable for $\alpha = 2$. Using our analysis, however, we have proved that providing uniform energy consumption of sensors in all the bands (i.e., *perfect uniform energy depletion*) is possible although hard to achieve, especially for $\alpha = 2$. Indeed, the number of bands cannot exceed 2, which imposes a severe restriction on the size of the field. This is due to the gap in the energy consumed by sensors in the first and $k^{th}$ bands. Indeed, most of the sensors in the $k^{th}$ band never forward data on behalf of others regardless of their transmission distance $D$. Thus, our result is in sharp contrast with the one reported in [164]. Also, while our analysis shows that the number of bands has a certain upper bound that does not depend on the size of the field but depends only on the transmission distance $D$ and the propagation model considered. Olariu and Stojmenovic [164] found that the number of coronnas (or bands) is based on the size of the field.

Next, we consider the following relaxed version of the above-mentioned problem.

#### 11.2.3.2 Partial Uniform Energy Depletion

Here, we investigate the value of $i$, where $1 < i < k$ and $\Re = k\,R$, such that a uniform energy depletion could be achieved amongst all sensors located in the lower $i$ bands (1 to $i$). Hence, our objective is to compute the value of $i$ such that sensors located in the first and $i^{th}$ bands have the same energy consumption rate. Guaranteeing uniform energy depletion of sensors in the first and $i^{th}$ bands will definitely yield uniform energy depletion of sensors located in the $j^{th}$ and $i^{th}$ bands, for any $j < i$. Thus, we focus only on the lower $i$ bands instead of all bands in the field and investigate the achievement of *partial energy depletion uniformity*.

Let $s_i$ be an arbitrary sensor located in the $i^{th}$ band and $ER(D_i,i)$ be the average energy consumption rate of $s_i$, where $D_i$ is the transmission distance of $s_i$. If we replace $R$ by $D_i$ in Eq. 1, we obtain

$$ER(D_i,i) = \frac{k^2-(i-1)^2}{2i-1} b\,\varepsilon\, D_i^{\alpha} + \left(\frac{2k^2-2i^2}{2i-1}+1\right) b\, E_{elec} \tag{5}$$

To achieve partial uniform energy depletion in the first and $i^{th}$ bands, their corresponding sensors must have the same energy consumption rate, i.e., $ER(d_1,1) = ER(D_i,i)$. Hence, by equating Eqs. 2 and 5, the transmission distance $d_1$ is computed as

$$d_1 = \frac{1}{k^{2/\alpha}}\left(\frac{k^2-(i-1)^2}{2i-1} D_i^{\alpha} + \left(\frac{k^2-i^2}{2i-1} - k^2 + 1\right)\frac{2\,E_{elec}}{\varepsilon}\right)^{1/\alpha} \tag{6}$$

where $0 < D_i \le R$ and $0 < d_1 \le R$. The value of $i$ should be chosen in a way such that the transmission distance $d_1$ is lower-bounded by $d_{\min}$. Hence, a value of $d_1$ in Eq. 6 exists if and only if $d_1 \ge d_{\min}$. Hence, Eq. 6 and Lemma 11.3 imply

$$\frac{k^2-(i-1)^2}{2i-1} D_i^{\alpha} + \left(\frac{2k^2-2i^2}{2i-1} - 3k^2 + 2\right)\frac{E_{elec}}{\varepsilon} \ge 0$$

Let $F(i,k)$ be a function defined as follows:

$$F(i,k) = \frac{k^2-(i-1)^2}{2i-1} D_i^{\alpha} + \left(\frac{2k^2-2\,i^2}{2i-1} - 3k^2 + 2\right)\frac{E_{elec}}{\varepsilon}$$

Let us evaluate the value of $i$ such that the function $F(i,k)$ is positive. For this purpose, we solve the following two equations:

$$\begin{cases} \dfrac{\partial F(i,k)}{\partial i} = \left(D_i^{\alpha} + \dfrac{2\,E_{elec}}{\varepsilon}\right)(k^2+i^2-i) = 0 & (7) \\[2ex] \dfrac{\partial F(i,k)}{\partial k} = k\left(D_i^{\alpha} + \dfrac{E_{elec}}{\varepsilon}(5-6\,i)\right) = 0 & (8) \end{cases}$$

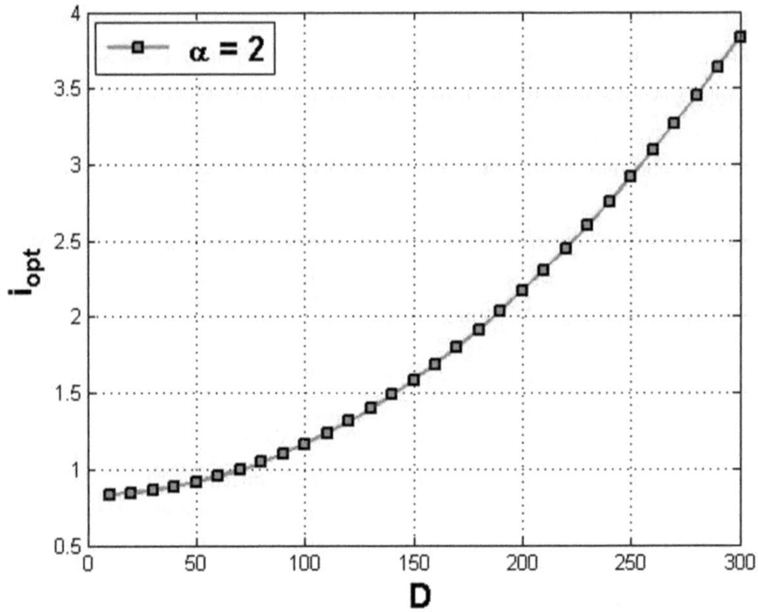


**Fig. 11.6** Plot of $i_{opt}$ for $\alpha = 2$

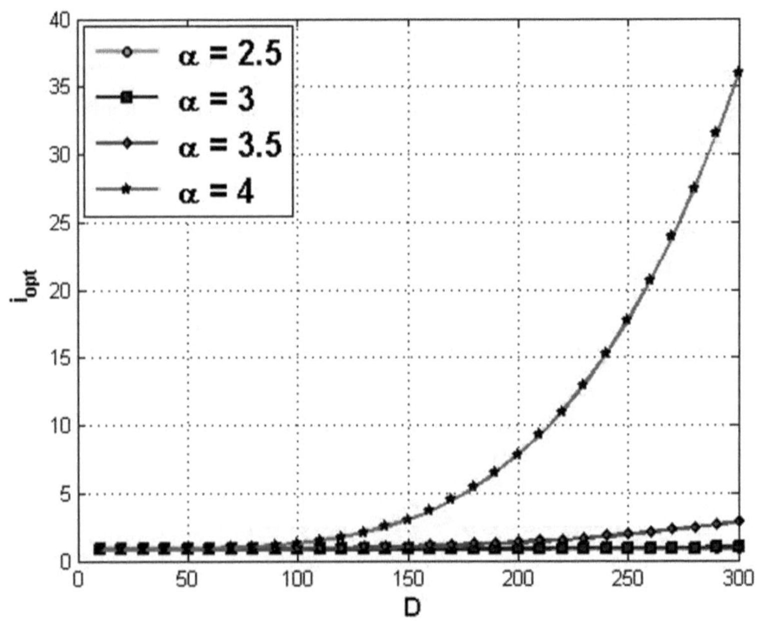


**Fig. 11.7** Plot of $i_{opt}$ for $2 < \alpha \leq 4$

Notice that the value of $k$ is immaterial because we are only interested in achieving partial uniform energy depletion amongst the first $i$ bands, where $i < k$. Eq. 8 provides a physical solution $i_{opt}$ given by

$$i_{opt} = \frac{\varepsilon}{6\,E_{elec}} D_i^{\alpha} + \frac{5}{6}$$

Figure 11.6 shows that partial uniform energy depletion can be achieved with regard to the first and the $3^{rd}$ bands when $\alpha = 2$, where the sensors in this band use their nominal communication range. Figure 11.7 shows that for $\alpha = 3.5$, we get $i = 2$; whereas for $\alpha = 4$, its value varies between 2 and 35. The number of bands does not depend on the size of the field.

Therefore, the energy sink-hole problem can be solved provided that sensors adjust the radii of their communication ranges. However, this solution imposes a severe restriction on the size of the field in terms of the number of its bands, especially for the free-space propagation model $(\alpha = 2)$. Next, we propose a scheme to overcome this shortcoming with the help of heterogeneous sensors, and evaluate its performance.

## 11.3 Using Heterogeneous Sensors

We consider heterogeneous sensors that are equipped with batteries with different initial energy. First, we describe our architecture for such a network with a goal to guarantee that all the sensors deplete their initial energy at the same time. Then, we evaluate its performance.

### *11.3.1 Multi-tier Architecture*

In this approach, heterogeneous sensors are assigned to the bands of the field in such a way that they all deplete their initial energy at the same time. In real-world scenarios, wireless sensor networks can have *heterogeneous* sensors with different capabilities, thus increasing the network reliability and lifetime. For tractability of the problem, we assume that *energy* is the only criterion that decides whether sensors are homogeneous or heterogeneous. Now, the bands do not contain the same type of sensors regarding their initial battery power. Precisely, each band has homogeneous sensors (i.e., sensors having the same amount of initial energy). But, any two bands have heterogeneous sensors (i.e., sensors with unequal amounts of initial energy). Sensors are supposed to use their nominal communication range when they forward data to the sink. Our goal is to guarantee uniform energy depletion of *all* sensors regardless of their bands.

Yarvis et al. [209] proposed a three-tier architecture for heterogeneous wireless sensor networks. The top layer contains only one sink that receives sensed data and analyzes them. The second layer includes sensors with no energy constraint. These sensors, called *line-powered* sensors, have unlimited energy resources by connecting them to a wall outlet. The third layer contains battery-powered sensors that are one-hop away from line-powered sensors. The rationale behind this architecture is that sensors closer to the sink in multi-hop sensor network with many-to-one delivery, consume more energy than all other sensors in the network. Thus, those sensors should be line powered. As can be observed, this three-tier architecture forms a dominating tree where each battery-powered sensor communicates with the sink via only line-powered sensors to transmit its sensed data. There is no communication amongst battery-powered sensors to save their energy, and hence no battery-powered sensor can play the role of data forwarder on behalf of other sensors. Definitely, there should be a sufficient number of line-powered sensors.

Now, we propose a multi-tier wireless sensor network architecture, where each band represents a tier and no sensor is supposed to be line powered. The data forwarding algorithm used by sensors is called *next band-based data forwarding* (NEAR) and is mainly based on the following premise: each sensor selects a neighbour in its adjacent band as its next hop forwarder the one whose remaining energy is the highest amongst all sensors in that band. Also, their transmission distance $R$. We prove that with such a sensor distribution, *all* sensors in the network deplete their initial energy at the same time. Consider the $k^{th}$ band. As discussed earlier, the average energy consumption rate per sensor in this band is $ER_{tx}(R) = (\varepsilon R^{\alpha} + E_{elec})\, b$, where $b$ (in bits/s) is the sensor's data rate. According to [88], the average lifetime of sensors of the $k^{th}$ band is given by

$$T(k) = \frac{e_0(j,k)}{ER_{tx}(R)} = \frac{e_0(k)}{ER_{tx}(R)}$$

where $e_0(j,k)$ is the *total initial energy* of the $j^{th}$ sensor in the $k^{th}$ band whose total number of sensors is $N(k)$. Since a given band is homogeneous in terms of initial energy,

$$e_0(1,k) = \ldots = e_0(N(k),k) = e_0(k)$$

Similarly, for the $i^{th}$ band, the average energy consumption rate per sensor in this band is given by

$$ER_{tx}(R) + \frac{\sum_{j=i+1}^{k} N(j)}{N(i)} (ER_{tx}(R) + ER_{rx})$$

where the first term is due to a sensor transmitting its own data and the second term is due to data forwarding on behalf of sensors in the $(i+1)^{th}, \ldots, k^{th}$ bands. Similarly, the average lifetime of sensors of the $i^{th}$ band is given by

$$\begin{aligned} T(i) &= \frac{N(i)\, e_0(i)}{\left(\sum_{j=i+1}^{k} N(j)\,(ER_{tx}(R) + ER_{rx})\right) + N(i)\, ER_{tx}(R)} \\ &= \frac{(2i-1)\, e_0(i)}{(k^2 - i^2)\,(ER_{tx}(R) + ER_{rx}) + (2i-1)\, ER_{tx}(R)} \end{aligned}$$

Note that the above equation holds for all $1 \le i \le k$. Sensors in all bands will deplete their initial energy at the same time if

$$T(1) = ... = T(i) = ... = T(k)$$

Thus, $T(i) = T(k)$, for all $1 \le i \le k-1$, which implies the following relationship between the initial energy $e_0(i)$ and $e_0(k)$ of the $i^{th}$ and $k^{th}$ bands, respectively,

$$e(i) = g(i,k)\, e_0(k) \text{ for } 1 \le i \le k-1 \tag{9}$$

where their ratio $g(i,k)$ is given by

$$g(i,k) = \frac{(k^2 - i^2)(ER_{tx}(R) + ER_{rx}) + (2i-1)\, ER_{tx}(R)}{(2i-1)\, ER_{tx}(R)} \tag{10}$$

Theorem 11.2 states the condition under which uniform energy depletion of all sensors can be guaranteed.

*Theorem 11.2:* Consider a deployment strategy where a circular field is sliced into $k$ concentric bands of constant width $R$ (i.e., nominal communication range of sensors) and the sensor spatial density is constant. Uniform energy depletion of sensors is guaranteed if each band is homogeneous and all bands are mutually heterogeneous in such a way that the ratio of the initial energy $e_0(i)$ of a sensor in the $i^{th}$ band to the initial energy $e_0(k)$ of a sensor in the $k^{th}$ band satisfies

$$\frac{e_0(i)}{e_0(k)} = g(i,k) \text{ for } 1 \le i \le k-1$$

where

$$g(i,k) = \frac{(k^2 - i^2)(ER_{tx}(R) + ER_{rx}) + (2i-1)\, ER_{tx}(R)}{(2i-1)\, ER_{tx}(R)} \qquad \blacksquare$$

Figures 11.8 and 11.9 plot the function $g(i,k)$ for different values of $i$, $k$, and $\alpha$ assuming a data rate of 1024 bits/s and $R = 300\,\text{m}$. We use the same values of the other constants listed in Table 1. Both figures show that $g(i,k)$ is high for the first inner bands and decreases continuously for the outer bands that tend to have comparable data forwarding load.

### 11.3.2 NEAR Performance Evaluation

We assume that $R = 300\,\text{m}$, $\alpha = 2$, and $\Re = 1500\,\text{m}$, i.e., $k = 5$ (there are 5 concentric circular bands in the circular field). The corresponding values of the constants $\varepsilon_{fs}$ and $E_{elec}$ are given in Table 11.1. Moreover, we assume that every sensor continuously generates constant bit rate (CBR) data of 1024 bits/s, i.e., 4 data packets per second ($b$ = 1024 bits/s). Sensors are randomly deployed in the circular field and their initial energy is determined based on their bands following Eqs. 9 and 10. The simulation is run for multiple times and the simulated time for each run is equal to 1000 s. Then, we average the results of all those runs.

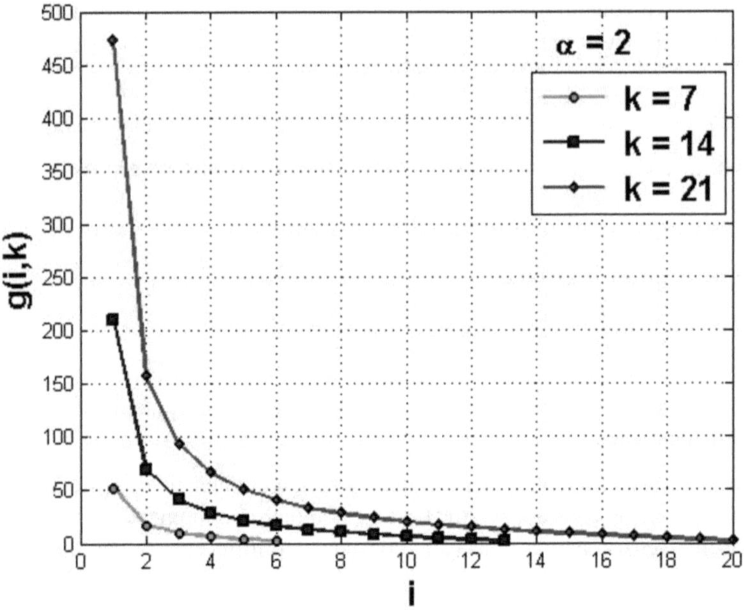


**Fig. 11.8** Plot of $g(i,k)$ for $\alpha = 2$

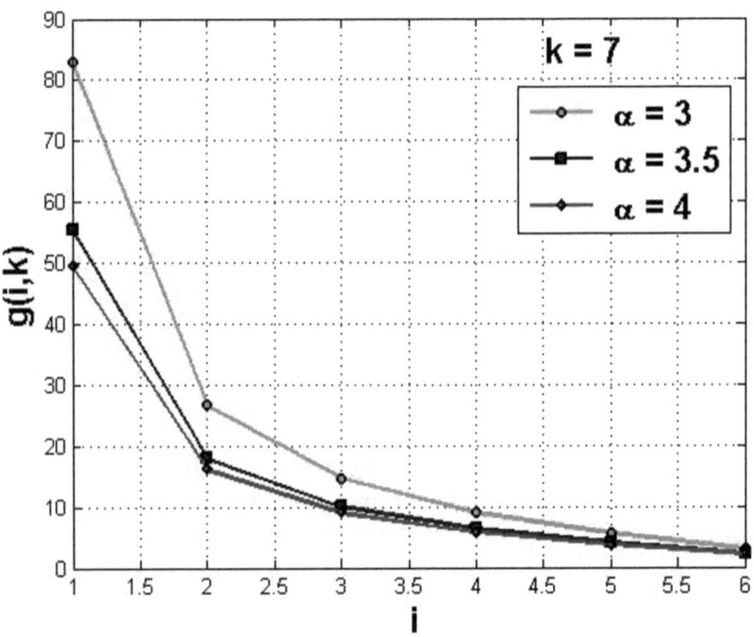


**Fig. 11.9** Plot of $g(i,k)$ for $2 < \alpha \leq 4$

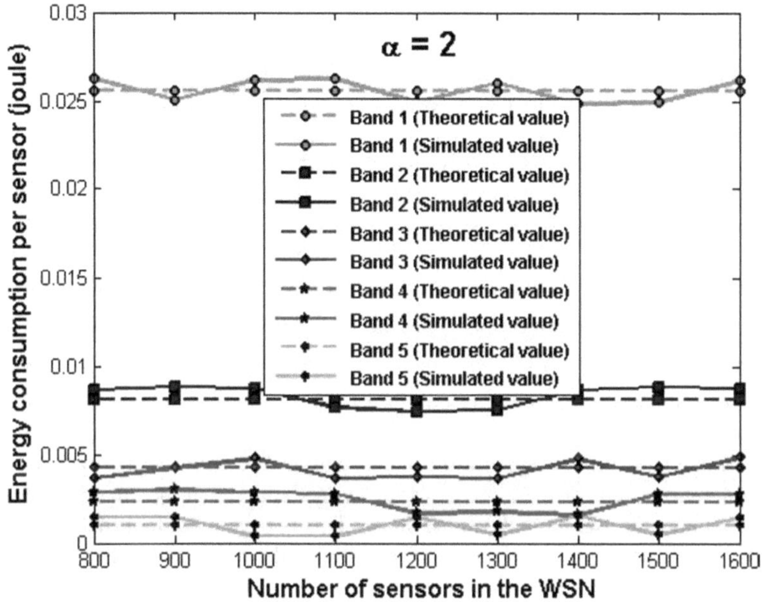


**Fig. 11.10** Average energy consumption of NEAR

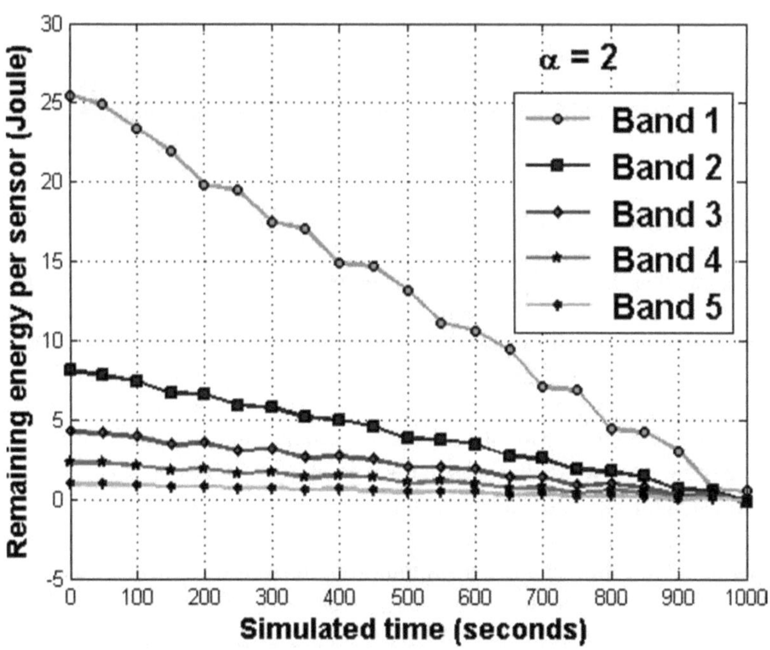


**Fig. 11.11** Uniform energy depletion of all sensors

In the first simulation, the number of sensors varies from 800 to 1600 (i.e., we vary the sensor spatial density). Figure 11.10 shows that sensors in all bands do not consume the same amount of energy. This remains true given that sensors are randomly and uniformly deployed. Indeed, sensors located in the bands closer to the sink still consume more energy than those placed in higher bands. However, the average energy consumption of sensors in a given band stays almost constant as we vary the total number of sensors deployed in the field. This implies that the energy sink-hole problem cannot be solved by simply increasing the number of sensors.

In the second simulation, the number of sensors is fixed to 1600. Assume that the initial energy of sensors in the $k^{th}$ band is $e_0(k) = 0.97$ J By Theorem 11.2, the initial energy of sensors in all other bands are computed based on $e_0(k)$. Figure 11.11 shows that the total remaining energy of sensors in all bands decreases as time advances since those sensors are continuously sending or forwarding sensed data to the sink. The simulation results almost totally agree with the theoretical results, which are omitted for the clarity of Fig. 11.11. Notice that the curves related to those five bands do not have the same slope. This is due to the fact that sensors in those bands do not have the same load, and hence the remaining energy of the inner bands depletes faster than that of the outer bands. More importantly, all sensors in all bands deplete their initial energy at the same time. This result confirms with the analysis of our sensor deployment strategy that assigns initial energy to sensors based on their load in data forwarding to the sink. Definitely, sensors closer to the sink have the highest initial amount of energy given that any sensed data should go through them before reaching the sink.

Next, we propose a data forwarding protocol for homogeneous wireless sensor networks, which uses sink mobility and a new variant of Voronoi diagram whose structure is time-varying.

## 11.4 Sink Mobility and Energy Aware Voronoi Diagram

Here, we assume that the sink is mobile so its neighbours change over time. Moreover, all sensors are homogeneous and randomly and uniformly distributed in a circular field of radius $\Re$ with sensor density $\lambda$. We also assume that the mobility trajectory of the sink follows the *random waypoint* (RWP) mobility model [116], thus covering the entire field (i.e., all locations in the field are equally likely to be visited). Initially, the sink randomly selects a waypoint and a speed between 0 and $v_{\max}$, and moves towards the selected waypoint at this constant speed. When it reaches a waypoint, the sink stays for a pause time and randomly selects new waypoint and speed. The sink repeats this process during the network lifetime. We assume that data collection continues via multi-hop forwarding wherever the sink stays. Thus, only the pause time has an impact on data dissemination.

To answer both questions raised in Chap. 1 (see Sect. 1.4), we propose a new concept, called *energy aware Voronoi diagram*, where all sensors in $SCN(s_i,s_m)$ act as candidate forwarders. When the sink arrives at a waypoint, it randomly selects its next waypoint and broadcasts it along with its current one and pause time in a single *info* packet, i.e., $info = \{rwp_{cur}, rwp_{fut}, ptime\}$. Each sensor that receives this *info* packet decides whether it would transmit its data to either the current or future waypoint of the sink based on its location and the average transmission delay. For a field whose size is in the order of a few miles, the average propagation delay is negligible compared to the average transmission delay [120]. Next, we discuss the concept of energy aware Voronoi diagram and compare it to weighted Voronoi diagram.

## 11.4.1 Why Energy Aware Voronoi Diagram?

The newly proposed concept of *energy-aware Voronoi diagram* differs from *weighted Voronoi diagram* [36]. While the Voronoi edges of the latter are not straight segments, the ones generated by the former are, indeed, straight segments. Also, energy aware Voronoi diagram depends on both the locations of sensors and their remaining energy. Hence, the structure of energy aware Voronoi diagram is dynamic in nature. As will be seen later, energy aware Voronoi diagram is an appropriate structure for data forwarding in wireless sensor networks for the following three reasons: First, it helps us design a localized routing protocol for wireless sensor networks in the sense that each sensor builds its energy aware Voronoi diagram based on its neighbours' information. Second, most of the existing geographical routing protocols, except GeRaF [226, 227], consider the closest sensor to destination as candidate forwarder. However, in GeRaF, a sensor may have the same subset of sensors alive as candidate forwarders. The concept of energy aware Voronoi diagram enables a sensor to have a subset of candidate forwarders from which it chooses the best one with respect to some metric. Third, it gives an equal chance to all sensors in the network to act as candidate forwarders on behalf of others to the sink. This is due to the randomness caused by the remaining energy metric to construct such a Voronoi diagram.

Next, we describe our protocol, evaluate its performance through simulations, and compare it with existing ones.

## 11.4.2 EVEN Detailed Description

Our proposed protocol (Fig. 11.12), called *energy aware Voronoi diagram-based data forwarding* (EVEN), is composed of three phases, namely *computing relative positions*, *computing energy-aware Voronoi diagram*, and *selecting appropriate forwarder*.

### 11.4.2.1 Computing Relative Positions

First, a source $s_0$ identifies its reference sensor $s_{ref}$. All other neighbours will be positioned relatively to $s_{ref}$ based on their remaining energy. If there are multiple reference sensors, EVEN selects $s_{ref}$ with the smallest distance to the shortest path $[s_0, s_m]$. The relative $(x,y)$-coordinates of a neighbour $s_j$ of $s_0$ are computed as follows:

$$\begin{cases} x_{rel}(s_j) = x(s_0) + x_{move}(s_j, s_0, s_{ref}) \\ y_{rel}(s_j) = y(s_0) + y_{move}(s_j, s_0, s_{ref}) \end{cases}$$

where

$$\begin{cases} x_{move}(s_j, s_0, s_{ref}) = [x(s_{ref}) - x(s_0)]\dfrac{E_{rem}(s_j)}{E_{rem}(s_{ref})} \\ y_{move}(s_j, s_0, s_{ref}) = [y(s_{ref}) - y(s_0)]\dfrac{E_{rem}(s_j)}{E_{rem}(s_{ref})} \end{cases}$$

---

Protocol: EVEN<br>
Begin<br>
// Actions executed by a source sensor $s_0$<br>
1: If sink $s_m \in CN(s_0)$ Then<br>
2: Forward sensed data packet directly to $s_m$<br>
3: Else<br>
4: Identify a reference sensor $s_{ref}$ such that<br>
$E_{rem}(s_{ref}) = \max\{E_{rem}(s_j) : s_j \in SCN(s_0, s_m, \beta)\}$<br>
5: Compute the relative positions of the neighbours<br>
$s_j \in SCN(s_0, s_m)$<br>
6: Compute the energy-aware Voronoi diagram:<br>
$EAVor(\{s_0, s_{ref}, s_m\} \cup SCN(s_0, s_m))$<br>
7: Identify a subset of candidate forwarders $CF(s_0, s_m)$<br>
8: Select an appropriate forwarder $s_{af}$ in $CF(s_0, s_m)$<br>
such that $CE_r(s_{af}) = \max\{CE_r(s_k) : s_k \in CF(s_0, s_m)\}$<br>
9: Forward the sensed data packet to $s_{af}$<br>
// Actions executed by appropriate forwarders<br>
10: While (sensed data packet has not reached $s_m$) Do<br>
11: If sink $s_m \in CN(s_{af})$ Then<br>
12: Forward sensed data packet directly to $s_m$<br>
13: Break;<br>
14: Else Repeat Steps 1–9 by replacing $s_0$ with $s_{af}$<br>
15: EndIf<br>
16: EndWhile<br>
End

---

**Fig. 11.12** The EVEN Protocol

The intuition behind this computation is to "virtually" move sensors with low remaining energy away from the sink. Thus, these sensors will not be considered as "good" candidate forwarders until their available energy reaches a certain value. Notice that only $s_{ref}$ has the same relative and physical positions with respect to $s_i$ as the ratio is equal to 1.

#### 11.4.2.2 Computing Energy-Aware Voronoi Diagram

A source $s_0$ computes its energy-aware Voronoi diagram, $EAVor(\{s_0, s_{ref}, s_m\} \cup SCN(s_0, s_m))$ with respect to its own actual location and that of the sink $s_m$ as well as the relative positions of sensors in the subset $SCN(s_0, s_m)$. As the remaining energy of sensors varies with time, the obtained structure of energy aware Voronoi diagram computed by sources and all forwarding sensors is time-varying too.

*Selecting Appropriate Forwarder*: A source $s_0$ uses the *relative* $Closeness \times Energy$ metric defined by $CE_r(s_j) = w_r(s_j) \times \overline{w}(s_j)$ to choose its appropriate forwarder $s_{af}$, where $s_j \in CF(s_0, s_m)$, $w_r(s_j) = \dfrac{\delta(s_0, s_m)}{\delta_r(s_0, s_j) + \delta_r(s_j, s_m)}$ is the *relative closeness ratio*, $\overline{w}(s_j) = \dfrac{E_{rem}(s_j)}{\sum_{s_k \in CF(s_0, s_m, \beta)} E_{rem}(s_k)}$ is the *energy ratio* with $E_{rem}(s_j)$ being the remaining energy of sensor $s_j$, $\delta_r(s_0, s_j)$ is the relative Euclidean distance between $s_0$ and $s_j$, and $\delta_r(s_j, s_m)$ is the relative Euclidean distance between $s_j$ and $s_m$. The source $s_0$ selects $s_{af}$ such that $CE_r(s_{af}) = \max\{CE_r(s_k) : s_k \in CF(s_0, s_m)\}$. Then $s_0$ forwards its data to $s_{af}$. After

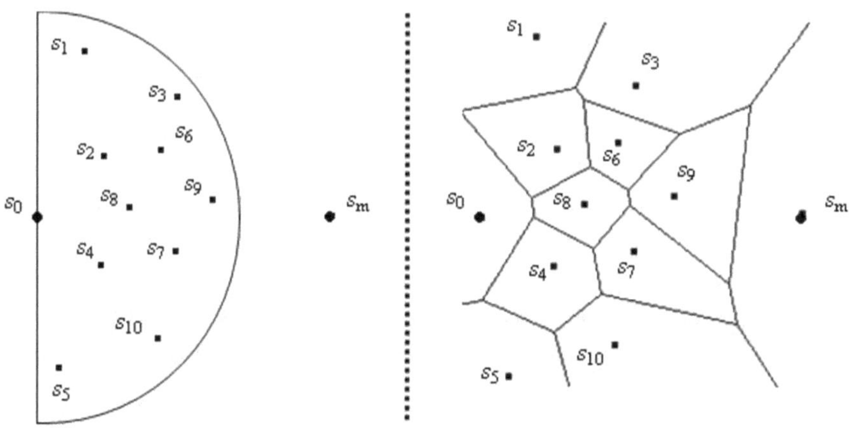


**Fig. 11.13** Voronoi diagram $Vor(\{s_0, s_m\} \cup SNS(s_0, s_m))$

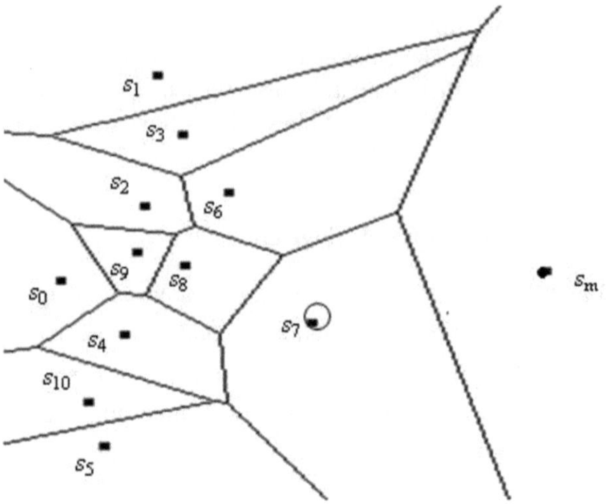


**Fig. 11.14** Energy-aware Voronoi diagram $EAVor(\{s_0, s_{ref}, s_m\} \cup SNS(s_0, s_m))$

receiving the data packet, $s_{af}$ performs the same above phases to identify its appropriate forwarder. This process repeats until the sink receives the data.

Using the concept of Voronoi diagram, only $s_3$, $s_7$, $s_9$, and $s_{10}$ can act as candidate forwarders for $s_0$ as shown in Fig. 11.13. However, the subset of candidate forwarders $CF(s_0, s_m)$ may vary depending on the remaining energy of the sensors when the concept of energy-aware Voronoi diagram is used. Figure 11.14 shows that $s_1$, $s_3$, $s_6$, and $s_7$ are candidate forwarders for $s_0$, which are computed based on the reference sensor $s_7$.

Lemma 11.4 states that all sensors in $SCN(s_i, s_m)$ are equally likely to be selected as appropriate forwarders for $s_i$.

*Lemma 11.4:* All sensors in $SCN(s_i, s_m)$ are equally likely to be selected as appropriate forwarders for $s_i$.

*Proof:* First, $s_i$ will pick sensor $s_{i1}$ having the maximum closeness ratio, i.e., $w(s_{i1}) = \max\{w(s_k) : s_k \in SCN(s_i, s_m)\}$, as a reference sensor and also as an appropriate forwarder. Obviously, $s_i$ will not select $s_{i1}$ as reference sensor in the next data dissemination as $s_{i1}$ will have less remaining energy than other sensors in $SCN(s_i, s_m)$. Let $s_{i2}$ be the second reference sensor, where $\overline{w}(s_{i2}) = \overline{w}(s_{i1}) + \overline{w}_2$ and $0 < \overline{w}_2 < 1$. However, $s_{i1}$ could be selected as an appropriate forwarder only if $CE_r(s_{i1}) > CE_r(s_{i2})$ or $w(s_{i1}) > w(s_{i2}) \times \frac{\overline{w}(s_{i2})}{\overline{w}(s_{i2}) - \overline{w}_2}$. Otherwise, either $s_{i2}$ or another sensor in $SCN(s_i, s_m)$ will be selected as an appropriate forwarder. As can be seen, any sensor could be successively selected as an appropriate forwarder a very few times. Because of the relative values of the metric $Closeness \times Energy$ and the notion of reference sensor, each sensor in $SCN(s_i, s_m)$ will be considered as an appropriate data forwarder for $s_i$. ∎

### 11.4.3 EVEN Performance Evaluation

In this section, we study the performance of EVEN based on simulation programs written in the C programming language. Our simulation set-ups consider 800 sensors randomly and uniformly distributed in a circular field of radius $\Re = 3000\,\text{m}$. The radius of the communication range of the sensors is $R = 300\,\text{m}$. Also, when mobility is considered, the maximum speed of a mobile sink is $v_{\max} = 5\,\text{m/s}$.

#### 11.4.3.1 Impact of Sink Mobility

In this experiment, we consider a *Voronoi diagram-based greedy forwarding* (VGF) protocol. VGF is similar to a greedy geographic routing algorithm, called *Bounded Voronoi Greedy Forwarding* (BVGF) [204]. On one hand, BVGF chooses the sensor that has the shortest distance to the sink and does not consider energy as a selection metric. On the other hand, VGF considers both the closeness of sensors to the sink and their remaining energy to build localized Voronoi diagram and select the best candidate forwarders. Notice that VGF and EVEN are quite identical except that the former uses the actual locations of sensors while the latter uses the virtual locations of sensors (computed in Sect. 11.3.2). We evaluate the performance of VGF for both cases of networks using a static sink and a mobile sink, respectively. We compute the average energy consumption of sensors as a function of their distance from the centre of the field. A static sink is positioned at the centre of the field for energy-efficient data gathering [148]. As can be seen in Fig. 11.15, VGF has better performance with a mobile sink than with a static sink. Indeed, sink mobility distributes the data forwarding load amongst all the sensors.

#### 11.4.3.2 Comparing EVEN with VGF

In this experiment, we consider a mobile sink and compare the performance of EVEN with that of VGF. EVEN leads to better load balance than VGF as shown in Fig. 11.16. While EVEN allows a sensor to select best forwarders amongst all of its neighbours, VGF considers only the closest sensors to the sink. As expected, EVEN distributes more evenly the data dissemination load amongst the neighbours of each sensor, and hence outperforms VGF. Figure 11.17 shows the impact of pause time of the mobile sink on the average energy consumption of sensors. As the pause time increases, more sensors will have the chance to transmit sensed data to the sink. Thus, sensors close to the sink become more active as they receive more data from sources. When the pause time has large value, those heavily used

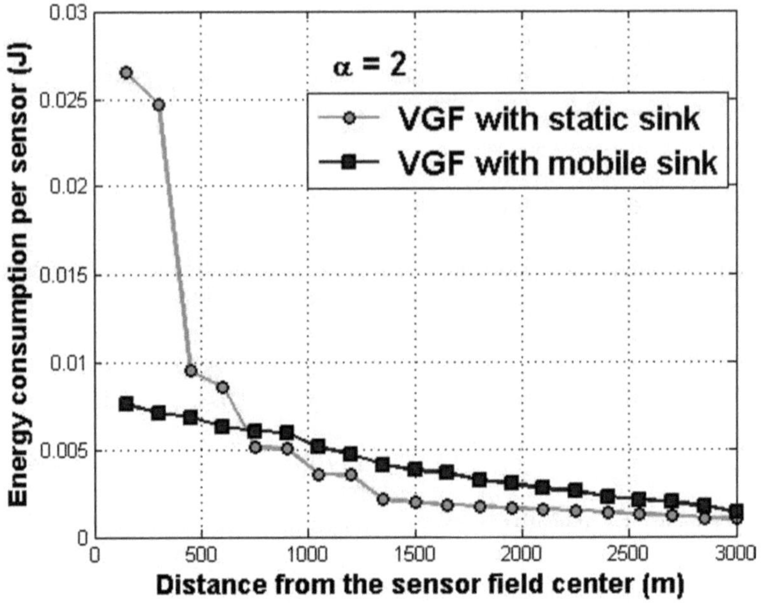


**Fig. 11.15** VGF – static sink vs. mobile sink

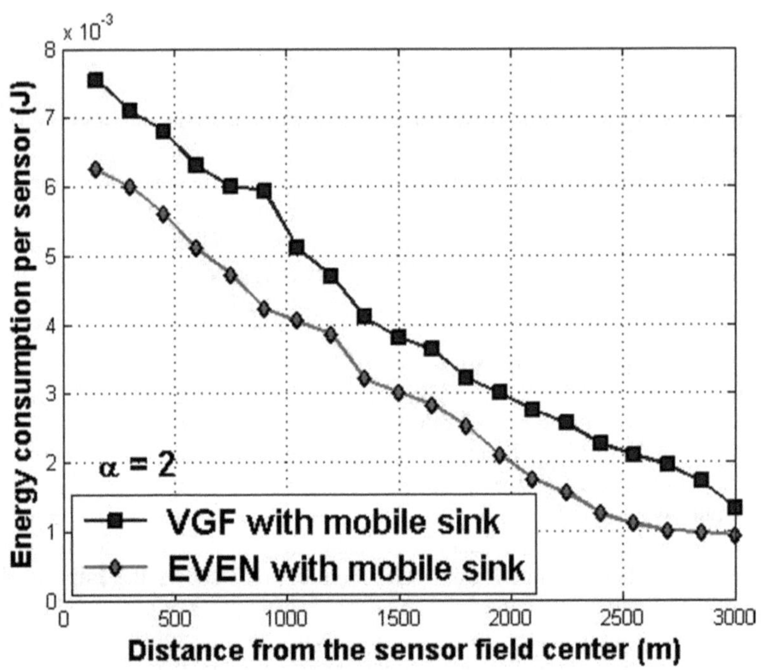


**Fig. 11.16** Comparing EVEN with VGF

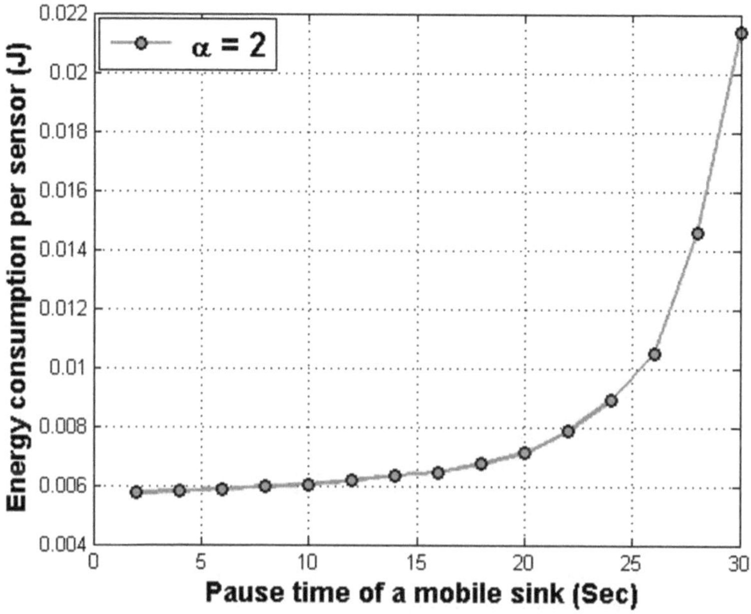


**Fig. 11.17** Impact of pause time on EVEN

sensors surrounding the sink deplete their energy more quickly than all other sensors. The performance of EVEN tends to that of VGF with a static sink. It is worth noting that no matter how energy efficient a data forwarding protocol is, the use of static sink significantly degrades its performance.

#### 11.4.3.3 Comparing EVEN with Another Protocol

Luo and Hubaux [148] discussed an energy efficient routing protocol for WSNs which exploits base station mobility and multi hop routing. Different mobility strategies of a mobile station have been studied to identify the optimum one in terms of balanced load distribution. Luo and Hubaux [148] showed through simulations that under the short-path routing strategy, a mobile sink reduces the average load of the sensors by about 75% compared to a static sink. This reduction implies approximately a 400% increase of the network lifetime. Luo and Hubaux [148] also showed by simulation that their joint mobility and routing strategy further reduces the network load by about 10% which corresponds to an overall improvement of the network lifetime of about 500% compared with the case of static sink. Their joint mobility and routing strategy is elegant and efficient in that it exploits the available energy at the sensors close to the border of the network, which are almost not used in data forwarding. Moreover, the sink moves on a circle of radius $R_m < R$, where $R$ stands for the radius of the field. Thus, the field is divided into two regions: the inner circle of radius $R_m$ and the annulus between

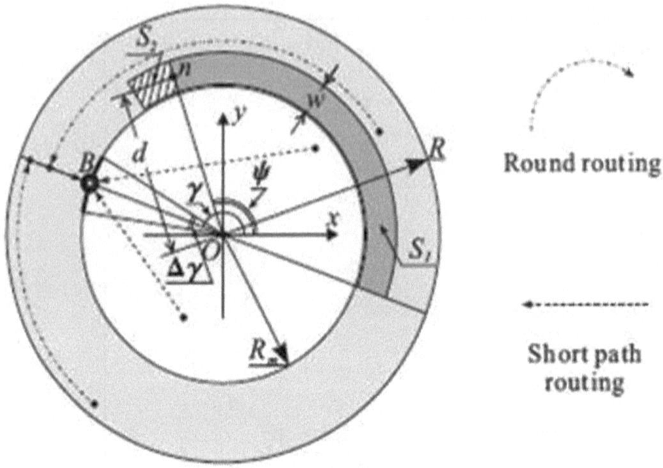


**Fig. 11.18** Joint mobility and routing strategy [148]

the periphery of the network and the trajectory of the sink as shown in Fig. 11.18. The sensors within the inner circle use short-path routing to transmit their sensed data to the sink, whereas the sensors in the annulus send their data to the sink using two steps. First, a sensor uses *round routing* around the centre of the network $O$ until the segment $OB$ is reached, where $B$ is the current position of the mobile sink. Then, the data is sent to the sink using a short path.

EVEN does not necessarily choose a short path. Using the concept of energy-aware Voronoi diagram, *all* neighbours of each sensor have the chance to participate in data forwarding towards the sink. Thus, a mixture of short-range and long-range data forwarding patterns would take place. This helps the sensors deplete their energy slowly and uniformly. Also, as the sink is moving randomly, all areas in the field are equally likely to be visited, including those on its periphery. Thus, sensors close to the border of the network would be selected to act as forwarders towards the sink as their remaining energy permits. According to [148], the network lifetime is inversely proportional to the maximum average energy consumption of the sensors. Thus, as can be seen from Fig. 11.16, EVEN improves the network lifetime by more than 430% compared with VGF that uses a static sink. Thus, EVEN has performance that is comparable to that of the joint mobility and routing strategy [148].

## 11.5 Related Work

Data dissemination protocols for WSNs can be categorized into two main classes. The first one consists of protocols that attempt to minimize the energy consumption of sensors. Unfortunately, these protocols do not consider some particular regions in the network whose sensors are heavily used compared to others, and hence deplete their initial energy very quickly. This behaviour may result in WSN disconnections although most of the rest of sensors have enough amount of energy to work correctly. The second class, however, includes protocols that attempt to balance the energy consumption amongst all sensors in the network. The next two subsections review protocols for the above classes.

### 11.5.1 Balancing Energy Consumption

The energy sink-hole problem in WSNs has gained relatively less attention in the literature. It is worth noting that this problem was originally addressed by Guo et al. [97]. They proposed an energy-balanced transmission scheme that adjusts the ratio between direct transmission to the sink and next-hop transmission. Precisely, sensors far away from the sink send larger percent of data to the next hop while sensors near the sink send more data directly to the sink. Zhang et al. [220] also exploited this combination of hop-by-hop transmission and direct transmission to find a trade-off between them. Efthymiou et al. [79] proposed a probabilistic data propagation algorithm for balancing energy consumption amongst all sensors. Powell et al. [170] used the probabilistic data propagation algorithm in [79] and proved that there is a relationship between energy balancing and lifespan maximization [113]. Leone et al. [134] considered non-uniform sensor distribution and proposed a *blind* algorithm that computes a solution to the energy-balancing problem on-line without prior knowledge on the occurrences of the events. Li and Mohapatra [137, 138] characterized the energy hole around the sink with an analytical model and investigated the effectiveness of a few approaches for mitigating the energy hole problem. Olariu and Stojmenovic [164] proved that energy-efficient routing can be guaranteed when the coronas of a circular field have the same width, but this would lead to uneven energy depletion of sensors. Hence, they computed the widths of coronas and their number to achieve even energy depletion of sensors. They also proved that uneven energy depletion is unavoidable for the free-space model but can be prevented for the multi-path model. Lian et al. [140] showed that up to 90% of the total initial energy is unused due to the static WSN model with uniformly distributed homogenous sensors and a stationary sink. They also proposed a non-uniform sensor distribution-based deployment strategy and showed by simulation that it can increase the total data capacity. Non-uniform node distribution was also considered in [199, 200] to achieve balanced energy depletion. Luo and Hubaux [148] proposed a data collection protocol for WSNs that makes use of multi-hop routing and base station mobility to solve the energy sink-hole problem.

### 11.5.2 Minimizing Energy Consumption

Chang and Tassiulas [57] proposed different approaches for maximizing the network lifetime based on finding the best link cost function. Intanagonwiwat et al. [112] proposed a data-centric paradigm, called *directed diffusion*, for sensor query dissemination and processing in static WSNs. Boukerche et al. [49] proposed power-efficient data dissemination protocols that combine sleep/awake and probabilistic forwarding techniques. Boukerche et al. [50] proposed a novel protocol, called *energy-aware data-centric* (EAD), which builds a virtual backbone composed of active sensors that are responsible for in-network data processing and traffic relaying. EAD attempts to construct a broadcast tree that approximates an optimal spanning tree with a maximum number of leaves, thus reducing the size of the backbone formed by active sensors. Antoniou et al. [35] and Boukerche et al. [48] proposed a protocol, called *Variable Transmission Range Protocol* (VTRP). VTRP contributes to solve the energy sink-hole problem by varying the transmission range of sensors to bypass sensors lying close to the sink and avoid their overuse. While EAD and VTRP assume that the sink is static, our protocol considers sink mobility. However, our protocol assumes that *all* sensors are active during forwarding, and hence do not consider *duty-cycling*. Our future work will address this issue. Gao and Zhang [89] proposed greedy forwarding algorithms for load-balanced routing in wireless networks when nodes lie either on a line or in a narrow strip. Sankar and Liu [175] proposed a distributed routing algorithm that checks whether it is possible to route flow in the network while satisfying all the demands in terms of total energy consumed by all nodes before threshold for lifetime expires. Zhou et al. [225] investigated the problem of finding energy-efficient paths to the sink using the notion of transmitter power control.

### *11.5.3 Mobility-Based Forwarding Protocols*

Kim et al. [120] proposed a Scalable Energy-efficient Asynchronous Dissemination (SEAD) protocol for WSNs, which is based on dissemination trees that are built to disseminate data to mobile sinks. Every mobile sink is supported by a special node, called access node, which acts as the relay between the mobile sink and source sensors. Shah et al. [176] proposed a three-tier architecture for data dissemination in sparse WSNs using the concept of MULES. Wang et al. [191] proposed distributed self-deployment protocols to discover coverage holes and cover those using mobile sensors that move from densely to sparsely areas to maximize coverage. Wang et al. [190] proposed a proxy-based sensor deployment protocol for mobile wireless sensor networks. A proxy sensor is a static sensor that is closest to the logical position of its delegated mobile sensor. As can be seen, checkpoints in our protocol differ from proxy sensors in Wang et al.'s protocol [190]. However, both of them are used for energy efficiency purpose. Proxy sensors are introduced to help mobile sensors move only when needed so that they save their energy, whereas our checkpoints are introduced to shorten data forwarding paths between source sensors and the sink, and hence minimize the total energy consumption. Zhang et al. [215] proposed a dynamic proxy tree-based data dissemination framework for mobile wireless sensor networks. Mobile sources and mobile sinks are associated with stationary source proxies and sink proxies, respectively, and proxies related to the same source form a proxy tree. The latter is used to multicast data from the source proxy to the sink proxies. When the distance between sources or sinks and their proxies do not change beyond the threshold distance, the sources and sinks will keep the same proxies. This situation could lead to a battery power depletion of the associated proxies. In our protocol, checkpoints dynamically change based on both their closeness to the shortest path between the senders and receivers and their remaining energy. Zhang et al. [215] protocol introduces an overhead in reconfiguring the proxy tree due to source and sink mobility. The overhead introduced by our protocol is due to the construction of localized Delaunay triangulation which occurs only once as the wireless sensor network is static, thus yielding little overhead. Our protocol, however, assumes that *all* sensors are always-on. Our future work will address duty-cycling. Zorzi and Rao [226, 227] proposed a geographic random forwarding (GeRaF) technique for wireless sensor networks, where relay nodes are decided only after the transmission has started. Biswas and Morris [43] proposed an integrated and MAC protocol for multi-hop wireless networks, where a source sends a batch of packets destined to the same destination. Ye et al. [210] proposed an efficient data delivery to multiple mobile sinks using a two-tier data dissemination model (TTDD). When a data source detects a stimulus, it builds a data dissemination grid structure over the deployment field and sets up the forwarding information at sensors closest to grid points. Intanagonwiwat et al. [112] proposed a data-centric paradigm for sensor query dissemination and processing in static WSNs, called directed diffusion (DD), which uses attribute-based naming to match data to sensors. The DD paradigm provides robust multi-path delivery and achieves energy saving when intermediate nodes aggregate responses to queries. Ammari and Das [32] proposed an information theory-based approach for data dissemination in wireless sensor networks with a mobile sink.

## 11.6 Summary

In this chapter, we have studied the energy sink-hole problem in static always-on wireless sensor networks, where the sensors around a sink are heavily used in data forwarding, thus depleting their energy quickly [17]. We proved that uniform energy depletion of all sensors can be achieved in uniformly distributed wireless sensor networks provided that the sensors adjust their communication ranges when forwarding data to a static sink. We have also proposed a deployment strategy using heterogeneous sensors that guarantees uniform depletion of sensors' initial energy. Sensors are placed in their bands based on their data forwarding activity and initial energy. Precisely, the inner bands contain sensors with large amount of energy as they are heavily loaded with data forwarding to the sink. We have found that our simulation results agree almost perfectly with our theoretical results. A critical factor for extending the lifetime of wireless sensor networks is load balancing in data dissemination, which depends on the nature of the sink and the selection scheme of appropriate forwarders. We have proposed a data dissemination protocol, called EVEN, which uses the concept of energy-aware Voronoi diagram to evenly distribute the load of data forwarding on all the neighbours of a given sensor, and exploits sink mobility to update the neighbours of the sink. The design goal of EVEN is to balance the load amongst the sensors so they deplete their energy as uniformly and slowly as possible, thus extending the network lifetime. Our results demonstrate that energy-aware Voronoi diagram and sink mobility lead to more than 430% improvement of the network lifetime compared to the case of static sink.

# Chapter 12
# Geographic Forwarding on Duty-Cycled Sensors in Two-Dimensional and Three-Dimensional Deployment Fields

This chapter focuses on the problem of joint $k$-coverage, sensor scheduling, and geographic forwarding in two-and three-dimensional wireless sensor networks. Specifically, it studies geographic forwarding in duty-cycled, $k$-covered wireless sensor networks and proposes the first design of protocols for geographic forwarding on duty-cycled sensors with and without data aggregation. The proposed protocols exploit the geometry of the configurations, which result from our connected $k$-coverage protocols, and the concept of virtual potential fields.

## 12.1 Introduction

The design of protocols for wireless sensor networks is challenging due to the scarce battery power of the sensors. On the one hand, it is well known that sensor *duty-cycling* is an important mechanism that helps densely deployed wireless sensor networks save energy so the sensors remain operational for as long as possible. On the other hand and as discussed earlier, geographic forwarding is an energy-efficient and practical scheme for wireless sensor networks in that the sensors need only maintain local knowledge on their one-hop neighbours with respect to their geographic location information. Although there is a dependency between sensor duty-cycling and data forwarding in the sense that data should be forwarded to a central gathering point, known as the *sink*, via active sensors, only a few works jointly consider them. Indeed, most of the geographic forwarding protocols assume that *all* sensors are *always on* during forwarding. However, such an assumption is not realistic in densely deployed wireless sensor networks [179], where sensors are *duty-cycled*, i.e., switched *on* or *off* to save energy. While most efforts focused on only single aspect of the problem (coverage, duty-cycling, routing), this chapter offers clarity into the issues that must be addressed for joint protocol development, where $k$-coverage, duty-cycling, and geographic forwarding are discussed in a unified framework.

In this chapter, we focus on geographic forwarding in two-and three-dimensional duty-cycled, $k$-covered wireless sensor networks, where each point in a field is *covered* by at least $k$-active sensors while *all* active sensors are *connected*. Geographic forwarding in $k$-covered wireless sensor networks, however, faces three major challenges. The first challenge is *how to determine the number of active sensors required to fully k-cover a field*. As of writing this chapter, there is no exact bound on

H.M. Ammari: Challenges and Opportunities of Connected $k$-Covered WSNs, SCI 215, pp. 271–291.
springerlink.com 

the sensor spatial density for $k$-coverage of a field. Besides $k$-coverage characterization, our approach *quantifies* a tight bound on the sensor spatial density that is required to achieve $k$-coverage. The second challenge is *how to design a minimum-energy duty-cycling protocol for a k-covered wireless sensor network that deploys as minimum number of active sensors as possible so all sensors deplete their energy slowly and uniformly.* Our goal is to prolong the lifetime of the sensors in order to extend the network lifetime. Although several $k$-coverage protocols for wireless sensor networks have been proposed, none of them provided a guarantee of using a minimum number of active sensors. The third challenge is *how to design energy-efficient geographic forwarding protocols running on top of a duty-cycled k-covered wireless sensor network with a specific requirement in terms of data aggregation.* A few works on joint duty-cycling and forwarding exist in the literature [43, 161, 226]. The work in [161] jointly considers duty-cycling and opportunistic routing at the network layer. Our joint protocols are complementary to [161] and are more general in that they jointly consider geographic forwarding on a duty-cycled $k$-covered wireless sensor network along with different data aggregation levels [23, 24].

In the previous chapters, we have addressed the first two challenges through the design of energy-efficient $k$-coverage and duty-cycling protocols. Although several elegant protocols have been proposed to solve the problem of $k$-coverage in wireless sensor networks as discussed earlier, the problem of routing on duty-cycled wireless sensor networks has received little attention in the literature. In particular, joint coverage and geographic forwarding in wireless sensor networks have been overlooked intentionally. This is due to the fact that *all* sensors are assumed to be always *on* during data forwarding. However, this assumption is not valid in real-world applications, where all sensors should not stay *on* all the times to save energy. This work is an effort complementing previous ones [43, 226], and particularly the one by Nath and Gibbons [161]. Precisely, we focus on the design of energy-efficient geographic forwarding on a minimum-energy duty cycled $k$-covered wireless sensor network, where every point in a field is covered by at least $k$ sensors. Similar to [161], our joint sleep–wakeup scheduling and geographic forwarding in a $k$-covered wireless sensor network is done at the network layer. We believe that our joint protocols could be useful for several applications and particularly those requiring data aggregation at intermediate sensors along paths to the sink. To the best of our knowledge, this is the first study of geographic forwarding on a duty-cycled $k$-covered wireless sensor network.

The remainder of this chapter is organized as follows: Section 12.2 presents our geographic forwarding protocols for two-dimensional duty-cycled $k$-covered wireless sensor network. Section 12.3 proposes our geographic forwarding protocols for three-dimensional duty-cycled $k$-covered wireless sensor network. Section 12.4 reviews related work. Section 12.5 concludes the chapter.

## 12.2 Two-Dimensional Sensor Deployment

In this section, we describe our geographic forwarding protocols on duty-cycled sensors in two-dimensional deployment fields. Our protocols are based on the clustered connected $k$-coverage protocol described in Chap. 5. Specifically, we present our first *potential field* [103] based solution for geographic forwarding on a duty-cycled $k$-covered wireless sensor network, called *Geographic Forwarding through Fish Bladders* (GEFIB), where data are forwarded through *fish bladders* (or *lenses*). Precisely, we discuss three geographic forwarding protocols with different levels of data aggregation.

*Assumption 12.1:* We assume that in each round, every active sensor has data to report to the sink. Also, only sensors selected to $k$-cover a field act as *relays*. ■

### 12.2.1 Potential Fields Based Modeling Approach

Sensors can be viewed as *particles*, and hence are subject to *virtual forces*, which attract sensors to each other. These virtual attractive forces are due to the *remaining energy* of the sensors and their *geographic locations*. Indeed, the sensors with highest remaining energy are preferred to act as relays in order to avoid *energy holes* (i.e., regions whose sensors have depleted their energy) that may disconnect the network. Also, as the energy spent in data transmission is proportional to the transmission distance [102], the sensors prefer closer ones to act as relays, thus forwarding data over short distances.

Using potential field terminology, each active sensor is subject to at least one *attractive force*, called *energy-location based force* and denoted by $F_{el}$, which is exerted by some active sensor and defined as the gradient of a unique scalar potential field, called *energy-location based potential field* and denoted by $U_{el}$, i.e., $F_{el} = -\nabla U_{el}$. We should mention that this notion of attractive force is *symmetric*. That is, if a sensor $s_i$ exerts a force on sensor $s_j$, the latter also exerts on the former a force with the same magnitude. Also, only active sensors can exert forces on each other. Our approach to modeling the resultant force that a sensor $s_i$ exerts on its sensing neighbour $s_j$ is borrowed from electromagnetism theory [186]. Using Coulomb's law [186], the magnitude of the electrostatic force $F(i, j)$ between two points electric charges $q_i$ and $q_j$ depends on their magnitudes and the Euclidean distance $d(i, j)$ between them and the permittivity $\varepsilon_0$ of free space, and is computed as

$$F(i,j) = -\nabla U(i,j) = \frac{1}{4\pi\,\varepsilon_0}\frac{|q_i||q_j|}{d^2(i,j)}$$

In our model, the charge of a sensor is its *remaining energy* and permittivity is the *transmitter amplifier* [102] in the free-space $(\varepsilon_{fs})$ model $(\alpha = 2)$ or the multipath $(\varepsilon_{mp})$ model $(2 < \alpha \leq 4)$, where $\alpha$ is the path-loss exponent. Thus, the magnitude of the force $F_{el}(i, j)$ that a sensor $s_i$ exerts on its sensing neighbour $s_j$ is proportional to the product of their remaining energy, $\pi(s_i)$ and $\pi(s_j)$, and inversely proportional to the Euclidean distance $d(i, j)$ between them. Moreover, it is important that $F_{el}(i, j)$ account for the type of model being used, i.e., free-space model $(\varepsilon = \varepsilon_{fs})$ vs. multi-path model $(\varepsilon = \varepsilon_{mp})$. Therefore, the attractive force $F_{el}(i, j)$ is computed as

$$F_{el}(i, j) = -\nabla U_{el}(i, j) = \frac{1}{4\pi\,\varepsilon} \frac{\pi(s_i)\,\pi(s_j)}{d^{\alpha}(i, j)}$$

Similarly, there is an interaction between one sensor and a set of sensors. Precisely, the resultant force exerted by sensor $s_i$ on a set of sensors $S$ is given by

$$F_{el}(i,S) = \sum_{s_j \in S} F_{el}(i, j) = -\sum_{s_j \in S} \nabla U_{el}(i, j)$$

Next, we present three geographic data forwarding protocols for a duty-cycled $k$-covered wireless sensor network based on potential fields.

### 12.2.2 Data Forwarding without Aggregation

In this section, we propose a simple approach for geographic forwarding without data aggregation on a duty-cycled $k$-covered wireless sensor network based on artificial potential fields that were described in Sect. 12.2.1. Let us first define the notion of *best relay* of a sensor.

*Definition 12.1:* A *best relay* of a sensor $s_i$ is a sensing neighbour $s_l$ of $s_i$ located between it and the sink such that $F_{el}(i,b)$ is the maximum over all resultant forces exerted by $s_i$ on its sensing neighbours, i.e., $F_{el}(i,l) = \max\{F_{el}(i, j) : s_j \in SN(s_i)\}$. ■

When a sensor receives data for which it is its *relay*, it selects a best relay from its sensing neighbour set and forwards the data to it. Otherwise, it just ignores it. This process repeats until the sink receives the data. The pseudo-code of our geographic forwarding without data aggregation on a duty-cycled $k$-covered wireless sensor network, called GEFIB-1, is given in Fig. 12.1.

---

ALGORITHM 2: JOINT-$k$-COVERAGE-FORWARDING (GEFIB-1)
(* This code is run in each round. It is called by every sensor $s_i$ holding data to be forwarded to the sink. It is assumed that our connected $k$-coverage protocol has been already called to $k$-cover a field *)

1. Sort all active sensing neighbours in a list, *LIST*, based on their resultant forces $F_{el}(i, j)$ where $s_j \in SN(s_i)$
2. Select a best relay $s_l$ from the sorted list, *LIST*
3. Forward sensed data to $s_l$
4. Return

---

**Fig. 12.1** Joint $k$-coverage and forwarding (GEFIB-1)

**Fig. 12.2** Communication between adjacent cluster-heads

---

ALGORITHM 2: JOINT-*k*-COVERAGE-FORWARDING (GEFIB-2)

(* This code is run in each round. It is called by every sensor $s_i$ holding data to be forwarded to the sink. It is assumed that $CSW_k$ has been already called to *k*-cover a field. We assume that each sensor already sent its sensed data directly to its cluster-head *)

Procedure Identify_Best_Relay (*id1*, *id2*, *id3*: integer)

1. Determine a best slice of $s_{id1}$ with respect to $s_{id2}$
2. Sort the three lenses $L_1, L_2, L_3$ of this slice based on the resultant forces $F_{el}(id1, L_j)$ of their sensors $(1 \le j \le 3)$
3. Select a best lens $L_b, 1 \le b \le 3$
4. Select a best relay $s_{id3} \in L_b$
5. Return(*id3*)

EndProcedure

/* The following code is run by a best relay */

1. If $s_i$ is a relay Then /* *i* is the *id* of $s_i$ */

1.1. Forward sensed data to the closest cluster-head

2. Else /* The following code is run by a cluster-head */

2.1. Identify_Best_Relay(*i*,*m*,*id*) /* *m* is the sink's *id* */

2.2. Forward sensed data to $s_{id}$ /* *id* is a best relay's *id* */

EndIf

3. Return

---

**Fig. 12.3** Joint *k*-coverage and forwarding (GEFIB-2)

### *12.2.3 Data Forwarding with Aggregation*

In this section, we present two geographic forwarding protocols with data aggregation on a duty-cycled $k$-covered wireless sensor network.

All data originated from sensors in a cluster are received by their corresponding cluster-head, which aggregates them with its own data into only one single data. Precisely, each sensor sends its data directly to its cluster-head, where data are aggregated. We distinguish two types of aggregation. In the first scenario, referred to as *local data aggregation*, aggregation occurs only within clusters and all data aggregated by cluster-heads are forwarded to the sink without further aggregation. Thus, the sink receives data from each cluster-head in each round. In the second scenario, referred to as *global data aggregation*, the sink receives only *one* data packet in each round that represents the aggregation of *all* data aggregated by cluster heads. Precisely, each cluster-head also aggregates its own aggregated data with the aggregated data it has received from a cluster-head and forwards the result to another cluster-head. Before we discuss our geographic data forwarding protocol with data aggregation on a duty-cycled $k$-covered wireless sensor network, namely GEFIB-2 (Sect. 6.2.3.1) and GEFIB-3 (Sect. 6.2.3.2), we define *best slice*, *best lens*, and *best relay* of a cluster-head with respect to a destination.

*Definition 12.2:* A *best slice* of a cluster-head $s_{ch}$ with respect to a destination $Dest$, which could be a cluster-head or the sink, is a slice that is crossed by a line segment connecting $s_{ch}$ and $Dest$ (Fig. 12.2). A *visible lens* $L_v(i,Dest)$ of $s_{ch}$ with respect to $Dest$ is a lens that belongs to the best slice of $s_{ch}$ with respect to $Dest$, where $1 \le v \le 3$ (Fig. 12.2). A *best lens* $L_b(i,Dest)$ of $s_{ch}$ with respect to $Dest$ is a visible lens such that $F_{el}(i,L_b(i,Dest))$ is the maximum over all resultant forces exerted by $s_{ch}$ on all its visible lenses $L_v(i,Dest)$, where $1 \le b,v \le 3$. A *best relay* $s_l$ of a cluster-head $s_{ch}$ with respect to a destination $Dest$ is a sensing neighbour of $s_{ch}$ selected from a best lens $L_b(i,Dest)$ such that $F_{el}(i,l)$ is the maximum over all resultant forces exerted by $s_i$ on its neighbours located in $L_b(i,Dest)$. That is, $F_{el}(i,L_b(i,Dest)) = \max\{F_{el}(i,L_v(i,Dest)) : 1 \le v \le 3\}$ and $F_{el}(i,l) = \max\{F_{el}(i,j) : s_j \in L_b(i,Dest)\}$. ■

#### 12.2.3.1 Locally Aggregated Data Forwarding

First, each sensor sends its data *directly* to its cluster-head without relaying them through other intermediate sensors. When a cluster-head $s_i$ receives data from all sensors belonging to its cluster, it aggregates them with its own data and forwards the result, called *locally aggregated data* (LAD), towards the sink. In each round, the sink receives as many LAD packets as cluster-heads. Indeed, when a cluster-head receives LAD packets initiated from other cluster-heads, it just forwards them without any update. Precisely, a cluster-head finds the best slice with respect to the sink (Definition 12.2) and chooses the best lens in terms of attractive force out of the three visible lenses (Fig. 12.2). From this lens, it selects the best relay based on the potential field-based force and forwards data to it. However, when a relay receives the data, it forwards it directly to the closest cluster-head. This forwarding process between cluster-heads using relays takes place through fish bladders (or lenses) and repeats until data arrive at the sink. All cluster-heads and relays apply the algorithm GEFIB-2 whose pseudo-code is given in Fig. 12.3.

---

ALGORITHM 3: DAT-CONSTRUCTION (DAT-C)

(* This code is run in each round by the sink and cluster-heads *)
/* The sink initiates the DAT construction process */

1. Randomly select one cluster-head as *ring aggregator* and two adjacent cluster-heads as *aggregation initiators*
2. The sink assign a *ring_id* to the first ring (*ring_id* = 1) and advertizes it to its ring aggregator and aggregation initiators
3. The sink designates one of the aggregation initiators as an *Aggregation proxy* to designate two of its adjacent cluster-heads as *aggregation* initiators for the next ring
4. The aggregation proxy advertizes *ring_id* clockwise while the other aggregation initiator advertizes it counter-clockwise

/* Ring aggregators and aggregation initiators build the DAT */

5. *ring_id* = *ring_id* + 1
6. While (all cluster-heads are not being considered) Do
7. A ring aggregator selects one of its adjacent cluster-heads as a ring aggregator for the next ring whose id is *ring_id*
8. An aggregation proxy selects two of its adjacent cluster-heads as aggregation initiators for the next ring whose id is *ring_id*. One of these cluster-heads is designated as an aggregation proxy
9. An aggregation proxy advertizes *ring_id* clockwise while an aggregation initiator advertizes it counter-clockwise
10. *ring_id* = *ring_id* + 1

    EndWhile

11. Return

---

**Fig. 12.4** DAT construction algorithm (DAT-C)

#### 12.2.3.2 Globally Aggregated Data Forwarding

In each round, the sink would receive only one data, called *globally aggregated data* (GAD), which represents the result of aggregation of all locally aggregated data generated by their cluster-heads. This implies that at the end of each round, only one cluster-head would forward the data aggregated by all other cluster-heads after it has been aggregated with its own data. Therefore, it is necessary that all cluster-heads participate in the data forwarding process. It follows that each

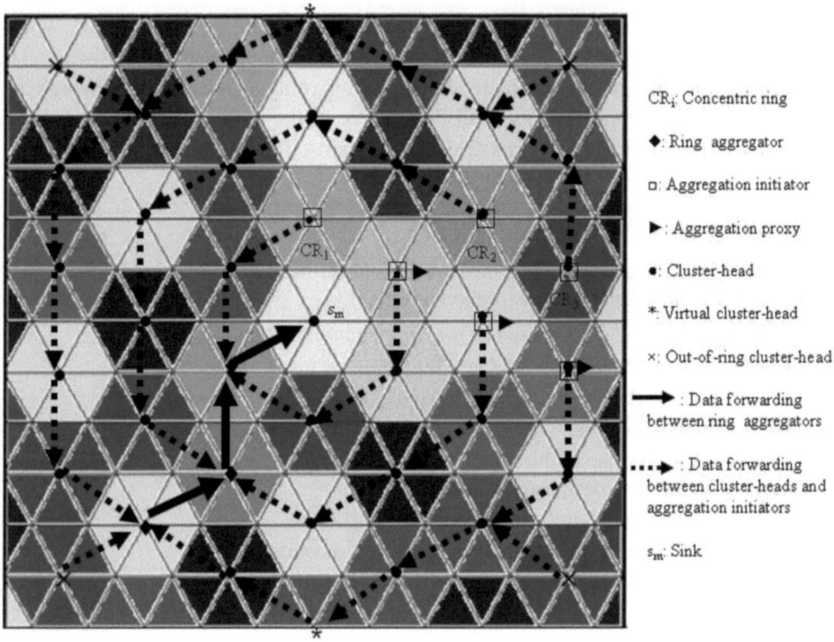


**Fig. 12.5** Data forwarding on a random data aggregation tree

**Fig. 12.6** Linear representation of a random data aggregation tree

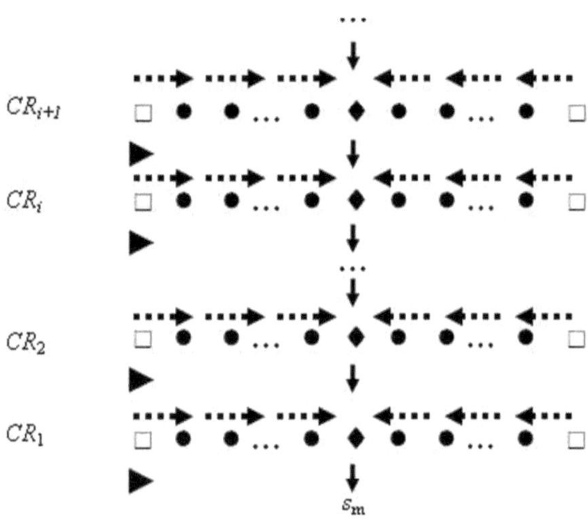


cluster-head would act as a relay on behalf of other cluster-heads until a GAD packet reaches the sink. Precisely, at any time, there is only one aggregated data packet that is forwarded by cluster-heads until it reaches the sink.

ALGORITHM 4: JOINT-$k$-COVERAGE-FORWARDING (GEFIB-3)

```
/* Run the first step at the start of each round. The sink and
   Cluster-heads participate in the DAT construction */
1. Call to DAT-Construction algorithm (ALGORITHM 3)
/* All aggregation initiators initiate the data forwarding process
   on their corresponding ring in parallel */
2. If s_i is an aggregation proxy Then
2.1. Identify_Best_Relay(i,j,id) /* s_id is a best relay */
2.2. Forward data clockwise to the immediate successor cluster-
       head s_j in the corresponding ring via the relay s_id
      (the destinations s_id and s_j are inserted in the data packet)
3. Else /* The other aggregation initiator */
3.1. Identify_Best_Relay(i,j,id)
3.2. Forward data counter-clockwise to the immediate successor
       cluster-head s_j in the corresponding ring via the relay s_id
      (the destinations s_id and s_j are inserted in the data packet)
   EndIf
/* Code is run by a best relay or a virtual cluster-head */
4. If s_i is a best relay or a virtual cluster-head Then
4.1. Forward data to the next cluster-head s_j
5. Else /* Cluster-heads and ring aggregators */
5.1. If s_i is a cluster-head Then /* Cluster-heads */
5.1.1. Identify_Best_Relay(i,j,id) /* s_id is a best relay */
5.1.2. Aggregate the received data with its own data
5.1.3. Forward data to the immediate successor cluster-head s_j
         in the corresponding ring via the relay s_id
         (s_id and s_j are inserted in the data packet)
5.2. Else /* Ring aggregator */
5.2.1. If s_i is a ring aggregator Then
5.2.1.1. Identify_Best_Relay(i,j,id) /* s_id is a best relay */
           /* s_j is a successor ring aggregator or the sink */
5.2.1.2. Wait until receipt of aggregated data from the
           immediate predecessor ring aggregator
5.2.1.3. Aggregate the received data from both neighbouring
           cluster-heads in its ring and the one received from a
           predecessor ring aggregator with its own data
5.2.1.4. Forward data to s_j via the relay s_id
           (s_id and s_j are inserted in the data packet)
       EndIf
     EndIf
   EndIf
6. Return
```

**Fig. 12.7** Joint $k$-coverage and forwarding (GEFIB-3)

Notice that we could enable forwarding of GAD packets only through cluster-heads. However, this solution would be very costly for cluster-heads [98, 99] as the energy spent in data transmission is proportional to the transmission distance, which would be around the nominal transmission range of the sensors. It is easy to check that the distance between the centres of two adjacent clusters is equal to $\sqrt{3}\,r$ (Chap. 5, Fig. 5.3), where $r$ is the radius of the sensors' sensing disks. Also, if the radius $R$ of the communication disks of the sensors does not satisfy $R \geq \sqrt{3}\,r$, then cluster-heads cannot directly communicate with each other, and hence other relays should forward aggregated data between adjacent cluster-heads. On the other hand, inserting several relays between two adjacent cluster-heads would incur high delay [31].

We believe that a more balanced approach should be used to account for both energy and delay. First, we show how to construct a data aggregation tree to enable data aggregation at cluster-heads.

*Data Aggregation Tree (DAT) Construction:* The process of forming a binary tree of cluster-heads rooted at the sink for data aggregation benefits from the distribution of cluster-heads in a field. The sink is supposed to be located at its optimum position in terms of energy efficient data gathering, which corresponds to the centre of a field [148]. Recall that cluster-heads are selected to be as close as possible to the centres of their clusters. The algorithm is initiated by the sink (Fig. 12.4). The sink starts by randomly selecting three cluster-heads: two of them, called *aggregation initiators*, are adjacent to each other and constitute the extreme points of an open ring that initiate data aggregation on the ring itself, while the third one, called *ring aggregator*, is located somewhere on the open ring and is responsible for the aggregated data on its ring. Each ring is associated with an *identification number*, called *ring_id*, which will be used by the cluster-heads to forward their aggregated data towards their corresponding ring aggregator. Furthermore, one of the aggregation initiators is designated as an *aggregation proxy*, which will select the aggregation initiators for the next ring and advertize the value of *ring_id* for this next ring. The first ring is the one whose centre is the closest one to the location of the sink (i.e., centre of a field) and has *ring_id* = 1. The value of *ring_id* is incremented by 1 from the sink out by aggregation proxies. Each cluster-head belongs to only one ring. While an aggregation proxy advertizes the value of *ring_id* clockwise to the cluster-heads members of its ring, the other aggregation initiator advertizes it counter-clockwise so all the cluster-heads that belong to the same ring receive the same value of *ring_id*. After the data aggregation tree construction, each cluster-head knows its immediate predecessor and successor cluster-head in its corresponding ring. Also, each ring aggregator knows its immediate predecessor and successor ring aggregators. Because of the boundary effect, the sink may designate a few sensors as *virtual cluster-heads* for clusters that are not complete (i.e., a portion of a disk) and which can be *k*-covered by adjacent cluster-heads provided that they select sensors to remain active from their boundary lenses, which they share with these non-complete clusters. Precisely, these virtual cluster-heads will not play the role of cluster-heads but will simply act just as relays between non-neighbouring cluster-heads on the same ring on the boundary of a field. Also, some cluster-heads located on the boundary of a field do not belong to any ring. They are called *out-of-ring cluster-heads*, which *connect* to the closest ring via their neighbouring cluster-heads to which they forward their data to be aggregated with theirs. Figure 12.5 shows a data aggregation tree and its components, whereas Fig. 12.6 is a linear representation of Fig. 12.5.

*Data Forwarding:* We assume that the concentric rings are numbered $CR_1, CR_2, CR_3, \ldots$ from the sink out. For a given ring $CR_i$, while the aggregation proxy initiates data aggregation in one direction of the ring, the other one initiates data aggregation in the other direction of the ring. The ring aggregator of $CR_i$ receives aggregated data originated from both of aggregation initiators of $CR_i$, and aggregated data from a ring aggregator of an outer adjacent ring $CR_{i+1}$. At the end, the ring aggregator of $CR_1$ aggregates its own data with those originated from both of the aggregation initiators of its ring and the one received from the ring aggregator of $CR_2$ and forward the *GAD* towards the sink. Specifically, when a cluster-head receives data from its predecessor cluster-head, it aggregates with its own data and forwards it to its successor cluster-head until it reaches the ring aggregator. Each ring aggregator would wait until it receives aggregated data from its immediate predecessor ring aggregator. Those out-of-ring cluster-heads would simply forward their data to the cluster-heads they registered with. Note that communication between adjacent cluster-heads follows the same scheme shown in Fig. 12.5. The pseudo-code of GEFIB-3 is given in Fig. 12.7. This pseudo-code is run in each round by a cluster-head or a relay $s_i$. It is assumed that the connected $k$-coverage protocol has been already called to $k$-cover a field and that each sensor already sent its data to its cluster-head.

## 12.2.4 Generalizability of GEFIB

The design of GEFIB framework for joint $k$-coverage, duty-cycling, and forwarding is based on the sensing and communication *disk*, and sensor *homogeneity* models. Although these assumptions are the basis for most of the coverage and forwarding protocols, they may not be valid in practice. In this section, we show how to relax them to promote the use of our protocols in real-world applications.

### 12.2.4.1 Convex Sensing and Communication Model

As mentioned earlier in Chap. 5, the communication range of MICA motes was found to be asymmetric and environment-dependent [221]. Furthermore, the communication range of radios was found to be highly probabilistic and irregular [224]. In this section, for problem tractability, we consider convex sensing and communication models, where sensors have the same *convex* sensing and communication ranges but not necessarily circular. Precisely, we consider the *largest enclosed disk* of the sensing range of the sensors whose radius is equal to $r_{led}$. In this case, the sink slices a field into overlapping Reuleaux triangles of width $r_{led}$.

However, all other processing remain the same for all the protocols, namely GEFIB-1, GEFIB-2, and GEFIB-3.

#### 12.2.4.2 Sensor Heterogeneity Model

It has been found that heterogeneity enhances reliability of the network and extends its lifetime [209]. In this section, we consider heterogeneous sensors with different yet convex sensing and communication ranges. Based on the notion of the largest enclosed disk of the sensing range of the sensors, the sink slices a field into overlapping Reuleaux triangles of width $r_{led}^{\min}$, the minimum radius of the largest enclosed disks of the sensing ranges of the sensors. However, a very small $r_{led}^{\min}$ could overestimate the required sensor density in the network. Indeed, with a single sensor with a very small radius, the network would be required to have a large sensor density. We believe that a more adaptive approach could be used to adapt the sensor density to the sensing ranges of the sensors in the area.

### *12.2.5 Performance Evaluation*

In this section, we evaluate GEFIB performance using a high-level simulator written in the C programming language. We consider a square deployment field of side length 1000 m where 16000 sensors are randomly and uniformly deployed. We use the energy model given in [211], where energy consumption in transmission, reception, idle, and sleep modes are 60 mW, 12 mW, 12 mW, and 0.03 mW, respectively. Following [218], we define *one unit of energy* as the energy required for a sensor to stay idle for 1 s. We assume that the initial energy of each sensor is 60 J enabling a sensor to operate about 5000 s in reception/idle modes [211]. All simulations are repeated 20 times and the results are averaged.

First, we compare GEFIB-1, GEFIB-2, and GEFIB-3. We assume that the energy consumption in data transmission and reception follows the model given in [102]. Recall that GEFIB-1, GEFIB-2, and GEFIB-3 use the same connected $k$-coverage protocol but *different* data collection protocols. Figures 12.8 and 12.9 show that GEFIB-3 outperforms GEFIB-1 and GEFIB-2 with respect to energy consumption and delay (i.e., average time for data sent by sensors to reach the sink), respectively. As expected, data aggregation yields significant energy savings as it reduces the amount of data communication in the network. Also, data aggregation improves on delay as relay forward data as they arrive without causing much delay overhead.

Second, we consider CCP [205] with a geographic forwarding protocol, such as BVGF [204], on top of it, denoted by CCP+BVGF, and compare it to GEFIB-1. We have slightly updated BVGF to consider remaining energy of the sensors that are candidate for data forwarding for a fair comparison. Here, the energy consumption is due to all activities of each sensor necessary to achieve $k$-coverage

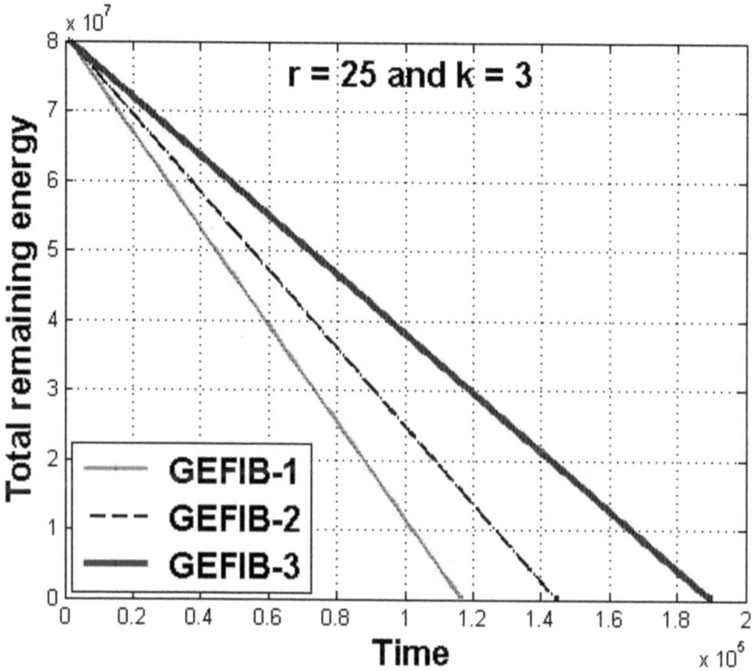


**Fig. 12.8** GEFIB-1 compared to GEFIB-2 (total remaining energy vs. time)

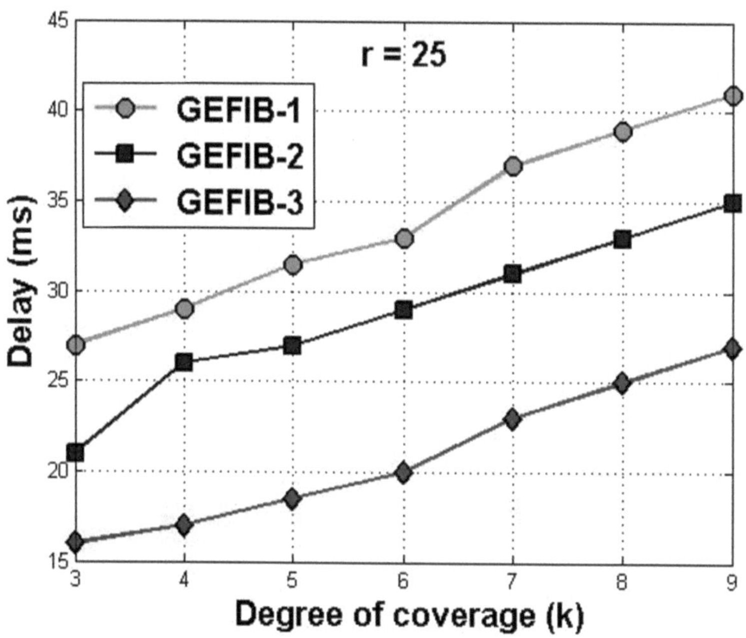


**Fig. 12.9** GEFIB-1 compared to GEFIB-2 (average delay vs. *k*)

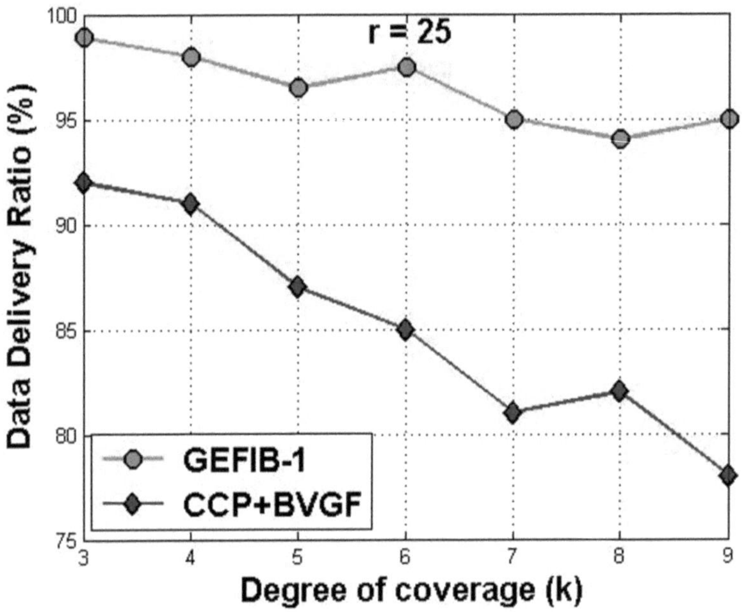


**Fig. 12.10** GEFIB-1 compared to CCP+BVGF (data delivery ratio vs. *k*)

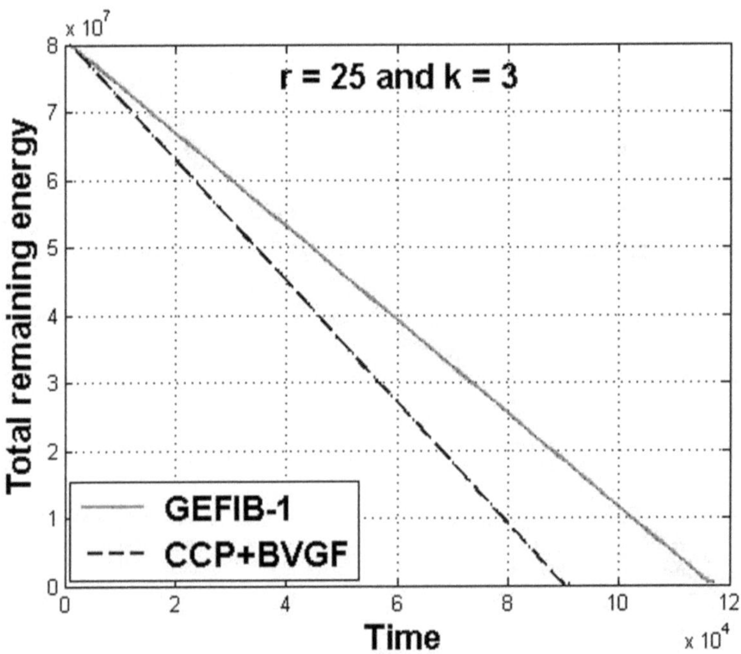


**Fig. 12.11** GEFIB-1 vs. CCP+BVGF (total remaining energy vs. time)

and to forward and/or send sensed data as well as control information. The plots in Figs. 12.10 and 12.11 show that the protocol GEFIB-1 yields higher percentage of data delivery (i.e., success rate) and less average energy consumption than CCP+BVGF. This is due in part to CCP, which uses higher number of active sensors for $k$-coverage than our connected $k$-coverage protocol. Also, BVGF forwards data over long distances. Hence, it consumes considerable energy. Those selected sensors would deplete their energy very quickly and die before expected, thus disconnecting the network. Moreover, data may reach sensors whose remaining energy is not enough to progress data towards the sink, which causes data to be dropped. Our potential field-based forwarding protocol, however, selects next forwarders based on their remaining energy and location, and hence all neighbours of a sensor are equally likely to be selected as next forwarders.

## 12.3 Three-Dimensional Sensor Deployment

In this section, we describe our geographic forwarding protocol for three-dimensional connected $k$-covered wireless sensor networks, which benefits from the advantages of deterministic and opportunistic forwarding schemes.

### 12.3.1 Hybrid Geographic Forwarding

In this section, we propose a protocol, called *hybrid geographic forwarding on a duty-cycled three-dimensional k-covered wireless sensor network* (HYGF-DC$_k$). Sensors that can forward data are called *relays*.

In *deterministic forwarding*, a sensor chooses a next best forwarder based on some metric and forwards data to it, i.e., the next forwarder is determined *a priori*. In *opportunistic forwarding*, however, a next best forwarder is decided on-the-fly and after the data are transmitted. Because of duty-cycling, sensors holding data to be forwarded to the sink and using deterministic forwarding are not totally certain that their currently awake sensing neighbours would remain awake after data are being forwarded. Clearly, duty-cycling introduces *uncertainty* at the sender side when selecting a next best forwarder, thus making opportunistic forwarding the most suitable approach. However, with opportunistic forwarding, several active sensors may hear the transmitted data, thus creating high contention at the receiver side to select a next best forwarder. Thus, it is important to find a *trade-off* between *uncertainty* due to duty-cycling with deterministic forwarding and *contention* due to opportunistic forwarding. Next, we describe our hybrid forwarding approach in detail.

Our hybrid forwarding protocol HYGF-DC$_k$ takes advantage of both deterministic and opportunistic forwarding approaches in order to achieve good data forwarding performance in terms of data delivery ratio, delay, and control overhead. First, we define the notion of *remaining energy* of a sensor.

As mentioned earlier, only sensors currently active to $k$-cover a three-dimensional field act as *relays*, i.e., can participate in data forwarding. Precisely, a sender $s_i$ specifies in its data packet the ids of the next best $p$ *candidate relays* with descending priorities (i.e., the first one has highest priority while the last one has the lowest priority) and broadcasts it using a transmission distance equal to $r$. Recall that the sensing sphere of a sensor has six slices, each of which has width equal to $r$. While the first sensor is the *primary relay*, the other $p-1$ sensors act as *backup relays*. The priorities assigned by $s_i$ to sensors are based on the knowledge it has about their *activity time*, i.e., the time interval a sensor has been recently awake (or active), where high priority means high small *activity time*. If there is a tie, $s_i$ would break it using the knowledge it has about their *remaining energy*. Thus, $s_i$ sorts its active sensing neighbours based on their priority defined earlier and chooses the first $p$ ones as candidate relays. The question we want to address now is: *Which slice,* called *target slice, the next best* $p$ *candidate forwarders are selected from*? First, a sensor $s_i$ randomly decomposes its sensing sphere into 12 slices (Chap. 7, Fig. 7.22). The *target slice* is the one that is traversed by a segment line between the locations of $s_i$ and the sink $s_m$. Given that all next best $p$ candidate *relays* belong to the same slice, they are guaranteed to be connected to each other. Precisely, they are within sensing range of each other, and hence can be aware of the status of each other (i.e., awake vs. asleep). Thus, the problem of multiple transmissions does not arise in our approach. When a sensor forwards a data packet towards the sink, it would also send an ACK to the sender from which it has received the packet. This would allow all potential relays to know that the underlying data packet has been successfully forwarded to the sink, i.e., no more action is needed by all other candidate relays.

When a sensor that is not located in a target slice receives data, it would simply ignore it. Otherwise, it checks whether it is one of the next best $p$ candidate relays. If so, it checks its priority and if it is the highest or the other next best $p-1$ candidate relays are not awake, it would forward the data packet towards the sink using the same approach as the original sender. However, when a sensor that is located in a target slice receives data and the next best $p$ candidate forwarders are not awake, it is considered as a *potential relay*. It would run the *opportunistic component* of our hybrid forwarding protocol. First, we define the *competition function* $\varphi(l,m)$ of a potential relay $s_l$ as the ratio of its remaining energy $\pi(s_l)$ to the Euclidean distance $\delta(s_l,s_m)$ between it and the sink $s_m$, i.e.,

$$\varphi(l,m) = \frac{\pi(s_l)}{\delta^{\alpha}(s_l,s_m)}$$

---

ALGORITHM 2: HYGF-DC$_k$

Procedure Official_Relay_Declared<br>
Begin<br>
1. $s_l$ forwards sensed data to the sink $s_m$<br>
2. $s_l$ sends an ACK to the sensed (i.e., previous relay)<br>
End<br>
Begin<br>
/* This code section is run by a sender */<br>
1. Sort all potential relays in a descending order of their most recent time activity and break their tie using their closeness to the sink $s_m$<br>
2. Select the first $p$ candidate relays as one *primary* and $p$ - 1 *back-up relays*, store them in a sensed data packet, and broadcast it using a transmission distance equal to $r$<br>
/* This code section is run by a candidate relay $s_l$ */<br>
3. If $s_l$ is active primary relay Then<br>
4. $s_l$ forwards sensed data to the sink $s_m$<br>
5. Else<br>
6. If $s_l$ is highest-priority, active back-up relay Then<br>
7. $s_l$ forwards sensed data to the sink $s_m$<br>
8. Else /* all other potential relays */<br>
9. Generate two random numbers in [1..6] and compute their sum<br>
10. Compute the competition function $\varphi(l,m)$<br>
11. Broadcast $BID(s_l) =< id_1, val, \varphi(l,m), id_2 >$<br>
/* Compare $BID(s_l)$ to all received $BID(s_j)$ broadcast by potential relays $s_j$ */<br>
12. If $val(s_l) > max\{val(s_j)\}$ Then<br>
13. Call Official_Relay_Declared<br>
14. Else<br>
15. If $val(s_l) = \max\{val(s_j)) \wedge \varphi(l,m) > \max\{\varphi(j,m)\}$<br>
16. Then Call Official_Relay_Declared<br>
17. Else<br>
18. If $val(s_l) = \max\{val(s_j)) \wedge \varphi(l,m) = \max\{\varphi(j,m)\}$ and $id(l) > max\{id(j)\}$ Then<br>
19. Call Official_Relay_Declared<br>
20. End<br>
21. End<br>
22. End<br>
23. End<br>
24. End<br>
25. Return<br>
End

---

**Fig. 12.12** Joint $k$-coverage and hybrid forwarding protocol

Intuitively, preference is given to potential relays that have higher remaining energy and are closer to the sink. The opportunistic component of our hybrid forwarding protocol is based on the following lemma [44, p. 64], which adds some randomness to the selection process of a potential relay.

*Lemma 12.1:* No matter what two loaded dices we have, it cannot happen that each of the sums 2, 3, …, 12 comes up with the same probability. ■

Each potential relay $s_l$ runs the following steps:

- Flips two loaded dices (i.e., randomly generates two numbers in [1…6]) and computes their sum *val*.
- Computes its competition function $\varphi(l,m)$.
- Builds a small packet $BID(s_l)$ containing the quadruplet $< id_1, val, \varphi(l,m), id_2 >$ and broadcasts it within its sensing range only, where $id_1$ is the id of a sender of a sensed data packet and $id_2$ is the id of $s_l$.

Any active sensor $s_a$ that receives $BID(s_l)$ will have to run the following sequence of steps:

- Checks whether it is a *potential relay* for the underlying data packet by looking at the first field $id_1$ of $BID(s_l)$. If not, it just drops the packet. Else, it runs the next step.
- Compares its random value (generated by flipping two loaded dices) to the second field *val* of all $BID(s_l)$. If it is smaller, $s_a$ cannot be a candidate relay. If both values are equal, it compares the value of its competition function with the third field $\varphi(l,m)$ of all $BID(s_l)$. If it is smaller, it does not consider itself as a candidate relay. If it is equal, it compares its *id* with the fourth field $id_2$ of all $BID(s_l)$. If it is smaller, $s_a$ is not a candidate relay.

As can be seen, the official relay of a sensed data packet is the one with the highest value of the second field *val*. The values of the other two fields, namely the value of competition function and *id*, are used to break ties. At the end of this selection process, the potential relay that has been designated as an *official relay* would forward the data packet to the sink and send back an ACK to the sender of the data.

Note that the opportunistic component of our hybrid approach requires a *little coordination* between potential relays. The pseudo-code of HYGF-DC$_k$ is given in Fig. 12.12.

## 12.3.2 Performance Evaluation

In this work, we propose the first analysis and solution to the problem of joint $k$-coverage and forwarding in duty-cycled three-dimensional wireless sensor networks. Thus, it is impossible to provide a fair quantitative comparison between our protocols and other existing ones, such as ExOR [43], CKN [161], and GeRaF [226], which were proposed for two-dimensional wireless sensor networks. Also,

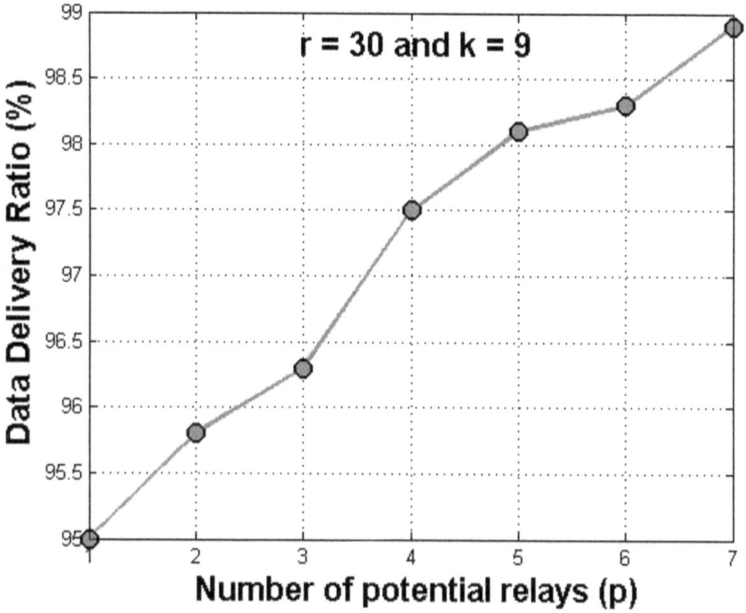


**Fig. 12.13** Data delivery vs. $p$

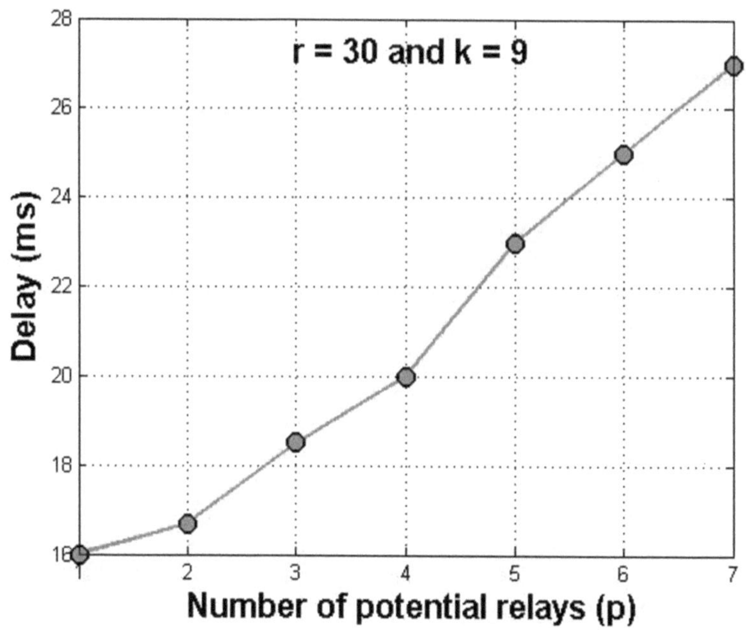


**Fig. 12.14** Delay vs. $p$

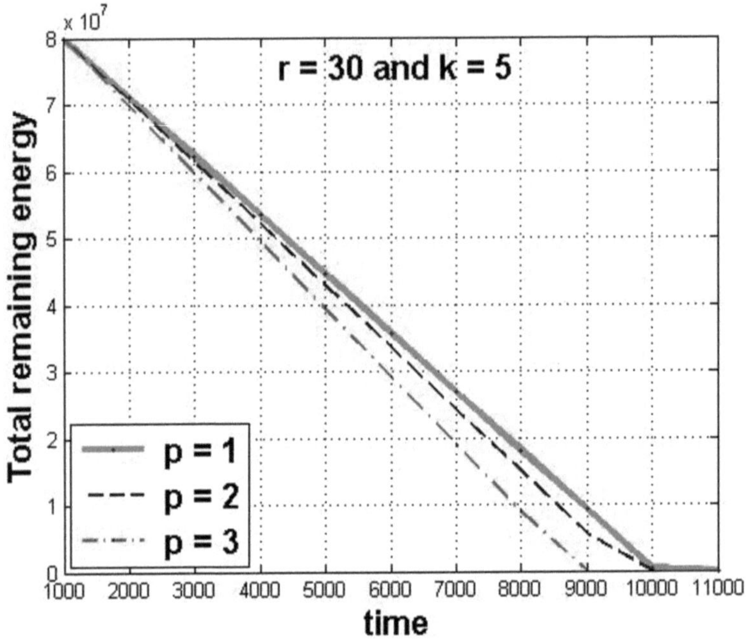


**Fig. 12.15** Remaining energy vs. *p*

CKN [161] considered only 1-coverage while Kao et al. [117] did not consider duty-cycling. In this section, we evaluate the performance of HYGF-DC$_k$ with a high-level simulator written in the C programming language.

Figures 12.13–12.15 show the impact of the number of *candidate relays* on the performance of our $k$-coverage and hybrid forwarding protocol HYFGF-DC$_k$. As we increase the number $p$ of *candidate relays*, the data delivery ratio (i.e., percentage of sensed data packets successfully received by the sink) increases (Fig. 7.13). Notice that the setting $p = 1$ corresponds to deterministic forwarding. When a single sensor is selected as a candidate relay $(p = 1)$, there is no guarantee that after data transmission this selected sensor would remain active. If it does not stay active, data packets destined to it would never reach the sink. Thus, the presence of multiple candidate relays increases the chance that sensed data would reach the sink. However, this improvement of data delivery ratio would lead to additional delay (Fig. 7.14) and more energy consumption (Fig. 7.15). Indeed, candidate relays need to exchange their local data to identify the best one of them as a relay. This process, however, needs more time and control overhead. These simulation results show that opportunistic forwarding helps improve data delivery ratio but requires more delay. However, deterministic forwarding yields better delay but does not guarantee high data delivery ratio. Our hybrid forwarding protocol on duty-cycled three-dimensional wireless sensor networks benefits from these nice features of opportunistic and deterministic schemes.

## 12.4 Related Work

The study of joint coverage and geographic forwarding, however, has received little attention. In particular, a few works addressed the problem of geographic forwarding on duty-cycled wireless sensor networks [43, 161, 226]. While Biswas and Morris [43] and Zorzi and Rao [226] assumed duty-cycling at the MAC layer, Nath and Gibson [161] considered both routing and duty-cycling at the routing layer. In traditional routing, a sender chooses the next forwarder before transmitting its data. However, when the link quality is poor, the probability that the selected forwarder receives the data is low. In contrast, using *opportunistic routing*, *any* node that overhears the transmission and is closer to the destination can participate in forwarding the packet. The packet duplication problem is solved using a scheme for contention among receivers. Zorzi and Rao [226] also gave a detailed description of a MAC protocol and an evaluation of the latency and energy performance. Zorzi and Rao [226] proposed an opportunistic data transmission scheme for wireless sensor networks, called *geographic random forwarding* (GeRaF), which uses geographic routing where a sensor acting as relay is not known *a priori* by a sender. In other words, relay nodes are decided only after the transmission has started. Thus, GeRaF does not guarantee that a sender will always be able to forward a message towards the sink, and hence GeRaF is said to be *best-effort* forwarding scheme. Biswas and Morris [43] proposed an integrated routing and MAC protocol, called ExOR, to enhance throughput in multi-hop wireless networks, where a source sends a batch of packets destined to the same destination. ExOR is also an opportunistic routing protocol that determines the next forwarder of a packet after the transmission of the packet. The node closest to the destination among all the candidate forwarders that receive the packet is selected in each hop. Nath and Gibbons [161] presented the first analysis of the performance of geographic routing on duty-cycled wireless sensor networks, where every sensor has $k$ awake neighbours. They also proposed a sleep–wakeup scheduling protocol for opportunistic geographic routing.

Kao et al. [117] proposed a heuristic for routing in three-dimensional space using the two-dimensional face routing algorithm. Pompili et al. [168] proposed different routing algorithms for applications in underwater sensor networks depending on whether delay is sensitive or not.

## 12.5 Summary

In this chapter, we have proposed energy-efficient geographic forwarding protocols for duty-cycled $k$-covered wireless sensor networks, which are deployed in two-and three-dimensional fields. We have considered both of the deterministic and opportunistic forwarding schemes in the design of our protocols. We have also relaxed two assumptions to promote the practicality of our joint protocol and evaluated its performance. Our joint $k$-coverage and geographic forwarding framework can be used for applications that demand high coverage degree, such as intruder detection and tracking. It is also useful for applications that require data aggregation as well as those applications where *all* data originated from sources should reach the sink without being aggregated or updated by intermediate forwarders.

## Related Work

# Chapter 13
# Conclusion and Future Work

This chapter gives a summary of the topics that have been discussed in this book. Precisely, it summarizes the contributions that we made in the areas of $k$-coverage, connectivity, duty-cycling, and geographic forwarding in two-and three-dimensional wireless sensor networks. All the proposed protocols are energy-efficient with a goal to minimize the total energy consumption of the sensors, thus extending their lifetime. Then, it presents our future research work in an attempt to solve some (open) problems more efficiently.

## 13.1 Contributions of This Book

The main challenge in the design of wireless sensor networks is the limited energy of the sensors and the difficulty of replacing and/or recharging their batteries due to the inherent characteristics of a field (hostile environments, for instance) and cost. To alleviate this problem, it is necessary that the sensors be densely deployed (i.e., *redundant sensor deployment*) and appropriate protocols be designed in order to exploit this redundancy while maximizing the operational lifetime of the network. One of the fundamental issues in the design of wireless sensor networks is sensor deployment. Indeed, several network configuration protocols for connected $k$-coverage have been proposed in the literature with a goal to save the energy of the sensors while meeting certain quality of service dictated by the critical application requirements in terms of coverage and connectivity. Another interesting problem is geographic forwarding in always-on wireless sensor networks and how to achieve a best trade-off between energy and delay for time-critical sensing applications. This problem of geographic forwarding becomes more challenging for duty-cycled $k$-covered wireless sensor networks. This is mainly due to the uncertainty introduced by sensor duty-cycling (or scheduling).

This book has four parts and focuses on the design of energy-efficient connected $k$-covered wireless sensor networks, and particularly *many-to-one* sensor network architectures, where *all* the sources sensors report their sensed data to a *single* sink. Nevertheless, the work published in this book could be easily extended to *many-to-many* sensor network architectures that include more than one sink node. The first part of this book reviews the major opportunities and challenges of wireless sensor networks and provides the background for the rest of the book. The second part investigates phase transitions in coverage and connectivity in wireless sensor networks from the perspective of percolation theory. The third part discusses the connected $k$-coverage problem while considering several real contexts of sensor deployment, such as sensor heterogeneity, stochastic sensing and communication models, and three-dimensional settings. The fourth part addresses the problem of geographic forwarding on always-on sensors while considering connectivity and coverage. It also deals with the problem of geographic forwarding on duty-cycled sensors under connected $k$-coverage constraint.

This book aims at exploring the design issues of energy-efficiency connected $k$-coverage, duty-cycling, and geographic forwarding protocols for two-and three-dimensional wireless sensor networks. It also explores several ways in order to jointly consider them in a unified framework. Our ultimate goal is to maximize the energy conservation, thus extending the operational network lifetime. In this book, we followed a bottom-up approach starting from sensor deployment to data gathering. This approach consists of a rigorous analysis of the design issues of each of the above-mentioned topics in Chaps. 3–11. Then, all these issues have been jointly considered in the same framework in Chap. 12. The contributions of this book can be summarized as follows:

- We studied coverage and connectivity in wireless sensor networks using percolation theory (Chaps. 3 and 4). Precisely, we computed a non-trivial value of the *covered area fraction* at critical percolation, called *critical covered area fraction*, and derived the critical sensor spatial density above which a field is *almost surely* covered. Furthermore, we proposed a model for percolation in wireless sensor networks, called *correlated disk model*, to study both coverage and connectivity in an integrated manner based on the relationship between the communication and sensing ranges of the sensors. We found that the value of the critical sensor spatial density depends on the ratio of the radius of the communication disks of the sensors to the radius of their sensing disks. This study helps network designer fully cover a deployment field with a minimum number of sensors and monitor it while maintaining connectivity between all the sensors.
- Motivated by the existence of different applications and environments with diverse requirements in terms of degrees of coverage and connectivity, we extended our above analysis for 1-coverage to redundant coverage (or $k$-coverage) so the network can be self-configured to support these applications. Precisely, we considered a deterministic approach to analyze and characterize $k$-coverage of a field, and compute the minimum active sensor spatial density required to full $k$-cover a field (Chap. 5). Our analysis is based on a fundamental theorem, namely Helly's Theorem, which characterizes the intersection of convex sets, as well as the geometric properties of the Reuleaux triangle. We found that this density depends only on the degree of coverage $k$ requested by a sensing application and the radius of the sensing disks of the sensors, thus reflecting the expected behaviour of the sensors. Moreover, our analysis requires that the communication range of the sensors be at least equal to their sensing range so that all active sensors in a $k$-covered wireless sensor network are guaranteed to be connected. Based on this analysis, we computed the unconditional and conditional connectivity of $k$-covered wireless sensor networks and proved that is higher than the degree of coverage $k$. Furthermore, we proposed four configuration protocols to achieve connected $k$-coverage in wireless

sensor networks. In the first one, called *centralized randomized connected k-coverage*, the sink is responsible for selecting a minimum number of sensors to guarantee $k$-coverage of a field while maintaining connectivity between active selected sensors. Each of the second and third protocols, called *Reuleaux triangle-based clustered randomized connected k-coverage* and *disk-based clustered randomized connected k-coverage*, is run under the control of the sink and a subset of sensors. Precisely, in each round, the sink selects a subset of sensors, called *cluster-heads*, each of which is responsible for selecting a subset of neighbouring sensors to $k$-cover its underlying cluster while remaining connected to each other. Both protocols consider different degree of network clustering. In the fourth protocol, called *distributed randomized connected k-coverage*, all the sensors are required to coordinate among themselves to $k$-cover a field in each round while being mutually connected. For our distributed connected $k$-coverage protocol, we designed two sensor scheduling schemes to guarantee full $k$-coverage of a field. In the first scheme, called *Self-Scheduling driven k-Coverage*, each sensor turns itself *on* based on the local information it has about its sensing neighbours in order to $k$-cover sensing range. In the second scheme, called *Triggered-Scheduling driven k-Coverage*, each sensor is allowed to trigger a necessary number of its sensing neighbours to become active in order to achieve $k$-coverage of its sensing range. We found that the latter outperforms the former in terms of the number of active sensors needed for $k$-coverage and network lifetime. Furthermore, we relaxed some commonly used assumptions for coverage configurations in wireless sensor networks, namely the sensing and communication disk model and sensor homogeneity model, to enhance the applicability of our connected $k$-coverage protocols in real-world sensing applications.

- We extended our connected $k$-coverage protocols to *mission-oriented wireless sensor networks*, where the region to be $k$-covered could change over time (Chap. 6). Moreover, the degree $k$ of coverage required for all the missions to be accomplished may not be the same. In this type of network, the sensors should be able to move towards a region of interest to be monitored when they are requested by a central entity, such as the sink. Our proposed protocols take into account the purposeful mobility of the sensors and the goals of the missions. Precisely, all the sensors in the network collaborate with each other to accomplish the missions successfully while minimizing the total energy consumption of the sensors due to their communication, sensing, and, particularly, mobility. In other words, our protocol for mobile connected $k$-coverage in mission-oriented wireless sensor networks is designed in such a way to find the best trade-off between the goals of the missions and the mobility of the sensors. Furthermore, we solved the problem of connected $k$-coverage for heterogeneous wireless sensor networks, where the sensors do not possess the same capabilities, and particularly, with regard to their sensing range, communication range, and initial energy. We identified the problems that arise in the design of connected *k*-coverage protocols for such networks and proposed multi-tier network architectures depending on the capabilities of the sensors, i.e., their sensing ranges, communication ranges, and initial energy.

- In order to account for the *stochastic* nature of the sensors, we adapted the analysis of the connected *k*-coverage problem in two-dimensional wireless sensor networks under a deterministic sensing model to solve the stochastic connected *k*-coverage problem under a stochastic sensing model (Chap. 7). This helps us develop a global framework for connected *k*-coverage in wireless sensor networks that considers both deterministic and stochastic sensing models. Our stochastic sensing model takes into account not only the distance between the sensors and the target locations but also the type of propagation model, i.e. free-space model or multi-path model. In addition, we focused on the problem of forwarding in duty-cycled three-dimensional *k*-covered wireless sensor networks, where *k*-coverage, duty-cycling, and data forwarding are discussed and addressed in a novel joint framework. Using *Helly's Theorem* [44], we find that the extension of the analysis of *k*-coverage in two-dimensional to three-dimensional wireless sensor networks is not straightforward. This is due to the fact that some properties that hold for two-dimensional space cannot generalize to three-dimensional space. Furthermore, we proposed new measures of connectivity of k-covered wireless sensor networks using a more realistic approach (Chap. 8). These measures take into account the inherent properties of wireless sensor networks, which are dense and heterogeneous in nature.
- We considered static wireless sensor networks, where all the sensors and the sink do not move. Moreover, the sensors are supposed to be *on* (or active) all the time. First, we addressed the problem of geographic forwarding on always-on sensors and proposed an energy-efficient protocol based on Delaunay triangulation (Chap. 9). This protocol enables the sensors to forward the sensed data through the shortest Delaunay edges, thus saving their energy and extending the network lifetime. We also solved the problem of finding a trade-off between energy and delay for time-critical sensing applications (Chap. 10). Our proposed protocol, called TED, divides the communication ranges of the sensors into concentric circular bands and classifies them depending on whether the underlying sensing application cares more about energy or delay. We formulated the trade-off as a multi-objective optimization problem whose solution is fed as the input to TED to select the band from which relay nodes have to be designated. Unfortunately, this type of architecture, which includes static always-on sensors with a static sink, suffers from a severe problem, called *energy sink-hole*, where the sensors near the sink suffer from severe battery power depletion problem. Indeed, the sensors close to the sink act as relays to the sink on behalf of *all* other sensors, and hence deplete their battery power more quickly, thus disconnecting the network. We proposed solutions that exploit heterogeneity, mobility, and our new concept of

energy-aware Voronoi diagram (Chap. 11). However, these solutions do not take coverage into consideration as the *energy sink-hole* is coverage-independent, but assume that the network is connected for data forwarding to take place. For homogeneous sensors in terms of their initial energy, we proposed EVEN, an *energy aware Voronoi diagram-based data forwarding* protocol. EVEN is a greedy, localized protocol that combines sink mobility with a new concept, called *energy aware Voronoi diagram* whose sites (sensors' locations) are time-varying as they depend on the remaining energy of the sensors. We find that EVEN yields a significant improvement in terms of network lifetime.

- We jointly considered connected $k$-coverage, duty-cycling, and geographic forwarding in two-and three-dimensional wireless sensor networks (Chap. 12). More specifically, on top of the connected $k$-coverage configuration protocols, we proposed several potential fields-based geographic forwarding protocols on duty-cycled sensors depending on whether data aggregation is considered. This effort constitutes the first design of geographic forwarding protocols for duty-cycled $k$-covered wireless sensor networks with and without data aggregation. Moreover, we proposed a hybrid forwarding protocol in duty-cycled three-dimensional $k$-covered wireless sensor networks, where *deterministic* and *opportunistic forwarding* approaches are considered. Our hybrid forwarding approach provides a trade-off between uncertainty and contention while achieving good data forwarding performance in terms of delay and control overhead.

## 13.2 Research Directions

In this section, we provide a few interesting broad directions for future work.

- The analysis of phase transitions in coverage and connectivity in wireless sensor networks considers homogeneous, two-dimensional wireless sensor networks. We plan to extend this analysis by considering heterogeneous sensors with different sensing and communication ranges capabilities. Real-world sensing applications may require heterogeneous sensors in order to enhance the reliability of the network and extend its lifetime. Moreover, even sensors equipped with identical hardware may not always have the same sensing model. We also plan to extend our work to irregular sensing and communication ranges of the sensors whose shape is not necessarily circular. In addition, we plan to extend the approach that we proposed study of coverage and connectivity in three-dimensional wireless sensor networks to two-dimensional wireless sensor networks and compare our results with the previously obtained ones in Chap. 3. That would help us compare between the two approaches.
- Our proposed framework considers static, connected $k$-covered wireless sensor networks, which still suffer from the energy-sink problem although it uses sensor duty-cycling. Indeed, this problem is inherent to static wireless sensor networks. We plan to benefit from the mobility that EVEN uses to solve the energy-sink problem. Precisely, we intend

to study joint connected $k$-coverage and geographic forwarding in mission-oriented wireless sensor networks, where sensors are mobile to accomplish a specific mission at some time that is requested by the sink. In particular, we will exploit our previous results for static, connected $k$-covered wireless sensor networks in order to investigate the problem of guaranteeing mobile $k$-coverage while maintaining network connectivity in mission-oriented wireless sensor networks [171].

- We are also interested in solving the following problem: given that the sensors may die or fail anytime due to their low battery power, it is necessary to determine the necessary minimum number of sensors that need to join the network to guarantee certain requirements in terms of coverage and connectivity, which should be satisfied for a sensing application. It is worth mentioning that heterogeneous wireless sensor networks require an adaptive approach in the sense that the sensor density should be defined based on the type of sensors deployed in a given area so as to ensure a coverage degree for a field with a minimum number of sensors.
- We believe that the problems of connected $k$-coverage and geographic forwarding in three-dimensional, static wireless sensor networks deserve deep investigation. In Particular, the problem of ensuring $k$-coverage of a three-dimensional field needs to be addressed using different tiling strategies of a sphere, assuming that the sensing range of the sensors is represented by a sphere. Indeed, the problems of $k$-coverage and tiling in three-dimensional space seem to be very inter-related. Different models of titling of a sphere could be adapted to solve the problem of $k$-coverage in three-dimensional wireless sensor networks with better bounds on the required sensor spatial density. We also plan to compute tight bounds on their unconditional and conditional connectivity and fault-tolerance. Furthermore, we plan to extend the analysis of our previous results to investigate the problem of stochastic $k$-coverage in three-dimensional wireless sensor networks using our stochastic sensing model. Similarly, the problem of connected $k$-coverage in heterogeneous wireless sensor networks should be investigated in depth. We believe that a more adaptive, energy-efficient approach needs to be studied to save more energy of the sensors.
- We also plan to extend our work [9, 10, 11] in order to build a hybrid network through the integration of wireless sensor networks, mobile ad-hoc networks, and the global IP Internet. The hybrid network would benefit from the advantages of each of these three networks, namely mobility of mobile ad hoc networks, high density of wireless sensor network, and continuous connectivity of the global IP Internet.
- Finally, the implementation of our proposed framework on a real sensor testbed is our ultimate goal. Our target framework should account for all the real characteristics of the sensors as well as the specific requirements of real-world sensing applications. We believe that all these features and design decisions will enhance the practicality and promote the applicability of our proposed framework to a variety of sensing applications running in different environments with different network setups.

# Appendix
# Network Connectivity and Fault-Tolerance Measures in Three-Dimensional Deployment Fields

This appendix computes the connectivity and fault-tolerance of three-dimensional homogeneous and heterogeneous $k$-covered wireless sensor networks. The proposed measures are based on the *Reuleaux tetrahedron* model, which is used to characterize $k$-coverage of a three-dimensional field. This choice is to make this computation problem more tractable. Moreover, based on the concepts of *conditional connectivity* and *forbidden faulty sensor set*, this appendix proposes conditional network connectivity and fault-tolerance measures for the above networks.

## 1 Introduction

Data accuracy depends on the size of the connected component that contains the sink. It reaches the highest value when the sink belongs to the largest connected component of the network. Thus, high-quality coverage requires *all* source sensors be connected to the sink. That is why we focus on the sink to compute the connectivity of three-dimensional $k$-covered wireless sensor networks [14]. In other words, connectivity of wireless sensor networks should be defined so as to take into consideration the inherent structure of this type of network. Indeed, sensors have neither the same role nor the same impact on the network performance. Thus, measuring the connectivity of wireless sensor networks should account for their specific morphology, where the sink is the most critical node in the network. Hence, we compute the connectivity of three-dimensional $k$-covered wireless sensor networks based on the *size of the connected component* that includes the *sink*.

The remainder of this appendix is organized as follows. Section 2 revisits $k$-coverage in three-dimensional deployment fields. Section 3 computes unconditional connectivity for three-dimensional homogeneous and heterogeneous $k$-covered wireless sensor networks while Section 4 computes their conditional counterparts. Section 5 discusses the results obtained in the previous sections. Section 6 shows how to relax the unit sphere model and account for a convex model for sensing. Section 7 discusses the results for underwater sensor networks while Sect. 8 concludes the appendix.

## 2 $k$-Coverage Characterization

In this section, we show how to guarantee $k$-coverage of a three-dimensional field and derive the corresponding sensor spatial density. Also, we provide some simulation results showing a perfect match with their theoretical counterparts.

# A Sensor Density for *k*-Coverage

Lemma A.1 characterizes the *breadth* of a three-dimensional *k-covered* convex region.

*Lemma A.1:* A three-dimensional convex region $C$ is guaranteed to be $k$-covered with exactly $k$ sensors if its breadth is less than or equal to $r$, the radius of the sensing spheres of sensors.

*Proof:* (By contradiction) Assume that the breadth of a three-dimensional convex region $C$ does not exceed $r$ and $C$ is not $k$-covered when exactly $k$ sensors are deployed in it. Notice that each of these $k$ sensors is located on the boundary or inside of $C$. Thus, there must be at least one location $\xi \in C$ that is not $k$-covered. In other words, there is at least one sensor $s_i$ located at $\xi'$ such that $|\xi - \xi'| > r$, which contradicts our hypothesis that the breadth of $C$ does not exceed $r$. ■

It is true that the deployment of $k$ sensors in a 3D convex region, say $C$, whose breadth is larger than $r$ cannot guarantee its $k$-coverage, where $r$ is the radius of the sensing range of the sensors. Let $p_i$ and $p_j$ be two points in $C$ such that one sensor $s_i$ is located at $p_i$ and $\delta(p_i,p_j) = b > r$, where $b$ is the breadth of $C$ and $\delta(p_i,p_j)$ is the Euclidean distance between $p_i$ and $p_j$. Given that $b > r$, it is impossible for $s_i$ to sense any event that occurs at $p_j$. Thus, there is at least one sensor (i.e., $s_i$) amongst those $k$ sensors, which cannot cover $p_j$, and hence $C$ cannot be $k$-covered.

Next, we compute the sensor spatial density required for guaranteeing $k$-coverage of a three-dimensional field. To this end, we compute the volume of a three-dimensional convex region $C$ that is guaranteed to be $k$-covered when exactly $k$ sensors are deployed in it. From *Helly*'s Theorem [44], we infer that given $k \geq 4$ sensors, a three-dimensional convex region $C$ is $k$-covered by those $k$ sensors if and only if $C$ is 4-covered by any four of those $k$ sensors. Given that the breadth of $C$ is $r$, the network induced by sensors located in $C$ is guaranteed to be connected if $R \geq r$, where $r$ and $R$ are, respectively, the radii of the sensing and communication spheres of sensors. Now, let us address the first question: *What is the sensor spatial density necessary to guarantee full k-coverage of a* three-dimensional *field?* Theorem A.1 computes this sensor density.

*Theorem A.1:* Let $r$ be the radius of the sensing spheres of sensors and $k \geq 4$. The sensor spatial density required to $k$-cover a three-dimensional field is computed as

$$\lambda(r,k) = \frac{k}{0.422\, r_0^3} \tag{1}$$

where $r_0 = r/1.066$.

*Proof:* Let $C_k$ be the intersection of $k$ sensing spheres and assume that their centres *do not* coincide in a three-dimensional field. From Lemma A.1, it follows that $C_k$ is guaranteed to be $k$-covered by exactly $k$ sensors if its breadth does not exceed the radius $r$ of the sensing spheres of sensors. Thus, the maximum volume of $C_k$ is obtained when its breadth is equal to $r$. From Helly's Theorem [44], it follows that the intersection of $k$ sensing spheres is not empty if the intersection of any four of these $k$ spheres is not empty. On the other hand, the intersection set operator requires that the *maximum* intersection volume of these $k$ sensing spheres be equal to that of four spheres provided that the maximum distance between any pair of these $k$ sensors does not exceed $r$. Let us focus on the analysis of four sensing spheres. The maximum overlap volume of four sensing spheres such that every point in this overlap volume is 4-covered, corresponds to the configuration where the centre of each sensing sphere is at distance $r$ from the centres of all other three sensing spheres. Precisely, the sensing sphere of each of the four sensors passes through the centres of the other three sensing spheres as shown in Fig. A.1. The edges between the centres of these four spheres form a *regular tetrahedron* and the shape of the intersection volume of these four spheres is known as the *Reuleaux tetrahedron* [234] and denoted by $RT(r)$. Unfortunately, it was proved that the Reuleaux tetrahedron does not have a constant breadth [234]. Indeed, while the distance between some pairs of points on the boundary of the Reuleaux tetrahedron $RT(r)$ is equal to $r$, the maximum distance between other pairs of points on the boundary of $RT(r)$ is equal to 1.066 $r$ [234], i.e., slightly larger than $r$. This implies that the Reuleaux tetrahedron $RT(r)$ cannot be $k$-covered with exactly $k$ sensors given that the distance between some pairs of points on the boundary of $RT(r)$ is larger than $r$, where $r$ is the radius of the sensing spheres of the sensors. Therefore, the Reuleaux tetrahedron that is guaranteed to be $k$-covered with exactly $k$ sensors should have a side length equal to $r_0 = r/1.066$. The volume of the Reuleaux tetrahedron $RT(r_0)$ is given by [234]

$$vol(r_0) = (3\sqrt{2} - 49\,\pi + 162\tan^{-1}(\sqrt{2})\,r_0^{\,3}/12 \approx 0.422\,r_0^{\,3}$$

Thus, $RT(r_0)$ is the maximum volume that can be $k$-covered by exactly $k$ sensors, where $k \geq 4$. We conclude that the *maximum* volume of $C_k$, denoted by $vol_{\max}(C_k)$, is equal to $vol_{\max}(C_k) = 0.422\,r_0^{\,3}$. Given that $vol_{\max}(C_k)$ has to contain $k$ sensors to $k$-cover $C_k$, we conclude that the sensor spatial density per unit volume required for full $k$-coverage of a three-dimensional field is computed as

$$\lambda(r,k) = k/vol_{\max}(C_k) = k/0.422\,r_0^{\,3} \qquad \blacksquare$$

Related to our work is the novel result discussed in [234], which proved that the breadth of the Reuleaux tetrahedron is not constant. This shows that the properties of 2D space cannot be directly extended to 3D space. Indeed, the Reuleaux triangle [233] (counterpart of Reuleaux tetrahedron in 2D space) has a constant width. Note that the Reuleuax tetrahedron is the symmetric intersection of four congruent spheres such that each sphere passes through the centres of the other three spheres. However, the Reuleaux triangle [233] corresponds to the symmetric intersection of three congruent disks such that each disk passes through the centres of the other two disks. Thus, its constant width is equal to the radius of these disks. Our work can be viewed as an extension of [8] by considering $k$-coverage in 3D WSNs. Moreover, existing works on coverage and connectivity in WSNs assumed the notion of traditional connectivity, whereas our work considers a more realistic measure, namely *conditional connectivity* [101], which is based on the concept of *forbidden faulty set* [81]. Also, our work exploits the result given in [234] with regard to the breadth of the Reuleaux tetrahedron discussed earlier. This helps us provide correct measures of connectivity and fault-tolerance of 3D WSNs based on an accurate characterization of $k$-coverage of 3D fields.

We should mention that the notion of the *arc length* discussed in [234] corresponds to the (maximum) breadth of the Reuleaux tetrahedron. Indeed, it is possible to find two parallel plans that bound the Reuleaux tetrahedron such that the maximum distance between these two plans is equal to $r$ only.

It is worth emphasizing that the value of $\lambda(r,k)$ is tight in the sense that it is minimum given that $k$ sensors should be located within $vol_{\max}(C_k)$, such that $C_k$ is guaranteed to be $k$-covered with exactly these $k$ sensors. Also, notice that $\lambda(r,k)$ depends only on the coverage degree $k$ dictated by a sensing application and the radius $r$ of the sensing range of sensors. Moreover, $\lambda(r,k)$ increases with $k$ and decreases as $r$ increases, thus reflecting the expected behaviour.

Lemma A.2 uses Theorem A.1 and states a *sufficient condition* for $k$-coverage of a three-dimensional field.

*Lemma A.2:* A three-dimensional field is guaranteed to be $k$-covered if any Reuleaux tetrahedron region of side $r_0$ in the field contains at least $k$ sensors, where $r_0 = r/1.066$ and $k \geq 4$. ■

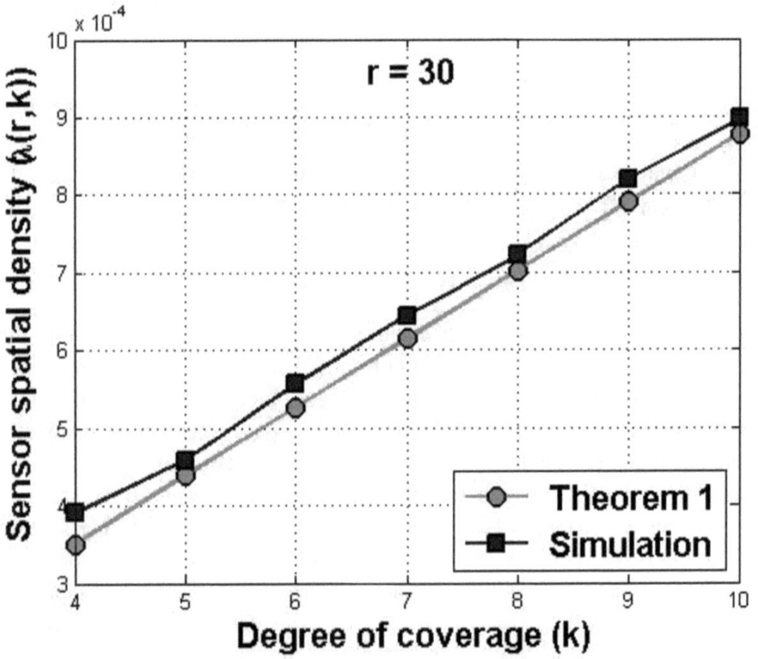


**Fig. A.1** Sensor spatial density $\lambda(r,k)$ vs. $k$

## B Simulation Results

In this section, we present the simulation results using a high-level simulator written in the C programming language. We consider a cubic field of side length 1000m, where all sensors are randomly and uniformly deployed. Moreover, we randomly decompose the cubic field into overlapping Reuleaux tetrahedra as follows. Two adjacent Reuleaux tetrahedra, say $RT_1$ and $RT_2$, overlap in a way such that one of the faces of the regular tetrahedron corresponding to $RT_1$ entirely coincides with one of the faces of the regular tetrahedron corresponding to $RT_2$. The volume of the intersection of two adjacent Reuleaux tetrahedra forms a three-dimensional *lens*. Each Reuleaux tetrahedron is adjacent to at most four other Reuleaux tetrahedra. In the first experiment, we fix the radius $r$ of the sensing range to 30m and vary $k$. In the second experiment, we fix $k$ to 4 and vary $r$. All simulations are repeated 100 times the results are averaged.

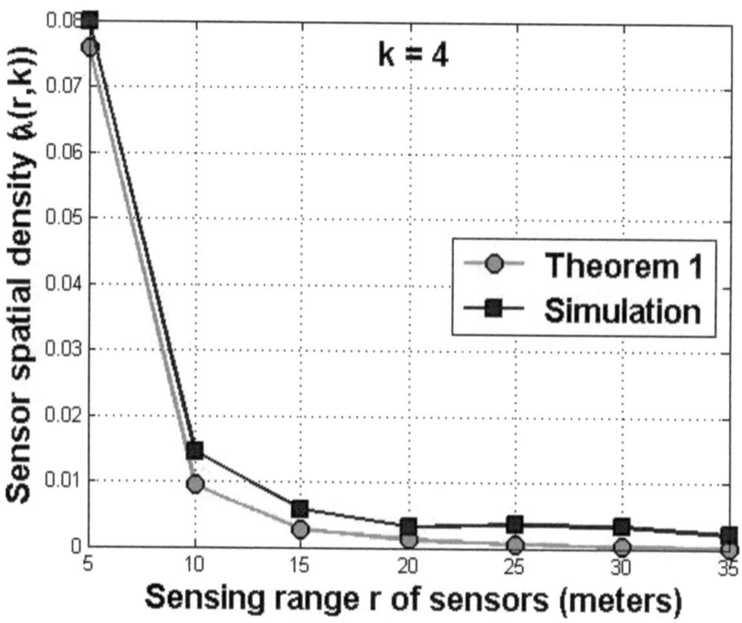


**Fig. A.2** Sensor spatial density $\lambda(r,k)$ vs. $r$

The plot in Fig. A.1 shows the analytical result of Theorem A.1 and simulations results of the first experiment. This figure demonstrates is a close match between the analytical and simulation results. Note that $\lambda(r,k)$ increases with the coverage degree $k$. Indeed, high coverage degree $k$ requires more sensors to be deployed, and hence a denser wireless sensor network is necessary. The plot in Fig. A.2 shows the analytical result of Theorem A.1 and simulations results of the second experiment with a close match. Notice that $\lambda(r,k)$ decreases when $r$ increases. When the sensing range gets larger, a fewer number of sensors is needed to fully $k$-cover a three-dimensional field. It is worth noting that the difference between analytical and simulation results is due to the *boundary effects*. It is impossible to decompose a three-dimensional field into *complete* Reuleaux tetrahedra such that the Reuleaux tetrahedra close to the border of a three-dimensional field lie entirely inside it. Hence, more than the required number of sensors is used to $k$-cover the border of the three-dimensional field.

Next, based on the minimum sensor spatial density (result of Theorem A.1) and another criterion to be specified in the following section, we compute the network connectivity of homogeneous and heterogeneous three-dimensional $k$-covered wireless sensor networks.

## 3 Unconditional Connectivity

In this section, we compute measures of *unconditional* (or traditional) connectivity for homogeneous and heterogeneous three-dimensional $k$-covered wireless sensor networks, where any subset of sensors can fail.

### A Homogeneous Sensors

Theorem A.2 deals with homogeneous wireless sensor networks.

*Theorem A.2:* Let $G$ be a communication graph of a homogeneous three-dimensional $k$-covered wireless sensor network deployed in a cubic field of volume $V$. The connectivity of $G$ is given by

$$\kappa_1(G) \leq \kappa(G) \leq \kappa_3(G) \tag{2}$$

where

$$\kappa_1(G) = 12.024\,\alpha^3\,k$$

$$\kappa_3(G) = \frac{R\,V^{2/3}\,k}{0.422\,r_0^3}$$

$r_0 = r/1.066$, $\alpha = R/r$, and $k \geq 4$.

*Proof:* The optimum position of the sink in terms of energy efficient data gathering is the centre of the cubic field [148]. Let $\xi_0$ be the location of the sink $s_0$. We consider the following three cases depending on the size of the connected component that includes the sink. Also, given that sensor failure is due to low battery power, we assume that the sink has infinite source of energy, thus excluding the possibility of a faulty sink.

*Case 1: Single-node connected component.* In this case, there are at least two components, one of them being the single-node component containing the sink. Finding the minimum number of nodes to disconnect the network requires that the disconnected network have only two components. To isolate the sink, all its neighbours must fail. Hence, at least the communication sphere of the sink should contain no sensor but the sink.

Let N be a random variable that counts the number of sensor failures to isolate the sink $s_0$. Given that sensors are randomly and uniformly deployed in a volume $V$ with density $\lambda(r,k)$ per unit volume, where $R << V^{1/3}$, the expected number of neighbours of the sink is given by

$$E[\mathrm{N}] = \lambda(r,k) \| B(\xi_0, R) \| \tag{2a}$$

where $\| B(\xi_0, R) \| = 4\pi R^3/3$ is the measure of the volume of the communication sphere $B(\xi_0, R)$ of the sink $s_0$ located at $\xi_0$. Thus, the expected minimum number of sensor failures to isolate $s_0$ is equal to $E[\mathrm{N}]$. Substituting Eq. 1 in Eq. 2a, we find that the network connectivity is given by

$$\kappa_1(G) = E[\mathrm{N}] = 12.024\,\alpha^3\,k \tag{2b}$$

where $\alpha = R/r$. Figures A.3 and A.4 plot the function $\kappa_1(G)$ in Eq. 2b. Clearly, $\kappa_1(G)$ increases with the ratio $\alpha$ and the degree of coverage $k$. More importantly, $\kappa_1(G)$ is much higher than $k$.

*Case 2: Non-trivial connected components.* As in the case of two-dimensional deployment (Chap. 8), two configurations of the disconnected network are of particular interest where the two connected components of the disconnected network

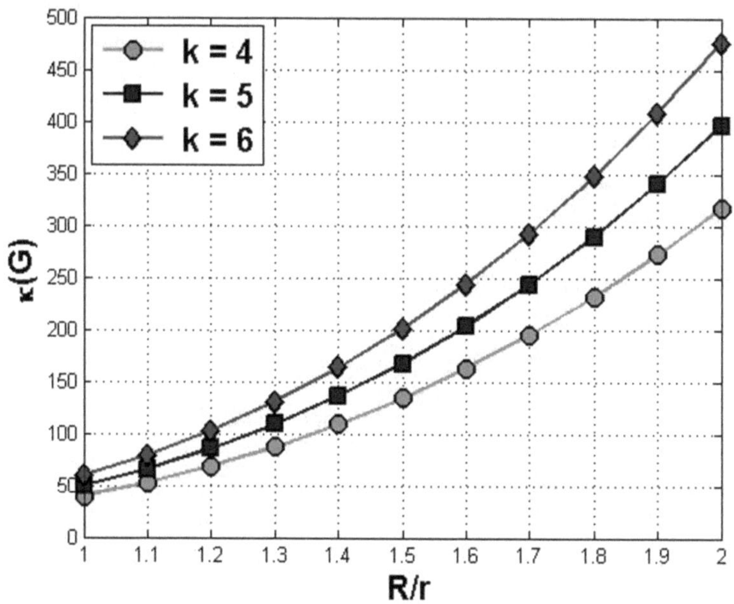


**Fig. A.3** Plot of the function $\kappa_1(G)$ (fix $k$ and vary $\alpha = R/r$)

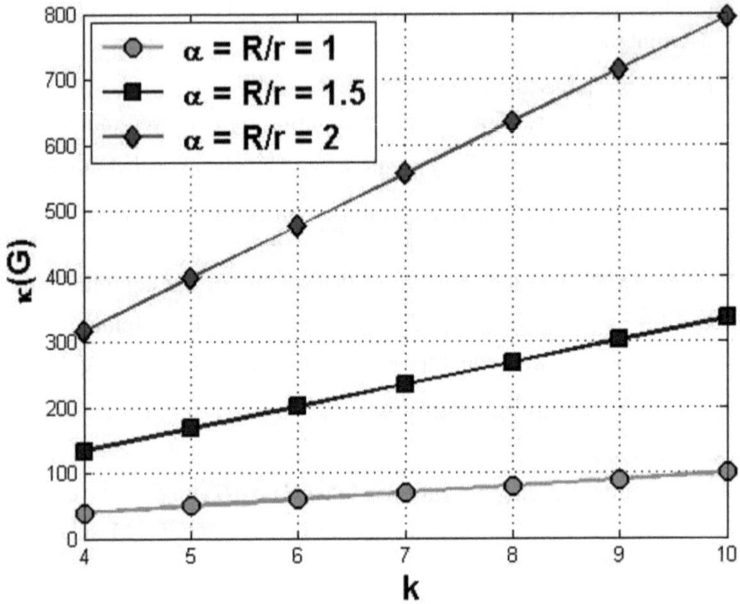


**Fig. A.4** Plot of the function $\kappa_1(G)$ (fix $\alpha = R/r$ and vary $k$)

are separated by a vacant region (or gap). Furthermore, any pair of sensors, one from each component, are separated by a distance at least equal to $R$ to prohibit any communication between the two components. In the first configuration, the component including the sink is reduced to its communication sphere. Thus, the volume of the vacant region, denoted by $gap(\xi_0, R)$, and surrounding the component of the sink, is given by

$$\begin{aligned} \| gap(\xi_0, R) \| &= 4\pi\,(2R)^3/3 - 4\pi\,R^3/3 \\ &= 9.333\,\pi\,R^3 \end{aligned} \tag{2c}$$

Thus, the expected minimum number of sensor failures to isolate the component of the sink is given by

$$E[\mathrm{N}] = \lambda(r,k) \| gap(\xi_0, R) \| \tag{2d}$$

Substituting Eq. 1 in Eq. 2d, we find that the network connectivity is equal to

$$\kappa_2(G) = E[\mathrm{N}] = 84.164\,\alpha^3\,k \tag{2e}$$

where $\alpha = R/r$.

In the second configuration, the original network is split into two components such that the vacant region forms a *cuboid*, denoted by $cub(R)$, and whose sides are $R$, $V^{1/3}$, and $V^{1/3}$. Now, this configuration corresponds to the smallest connected component containing the sink if the field has to be divided into two regions such that none of them surrounds the other. Thus, the expected minimum number of sensor failures to isolate the connected component containing the sink is given by

$$E[\mathrm{N}] = \lambda \| cub(R) \| \tag{2f}$$

where

$$\| cub(R) \| = R\, V^{2/3} \tag{2g}$$

Substituting Eqs. 1 and 2g in Eq. 2f, it follows that network connectivity is equal to

$$\kappa_3(G) = E[\mathrm{N}] = \frac{R\, \mathrm{V}^{2/3}\, k}{0.422\, r_0^3} \tag{2h}$$

It is easy to check that $\kappa_3(G) > \kappa_2(G)$ since $R << V^{1/3}$.

*Case 3: Largest connected component.* This configuration is totally opposite to the one given in Case 1 and has only one isolated sensor. Since we are interested in $k$-coverage of the entire field, such a network is considered as disconnected. The network connectivity is the same as in Case 1. Thus,

$$\kappa_1(G) \le \kappa(G) \le \kappa_3(G)$$

It is easy to check that $\kappa(G) > k$. ■

All these bounds and those to be derived in next sections are based on our fundamental result stated in Theorem A.1 related to the minimum sensor spatial density for full $k$-coverage of a three-dimensional field. Thus, these are lower bounds, and hence tight.

## B Heterogeneous Sensors

Achieving $k$-coverage of a three-dimensional field by heterogeneous sensors would depend on the least powerful ones in terms of their sensing capabilities. Lemmas A.3 and A.4 correspond to Lemma A.1 and Theorem A.1, respectively.

*Lemma A.3:* If the breadth of a three-dimensional convex region $C$ is at most equal to the minimum radius $r_{\min}$ of the sensing spheres of sensors, then $C$ is guaranteed to be $k$-covered if $k \ge 4$ sensors are deployed in it, where $r_{\min} = \min\{r_j / 1.066 : s_j \in S\}$. ■

From Lemma A.4, it follows that connectivity between sensors located in the Reuleaux tetrahedron of side $r_{\min}$ requires that $R_{\min} \ge r_{\min}$.

*Lemma A.4:* Let $r_{min}$ be the minimum radius of the sensing spheres of sensors and $k \geq 4$. The minimum sensor spatial density needed for *k*-coverage of a three-dimensional field by heterogeneous wireless sensor networks is given by

$$\lambda(r_{min}, k) = \frac{k}{0.422\, r_{min}^{\;3}} \tag{3}$$

where $r_{min} = \min\{r_j / 1.066 : s_j \in S\}$. ■

Lemma A.5 computes the connectivity measures for heterogeneous three-dimensional *k*-covered wireless sensor networks.

*Lemma A.5:* Let $G$ be a communication graph of a heterogeneous three-dimensional *k*-covered wireless sensor network with $R_{min} \geq r_{min}$ and $k \geq 4$. The connectivity of the graph $G$ is given by

$$\kappa_4(G) \leq \kappa(G) \leq \kappa_3(G) \tag{4}$$

where

$$\kappa_3(G) = \frac{R_{max}\, V^{2/3}\, k}{0.422\, r_{min}^{\;3}}$$

$$\kappa_4(G) = 12.024\, \alpha_2^{\;3}\, k$$

$\alpha_2 = R_{min} / r_{min}$, $k \geq 4$, $r_{min} = \min\{r_j / 1.066 : s_j \in S\}$, $R_{min} = \min\{R_j : s_j \in S\}$, and $R_{max} = \max\{R_j : s_j \in S\}$. ■

## C Boundary Effects

Network connectivity is defined as the *minimum* number of sensors whose failure (or removal) disconnects the network. Given the cubic geometry of a three-dimensional field we considered, sensors located close to the border of the field are affected by the boundary effects. Indeed, the communication ranges of these sensors cover areas outside the deployment area, and hence have fewer number of communication neighbours compared to all other sensors (especially the ones whose communication ranges lay entirely inside the field). However, the proposed network connectivity measures consider the sink as the most critical node in the network and whose isolation would definitely kill the entire network. Because the optimum position of the sink in terms of energy efficient data gathering is the centre of the cubic field [148], the boundary effects do not exist at all, and thus our derived bounds on the connectivity are correct. Even when our approach for computing network connectivity measures considers all sensors as critical (see Appendix C for detailed discussion), and the boundary effects do not have any impact on the derived bounds on network connectivity. Indeed, we are interested in the *minimum number* of sensor failures to disconnect the network.

As far as sensor deployment to achieve three-dimensional *k*-coverage is concerned, we should mention that the boundary effects have an impact on the performance of the network. By Lemma A.2, a three-dimensional field is guaranteed to be fully $k$-covered if each Reuleaux tetrahedron region in the field contains at least $k$ sensors. However, it is impossible to randomly decompose a three-dimensional field into an integer number of Reuleaux tetrahedron regions because of the boundary of the field. Indeed, most of the Reuleaux tetrahedron regions close to the border of a three-dimensional field do not entirely lay inside the deployment area. Therefore, more than necessary sensors would be needed to achieve $k$-coverage of these Reuleaux tetrahedron regions on the border of the field. Simulation results reported in Sect. 2 show that the sensor spatial density necessary to fully $k$-cover a cubic field is a bit higher than the bound given in Theorem A.1, mainly due to the boundary effects.

Next, we introduce new measures of connectivity of three-dimensional $k$-covered wireless sensor networks by placing a specific constraint on a subset of sensors that would fail.

# 4 Conditional Connectivity

In this section, we compute measures of *conditional* connectivity for homogeneous and heterogeneous three-dimensional $k$-covered wireless sensor networks based on the concepts of *conditional connectivity* [101] and *forbidden faulty set* [81].

As we will see, our results prove that three-dimensional $k$-covered wireless sensor networks can sustain a larger number of sensor failures under the restriction imposed on the faulty sensor set. Similarly to our discussion in Chap. 8, it is easy to show that the traditional connectivity, which does not impose any restriction on the faulty sensor set, is not a useful metric for three-dimensional $k$-covered wireless sensor networks, which are highly dense networks and denser than their two-dimensional counterparts.

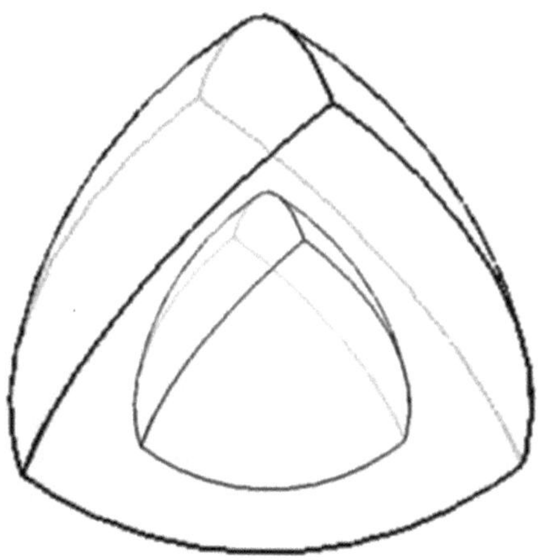

**Fig. A.5** Two nested concentric Reuleaux tetrahedra

## A Homogeneous Sensors

Theorem A.3 computes the conditional connectivity of homogeneous three-dimensional $k$-covered wireless sensor networks.

*Theorem A.3:* The conditional connectivity of a homogeneous three-dimensional $k$-covered wireless sensor network $(k \geq 4)$ is given by

$$\kappa(G:\mathrm{P}) = \frac{((r_0 + 2R)^3 - r_0^3)\,k}{r_0^3} \tag{5}$$

where $r_0 = r/1.066$.

*Proof:* We consider the following two cases based on the type of connected component that contains the sink.

*Case 1: Smallest connected component.* According to our conditional connectivity model, no sensor can be isolated and hence no trivial component can be part of the disconnected network. Under the assumption of forbidden faulty sensor set, the smallest connected component that is disconnected from the rest of the network and contains the sink can be determined as follows. To achieve $k$-coverage of the cubic field, every location must be $k$-covered, including the location $\xi_0$ of the sink $s_0$. Otherwise, the $k$-coverage property will not be satisfied. Therefore, the smallest connected component that includes the sink consists of $k$ sensors deployed in the Reuleaux tetrahedron of side $r_0$ and centered at $\xi_0$. To disconnect the sink under the forbidden faulty sensor set constraint, the Reuleaux tetrahedron should be surrounded by an empty annulus of width equal to $R$, i.e., sensors located in the annulus have failed (Fig. A.5). The Reuleaux tetrahedron together with this annulus forms a larger Reuleaux tetrahedron of side $r_0 + 2R$. The volume of the annulus, denoted by $A(\xi_0, R)$, is equal to

$$\| A(\xi_0, R) \| = 0.422\,(r_0 + 2R)^3 - 0.422\,r_0^3$$

Therefore, the expected conditional minimum number of sensor failures to disconnect the smallest component including the sink can be computed as

$$E[\mathrm{N}:\mathrm{P}] = \lambda(r,k) \| A(\xi_0, R) \| \tag{5a}$$

Substituting Eq. 1 in 3a, we find that the conditional connectivity is given by

$$\kappa_1(G:\mathrm{P}) = E[\mathrm{N}:\mathrm{P}] = \frac{((r_0 + 2R)^3 - r_0^3)\,k}{r_0^3} \tag{5b}$$

It is easy to check that the forbidden faulty set constraint is satisfied for both the faulty sensors (located inside the annulus and which have failed) and non-faulty sensors (located outside the annulus). Indeed, any sensor in the inner Reuleaux tetrahedron still has non-faulty neighbours in the inner Reuleaux tetrahedron. Besides, any sensor outside the outer Reuleaux tetrahedron still has non-faulty neighbours outside the outer Reuleaux tetrahedron. Also, any faulty sensor within the annulus $A(R)$ has non-faulty neighbours located in the inner Reuleaux tetrahedron and outside the outer Reuleaux tetrahedron.

*Case 2: Largest connected component.* This case is similar to the previous one. However, the sink belongs to the largest connected component. Hence, the disconnected network consists of two components: the one including the sink and a second component associated with sensors located in a Reuleaux tetrahedron of side *r*. Using the same analysis as in Case 1, we obtain the same conditional network connectivity:

$$\kappa_2(G:P) = \kappa_1(G:P) \tag{5c}$$

From both cases 1 and 2, the conditional connectivity of homogeneous three-dimensional *k*-covered wireless sensor networks is computed as
$\kappa(G:P) = \kappa_1(G:P)$ ■

## B Heterogeneous Sensors

We observe that computing the conditional connectivity of heterogeneous three-dimensional *k*-covered wireless sensor networks is not a straightforward generalization of the approach used previously for homogeneous three-dimensional *k*-covered wireless sensor networks. If we use the same process as previously, we either violate the forbidden faulty sensor set constraint or maintain network connectivity. Precisely, if the width of the annulus containing the faulty sensors is $R_{max}$, then sensors whose communication radii are less than or equal to $R_{\max}/2$, may be located in the annulus. Thus, the entire neighbour set of this type of sensors would fail at the same time and hence the property P would be violated. Now, if the width of the annulus containing the faulty sensors is less than $R_{\max}$, then the non-faulty sensors of one connected component will be able to communicate with the non-faulty sensors of the other connected component of the disconnected network. Hence, the network is still connected. In this case, it is impossible to find an exact value of conditional connectivity for heterogeneous three-dimensional *k*-covered wireless sensor networks. In the following, we compute their lower and upper bounds based on the types of sensors in and around the annulus.

*Lemma A.6:* The conditional connectivity of the heterogeneous three-dimensional *k*-covered wireless sensor networks is given by

$$\kappa_1(G:F_P) \leq \kappa(G:\mathrm{P}) \leq \kappa_2(G:F_P) \tag{6}$$

where

$$\kappa_1(G:\mathrm{P}) = \frac{((r_{\min}^0 + 2\,R_{\min})^3 - {r_{\min}^0}^3)\,k}{{r_{\min}^0}^3}$$

$$\kappa_2(G:\mathrm{P}) = \frac{((r_{miax} + 2\,R_{\max})^3 - r_{\max}{}^3)\,k}{r_{\min}{}^3}$$

$k \geq 4$, $r_{\min} = \min\{r_j / 1.066 : s_j \in S\}$, $r_{\max} = \max\{r_j / 1.066 : s_j \in S\}$, $R_{\min} = \min\{R_j : s_j \in S\}$, and $R_{\max} = \max\{R_j : s_j \in S\}$.

*Proof:* As above, we consider the following two cases depending on the size of the connected component that includes the sink.

*Case 1: Smallest connected component.* To compute a lower bound on the conditional connectivity of heterogeneous three-dimensional *k*-covered wireless sensor networks, we consider the Reuleaux tetrahedron centered at location $\xi_0$ of the sink $s_0$, which will be disconnected from the network. First, we assume that the annulus containing the faulty sensors as well as the volume surrounding it contains only least powerful sensors, and hence the width of this annulus is equal to $R_{\min}$. Also, to guarantee that the sink will not be isolated, which would violate the forbidden faulty sensor set constraint, the Reuleaux tetrahedron centred at $\xi_0$ should have a side equal to $r_{\min}$. These two conditions help disconnect the network while satisfying the forbidden faulty sensor set constraint. The volume of the annulus $A(\xi_0, R_{\min})$ is given by

$$\| A(\xi_0, R_{\min}) \| = 0.422\,(r_{\min} + 2\,R_{\min})^3 - 0.422\,r_{\min}{}^3 \tag{6a}$$

Hence, the expected conditional minimum number of sensor failures to disconnect the connected component including the sink (or the inner Reuleaux tetrahedron) from the rest of the network is computed as

$$\mathrm{E}[\mathrm{N}:\mathrm{P}] = \lambda(r_{\min}, k) \| A(\xi_0, R_{\min}) \| \tag{6b}$$

where $\lambda(r_{\min}, k) = \dfrac{k}{0.422\,r_{\min}{}^3}$ and $r_{\min} = \min\{r_j / 1.066 : s_j \in S\}$. Thus, the conditional network connectivity is given by

$$\kappa_1(G:\mathrm{P}) = \mathrm{E}[\mathrm{N}:\mathrm{P}] = \frac{((r_{\min} + 2\,R_{\min})^3 - r_{\min}{}^3)\,k}{r_{\min}{}^3} \tag{6c}$$

where $k \geq 4$ and $R_{\min} = \min\{R_j : s_j \in S\}$.

To compute an upper bound on the conditional connectivity of heterogeneous three-dimensional *k*-covered wireless sensor networks, we assume that sensors inside the annulus are the most powerful ones. Thus, the side of the inner Reuleaux tetrahedron should be equal to $r_{\max}$, while the width of the annulus surrounding it should be equal to $R_{\max}$. It is easy to check that this set-up will disconnect the network while satisfying the forbidden faulty set constraint. The upper bound on the conditional connectivity is given by

$$\kappa_2(G:\mathrm{P}) = \mathrm{E}[\mathrm{N}:\mathrm{P}] = \lambda(r_{\min},k)\| A(\xi_0, R_{\max}) \| = \frac{((r_{\max} + 2\,R_{\max})^3 - r_{\max}{}^3)\,k}{r_{\min}{}^3} \tag{6d}$$

where $\lambda(r_{\min},k) = \frac{k}{0.422\, r_{\min}{}^3}$, $k \geq 4$, $r_{\min} = \min\{r_j/1.066 : s_j \in S\}$, $r_{\max} = \max\{r_j/1.066 : s_j \in S\}$, and $R_{\max} = \max\{R_j : s_j \in S\}$.

*Case 2: Largest connected component.* In this case, the sink belongs to the largest connected component of the disconnected network. Hence, the previous analysis applies to any sensor in the network instead of the sink.

Thus, the conditional connectivity of heterogeneous three-dimensional *k*-covered wireless sensor networks satisfies

$$\kappa_1(G,P) \leq \kappa(G,P) \leq \kappa_2(G,P)$$ ■

Next, we relax the assumptions used in our previous analysis to enhance the practicality of our results.

# 5 Discussion

The analysis of the minimum sensor spatial density necessary for *k*-coverage of a three-dimensional field and network connectivity of three-dimensional *k*-covered wireless sensor networks is based on the *unit sphere model*, where the sensing and communication ranges of sensors are modelled by spheres. In other words, sensors are supposed to be typically isotropic. Although this assumption is the basis for most of the protocols for coverage and connectivity in wireless sensor networks, it may not hold universally and thus may not be valid in practice. In Appendix B, we show how to relax this assumption to promote the applicability of our results to real-world three-dimensional wireless sensor network scenarios, and summarize our results for the convex model, where the sensing and communications ranges of sensors are convex but not necessarily spherical. Moreover, we assumed that our results for network connectivity hold for a degree of sensing coverage $k$, where $k \geq 4$.

In this section, we show how to relax the latter assumption. Then, we discuss a sensor placement strategy for full *k*-coverage of a three-dimensional field.

## A Relaxing the Assumption of $k \geq 4$

The analysis of *k*-coverage and connectivity for three-dimensional *k*-covered wireless sensor networks are valid for all $k \geq 4$. Since the breadth of the Reuleaux tetrahedron is equal to $r$ (or $r_{\min}$), our results can also be used for $k \leq 3$. In other words, *k*-coverage of the cube can be met by deploying $k$ sensors in the Reuleaux tetrahedron, where $k \leq 3$. However, the network would be denser than necessary (especially for $k = 1$) and the coverage degree would be higher than that dictated by the application.

## B Sensor Placement Strategy

Notice that under the assumption of spherical model, it is impossible to achieve a degree of coverage exactly equal to $k$ in all the locations of the cube. Therefore, a sensor placement strategy to achieve $k$-coverage should be devised in such a way that every location in the cube is $k'$-covered, where $k'$ is very close to $k$. This placement strategy should benefit from the geometry of the Reuleaux tetrahedron. The sensor placement problem can be transformed into a problem of covering a cube with overlapping sets of congruent Reuleaux tetrahedra. An optimal covering consists to use a minimum number of Reuleaux tetrahedra by minimizing the overlap volume between them. More precisely, two adjacent Reuleaux tetrahedra overlap such that the faces of their corresponding regular tetrahedra are entirely coinciding with each other. Thus, the curved edges of the Reuleaux tetrahedra should overlap so the same subset of sensors deployed on these curved edges could participate in covering the space associated with both Reuleaux tetrahedra at the same time, thus minimizing the total number of sensors required to $k$-cover the cube. As in the case of two-dimensional deployment, the sensors located in a three-dimensional lens, which corresponds to the overlap volume of two adjacent Reuleaux tetrahedra of side length $r_0$, are at distance $r_0$ from any point in their volumes, and hence participate in $k$-covering both tetrahedra. The design of duty-cycling protocols to $k$-cover a three-dimensional field with a minimum number of sensors should select sensors based on this observation.

## C Sink-Independent Connectivity Measures

Although in centralized algorithms the concept of sink is well defined, it is likely that distributed algorithms, such as the consensus based algorithm, will be implemented for WSNs. In this case, the concept of (fixed) sink would lose value. It would be interesting to revise the definition of connectivity to take this concept into account. We suggest that all the nodes be considered as peer-to-peer. Thus, we define connectivity with respect to all sensors in the network. Given the geometry of the deployment field that we consider (cube), the *boundary sensors*, i.e., sensors located at the boundary of the cube, have small neighbour sets. In particular, the sensors located at the eight corners of the cube $s_{b1},\ldots, s_{b8}$ as shown in Fig. A.6 are highly likely to have the smallest neighbour set. Our measures of connectivity will be based on one of these boundary sensors to find the minimum number of sensors of its neighbour set that should fail in order to disconnect the network. It is easy to check that compared to the sink, the size of the neighbour set of a boundary sensor is equal to a *quarter* of that of the neighbour set of the sink. Thus, the connectivity measures computed in the previous sections with respect to the sink remain the same for a boundary sensor but are weighted by a coefficient equal to ¼. For more details, the interested reader is referred to Sect. 7.

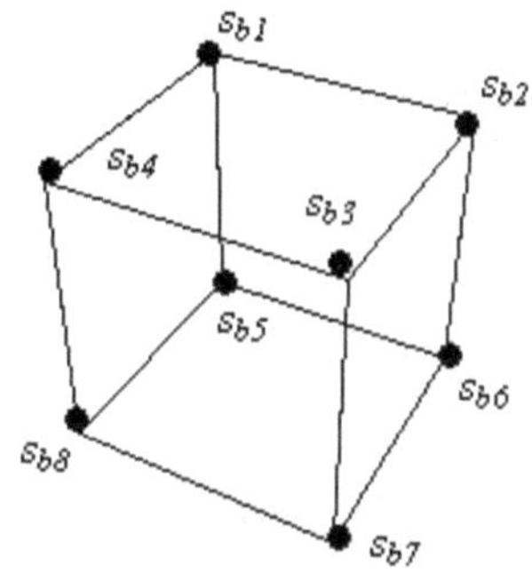


**Fig. A.6** Eight boundary sensors located on the corners of a cube

# D Stochastic Sensing and Communication Models

In this section, we exploit the results of Sect. 2 to characterize probabilistic $k$-coverage in 3D WSNs based on our stochastic sensing model. Theorem A.4 computes the *minimum k-coverage probability* $p_{k,min}$ such that every point in a field is probabilistically $k$-covered.

*Theorem A.4:* Let $r$ be the radius of the nominal sensing range of the sensors, $r_0 = r/1.066$, and $k \geq 4$. The minimum $k$-coverage probability so that each point in a 3D field is probabilistically $k$-covered by at least $k$ sensors under our stochastic sensing model defined in (5) is computed as

$$p_{k,\min} = 1 - \left(1 - e^{-\beta r_0^{\alpha}}\right)^k \quad (7)$$

*Proof:* First, we identify the least $k$-covered point in a 3D field so we can compute $p_{k,min}$. By Lemma A.2, $k$ sensors should be deployed in a Reuleaux tetrahedron region of side $r_0 = r/1.066$ in the field to achieve $k$-coverage of a 3D field with a minimum number of sensors. It is easy to check that the least $k$-covered point $\xi_{lc}$ is the one that corresponds to the configuration where all deployed $k$ sensors are located at a distance $r_0 = r/1.066$ from $\xi_{lc}$. In other words, $\xi_{lc}$ is the farthest point from all those $k$ sensors. Hence, the distance between $\xi_{lc}$ and each of these $k$ sensors is equal to $r_0 = r/1.066$. Thus, the *minimum k-coverage probability* for the least $k$-covered point $\xi_{lc}$ by $k$ sensors under the stochastic sensing model introduced in Chap. 2 is given by

$$p_{k,\min} = 1 - \prod_{i=1}^{k} (1 - p(\xi, s_i)) = 1 - \left(1 - e^{-\beta r^{\alpha}}\right)^k \quad \blacksquare$$

The stochastic $k$-coverage problem is to select a minimum subset $S_{min} \subseteq S$ of sensors such that each point in a 3D field is $k$-covered by at least $k$ sensors and that the minimum $k$-coverage probability of each point is at least equal to some given threshold probability $p_{th}$, where $0 < p_{th} < 1$. This helps us compute the stochastic sensing range $r_s$, which provides probabilistic $k$-coverage of a field with a probability no less than $p_{th}$. Lemma A.7 computes the value of $r_s$.

*Lemma A.7:* Let $k \geq 3$ and $2 \leq \alpha \leq 4$. The stochastic sensing range $r_s$ of the sensors that is necessary to probabilistically $k$-cover a 3D field with a minimum number of sensors and with a probability no lower than $0 < p_{th} < 1$ is given by

$$r_s = \left( -\frac{1}{\beta} \log (1 - (1 - p_{th})^{1/k}) \right)^{1/\alpha} \tag{8}$$

where $\beta$ represents the physical characteristic of the sensors' sensing units.

*Proof:* From Eq. 7, we deduce that $p_{k,\min} \geq p_{th} \Rightarrow r_s \leq \left( -\frac{1}{\beta} \log (1 - (1 - p_{th})^{1/k}) \right)^{1/\alpha}$.

Since we are interested in the *minimum* number of sensors to probabilistically $k$-cover a 3D field, we should consider the maximum value of $r_s$, i.e., the maximum stochastic sensing range of the sensors. This will allow the sensors to probabilistically $k$-cover as much space of the 3D deployment field as possible. Thus, $r_s = \left( -\frac{1}{\beta} \log (1 - (1 - p_{th})^{1/k}) \right)^{1/\alpha}$. ■

The upper bound on the stochastic sensing range $r_s$ of the sensors computed in Eq. 8 will be used to compute our measures of connectivity and fault-tolerance of 3D $k$-covered WSNs under the assumption of more realistic stochastic sensing and communication models. Figures A.7 and A.8 show $r_s$ for different values of $p_{th}$ and $k$ while considering the free-space model ($\alpha$=2) (Fig. A.7) and

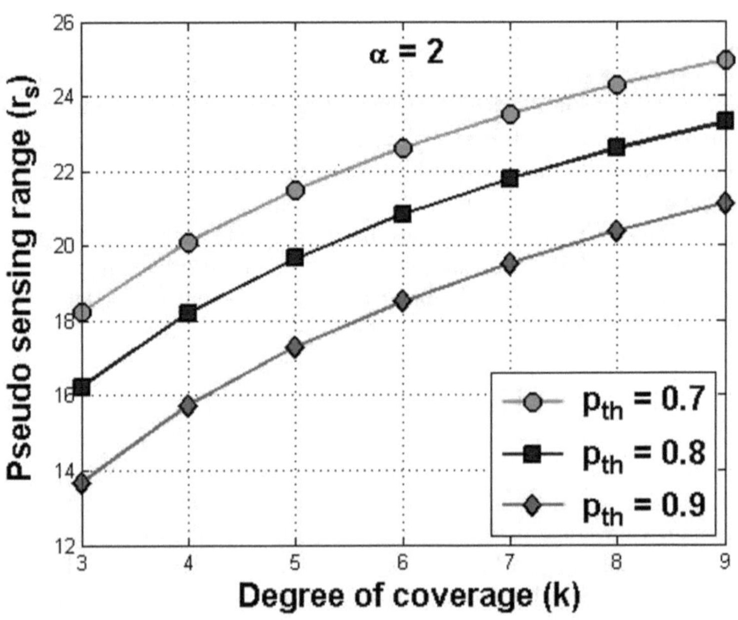


**Fig. A.7** Upper bound of $r_s$ vs. $k$ for $\alpha = 2$

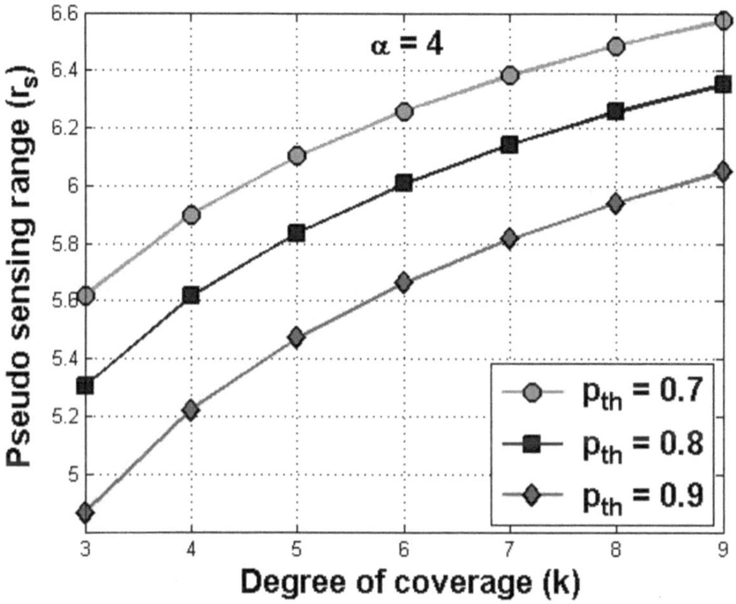


**Fig. A.8** Upper bound of $r_s$ vs. $k$ for α = 4

the multi-path model ($\alpha$=4) (Fig. A.8). Note that $r_s$ decreases as a function of $p_{th}$, $k$, and $\alpha$. This is due to the fact that the minimum probability $p_{k,min}$ of $k$-coverage of the same location by multiple sensors decreases as $p_{th}$, $k$, and $\alpha$ increase.

Lemma A.8 states a sufficient condition for probabilistic $k$-coverage of a 3D field based on our stochastic sensing model given in Chap. 2, the threshold probability $p_{th}$, and the degree $k$ of coverage.

*Lemma A.8:* Let $k \geq 4$. A 3D field is probabilistically $k$-covered with a probability no lower than $0 < p_{th} < 1$ if any Reuleaux tetrahedron of maximum breadth $r_s$ /1.066 in the field contains at least $k$ sensors. ■

Lemma A.9 states a sufficient condition for connectivity between sensors under our stochastic sensing model.

*Lemma A.9:* Let $k \geq 4$. The sensors that are selected to $k$-cover a 3D field with a probability no less than $0 \leq p_{th} \leq 1$ under the stochastic sensing model defined in Eq. 5 are connected if the radius of their stochastic communication range $R_s$ is at least equal to their stochastic sensing range $r_s$., $R_s \geq r_s$. ■

Given that network connectivity is defined as the *minimum* number of sensors of a neighbour set which need to fail to disconnect the network, we should consider the minimum stochastic communication range of the sensors, i.e., $R_s = r_s$, to minimize the size of the neighbour set of each sensor. Therefore, to find our

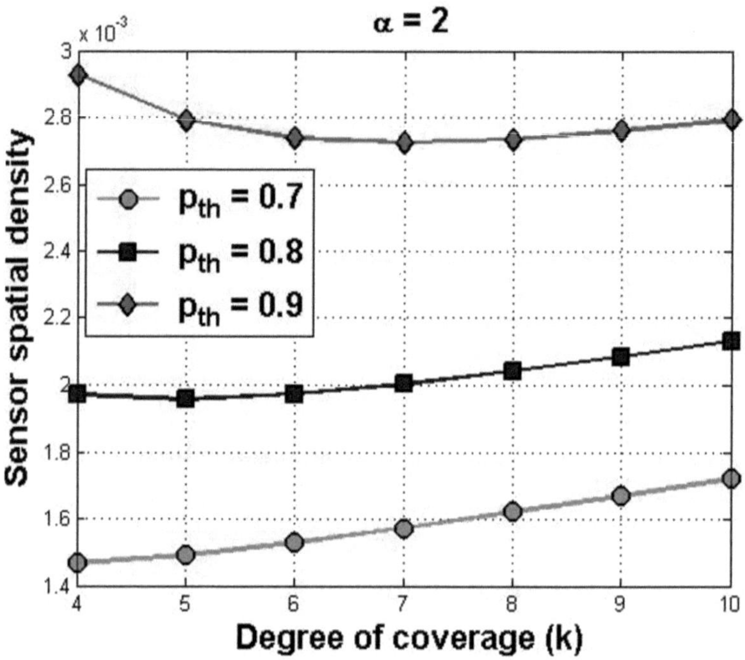


**Fig. A.9** Sensor spatial density vs. *k*

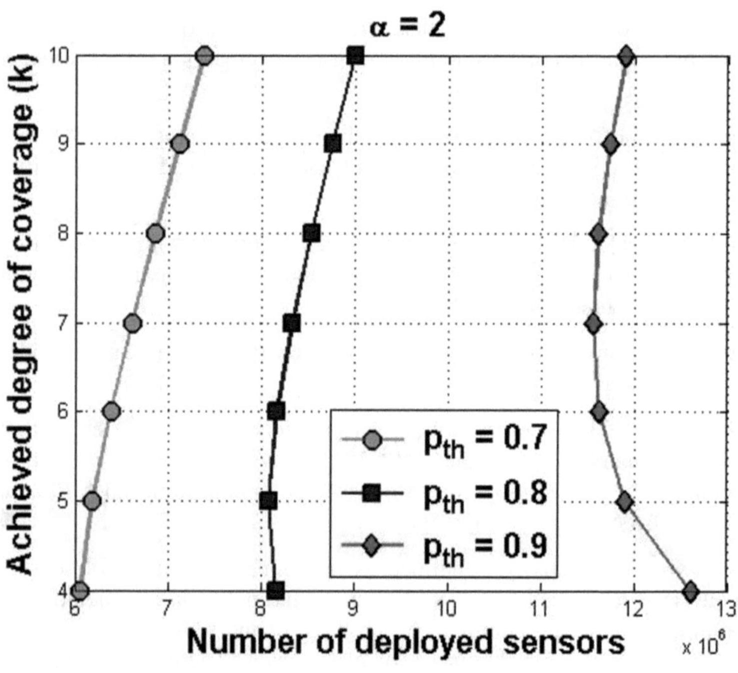


**Fig. A.10** *k* vs. number of deployed sensors

unconditional and conditional measures of network connectivity and fault-tolerance for 3D $k$-covered WSNs using more realistic scenarios, including stochastic models for sensing and communications, we need to replace $r$ by $r_s$ and $R$ by $r_s$ in Sects. 2, 3, 4, 5, and 7.

Now, we present the simulation results using a high-level simulator written in the C language. We consider a cubic field of side length 1000 m. All simulations are repeated 200 times and the results are averaged.

Figure A.9 plots the sensor spatial density as a function of the degree of coverage $k$ for different values of the threshold probability $p_{th}$ and for a path-loss exponent $\alpha = 2$. As expected, the density increases with $p_{th}$. Indeed, as we increase $p_{th}$, more sensors would be needed to achieve the same degree of coverage $k$. Recall that the breadth of the Reuleaux tetrahedron that is guaranteed to be covered with exactly $k$ sensors decreases as $p_{th}$ and $\alpha$ increase. Precisely, this breadth is equal to $r_s$ /1.066. However, for $p_{th}$=0.8, the sensor density tends to decrease when $k$ goes from 4 to 5 and increases afterwards. This behaviour is clearly noticeable for $p_{th}$=0.9. This is mainly due to the stochastic nature of the sensing range of the sensors, which depends on the logarithm of $p_{th}$, the threshold probability, $k$, and $\alpha$.

Figure A.10 plots the achieved degree of coverage $k$ vs. the total number of deployed sensors. Moreover, we vary both $p_{th}$ and we fix $\alpha$=2. Definitely, higher number of deployed sensors would yield higher coverage degree. Here also, any increase in $p_{th}$ would require a larger number of deployed sensors to provide the same degree of coverage. Notice that the same observation holds for $p_{th}$=0.8 and $p_{th}$=0.8 as in the previous experiment.

## E Three-Dimensional Sensing Applications

In the case of WSNs deployed on the trees of different heights in a forest, the sensors could be seen almost everywhere in the space. Pompili et al. [169] proposed different deployment strategies for 2D and 3D communication architectures for underwater acoustic sensor networks, where the sensors are anchored at the bottom of the ocean for the 2D design and float at different depths of the ocean to cover the entire 3D region. Indeed, oceanographic data collection, pollution monitoring, offshore exploration, disaster prevention, and assisted navigation are all typical applications of underwater sensor networks [4, 5]. For WSNs deployed in buildings with multiple floors, sensors are placed on the ground and/or the wall, but the networks seldom contain sensors floating in the middle of the room. The first examples show that our proposed 3D model is valid and can be applied to choose the sensor density in practical problems. The last example, however, shows the limited validity of our model due to the restriction imposed on the placement of sensors inside buildings or rooms.

# 6 Relaxing the Unit Sphere Model: Convex Model

The assumption of spherical sensing and communication ranges of the sensors may not hold in real-world wireless sensor network platforms. It has been observed in [209] that the communication range of MICA motes [230] is asymmetric and depends on the environments. It is also found in [224] that the communication range of radios is highly probabilistic and irregular.

In this appendix, for problem tractability, we consider a *convex model*, where the sensing and communication ranges of sensors are *convex* but not necessarily spherical.

First, we define the notion of *largest enclosed sphere* of a three-dimensional convex region $C$ as a sphere that lies entirely inside $C$ and whose diameter is equal to the minimum distance between any pair of points on the boundary of the region $C$.

# A Homogeneous Sensors

We consider homogeneous sensors that have the same convex sensing ranges and communication ranges. Lemmas A.10 and A.11 correspond to Lemma A.1 and Theorem A.1, respectively. Their proof is similar to that in Sect. 2 by using the largest enclosed sphere instead of the sensing sphere.

*Lemma A.10:* If $k \geq 4$ homogeneous sensors are deployed in a three-dimensional convex region $C$, then the region $C$ is $k$-covered if its breadth does not exceed $r_{led}$, the radius of the largest enclosed sphere of the sensing range. ■

*Lemma A.11:* The minimum sensor spatial density required to guarantee $k$-coverage of a three-dimensional field is given by

$$\lambda(r_{\text{led}}, k) = \frac{k}{0.422\, {r_{led}^{0}}^{3}} \tag{9}$$

where $r_{led}$ stands for the radius of the largest enclosed sphere of the sensing range, $r_{led}^{0} = r_{led}/1.066$, and $k \geq 4$. ■

Now, we discuss how those results can be derived using a convex model, where the sensing and communication ranges of the sensors may not necessarily be spherical.

# B Heterogeneous Sensors

Lemmas A.12 and A.13 correspond to Lemma A.1 and Theorem A.1, respectively. Their proof is also the same as that in Sect. 2 by using the notion of largest enclosed sphere instead of sensing sphere.

*Lemma A.12:* A three-dimensional convex region $C$ is guaranteed to be $k$-covered when exactly $k$ heterogeneous sensors are deployed in it, if the breadth of $C$ does not exceed $r_{led}^{\min}$, where $r_{led}^{\min} = \min\{r_{led}/1.066 : s_j \in S\}$ and $k \geq 4$. ■

*Lemma A.13:* The minimum sensor spatial density required to $k$-cover a three-dimensional field is given by

$$\lambda(r_{led}^{\min}, k) = \frac{k}{0.422\,(r_{led}^{\min})^2} \tag{10}$$

where $r_{led}^{\min} = \min\{r_{led}/1.066 : s_j \in S\}$ and $k \geq 4$. ∎

The measures of network connectivity can be derived using the same approach as in the previous sections. Thus, the assumption of unit sphere model for sensing and communication ranges of sensors can be relaxed with the aid of largest enclosed sphere of their sensing range.

# 7 Underwater Sensor Networks

The results in the previous sections are only applicable to the connectivity of sink node. Although the connectivity of sink is critical, in some scenarios, such as underwater wireless sensor networks [4, 5], any sensor may be critical due to the high cost, for instance.

In the following, we extend our network connectivity measures for three-dimensional $k$-covered wireless sensor networks to the case where any sensor in the network is critical. Specifically, we consider a *boundary sensor*, i.e., a sensor located at one corner of the cubic field. Such a boundary sensor has the minimum number of communication neighbours given that all sensors are located within the deployment region—the cube, and hence the actual communication range of a boundary sensor is only a quarter of its communication sphere. In [205], a boundary sensor is considered to compute the connectivity of 2D $k$-covered wireless sensor networks.

Theorem A.5 summarizes the connectivity measures with respect to a boundary sensor for homogeneous three-dimensional $k$-covered wireless sensor networks. The case of heterogeneous wireless sensor networks and the case of sensors with convex sensing and communication ranges can be treated similar to the previous sections. Thus, we omit the proof of Theorem A.5.

*Theorem A.5:* Let $G$ be a communication graph of a homogeneous three-dimensional $k$-covered wireless sensor network deployed in a cubic field, where the radii of the sensing and communication spheres of sensors are $r$ and $R$, respectively. The connectivity of $G$ is computed as

$$\kappa(G) = 3.02\,\alpha^3\,k \tag{11}$$

whereas the conditional connectivity of $G$ is given by

$$\kappa(G:\mathrm{P}) = \frac{3.02\,\left((r_0 + R)^3 - r_0^{\,3}\right)k}{r_0^{\,3}} \tag{12}$$

where $r_0 = r/1.066$, $\alpha = R/r$, and $k \geq 4$. ∎

## 8 Summary

In this appendix, we investigated coverage and connectivity in three-dimensional $k$-covered wireless sensor networks with emerging applications, such as underwater acoustic sensor networks that require three-dimensional design. We proposed the Reuleaux tetrahedron model to guarantee $k$-coverage of a three-dimensional field. Based on the geometric properties of Reuleaux tetrahedron, we derived the sensor spatial density for guaranteeing $k$-coverage of a three-dimensional space. We also computed the connectivity of homogeneous and heterogeneous three-dimensional $k$-covered wireless sensor networks. Our results on connectivity take into consideration an inherent characteristic of wireless sensor networks in that the sink has a critical role in terms of data processing and decision making, compared to the rest of the network. Therefore, we computed the connectivity of three-dimensional $k$-covered wireless sensor networks based on the size of the connected component that includes the sink. We conclude that the connectivity of three-dimensional $k$-covered wireless sensor networks is much higher than the degree of sensing coverage $k$ provided by the network. The traditional connectivity metric, however, is defined in an abstract way and does not consider the inherent properties of wireless sensor networks because it assumes that any subset of nodes can fail at the same time. This assumption is not valid for heterogeneous three-dimensional $k$-covered wireless sensor networks. To compensate for these shortcomings, we used proposed more realistic measures of connectivity based on the concept of forbidden faulty set. We found that three-dimensional $k$-covered wireless sensor networks can sustain a large number of sensor failures provided that the neighbour set of a sensor cannot fail at the same time.

We believe that our results have practical significance for sensor network designers to develop three-dimensional applications with prescribed degrees of coverage and connectivity. These connectivity measures can be exploited in the design of fault-tolerant topology control protocols for three-dimensional $k$-covered wireless sensor networks. Our future work will focus on the design of efficient sensor deployment strategies for three-dimensional $k$-covered wireless sensor networks. We are also interested in the design of data routing protocols on duty-cycled three-dimensional $k$-covered wireless sensor networks, which pose major challenges due to the time-varying connectivity of the network as sensors are turned *on* or *off* to save energy and extend the network lifetime.

# References


[1] Abrams, Z., Goel, A., Plotkin, S.: Set $k$-cover algorithms for energy efficient monitoring in wireless sensor networks. In: Proc. IPSN, Berkeley, California, USA, pp. 424–432 (April 2004)

[2] Adlakha, S., Srivastava, M.: Critical density threshold for coverage in wireless sensor networks. In: Proc. IEEE WCNC, pp. 1615–1620 (2003)

[3] Ai, J., Abouzeid, A.: Coverage by directional sensors in randomly deployed wireless sensor networks. Journal of Combinatorial Optimization 11(1), 21–41 (2006)

[4] Akyildiz, I.F., Pompili, D., Melodia, T.: Underwater acoustic sensor networks: research challenges. Ad Hoc Networks 3, 257–279 (2005)

[5] Akyildiz, I., Pompili, D., Melodia, T.: Challenges for efficient communications in underwater acoustic sensor networks. ACM SIGBED Review 1(2), 3–8 (2004)

[6] Akyildiz, I.F., Su, W., Sankarasubramaniam, Y., Cayirci, E.: Wireless sensor networks: A survey. Computer Networks 38(4), 393–422 (2002)

[7] Akyildiz, I.F., Su, W., Sankarasubramaniam, Y., Cayirci, E.: A Survey on Sensor Networks. IEEE Communications Magazine 40(8), 102–114 (2002)

[8] Alam, S.M., Haas, Z.J.: Coverage and connectivity in three-dimensional networks. In: Proc. ACM MobiCom, pp. 346–357 (2006)

[9] Ammari, H.M.: A survey of current architectures for connecting wireless mobile ad hoc networks to the Internet. International Journal of Communication Systems 20(8), 943–968 (2007)

[10] Ammari, H.M.: Lessons learned from the simulation experience of a three-tier multihop wireless Internet architecture. Information Sciences 177(8), 1806–1833 (2007)

[11] Ammari, H.M.: Using group mobility and multihomed mobile gateways to connect mobile ad hoc networks to the global IP Internet. International Journal of Communication Systems 19(10), 1137–1165 (2006)

[12] Ammari, H.M., Das, S.K.: Coverage and connectivity in three-dimensional wireless sensor networks using percolation theory. IEEE Transactions on Parallel and Distributed Systems 20(6), 872–885 (2009)

[13] Ammari, H.M., Das, S.K.: Fault tolerance measures for large wireless sensor networks. ACM Transactions on Autonomous and Adaptive Systems 4(1), 2:1–2:28 (2009)

[14] Ammari, H.M., Das, S.K.: A study of $k$-coverage and measures of connectivity in three-dimensional wireless sensor networks. IEEE Transactions on Computers (accepted for publication) (to appear, 2010)

[15] Ammari, H.M., Das, S.K.: Energy-efficient $k$-coverage configurations in wireless sensor networks. IEEE Transactions on Mobile Computing (under review) (May 2009)
[16] Ammari, H.M., Das, S.K.: Integrated coverage and connectivity in wireless sensor networks: A two-dimensional percolation problem. IEEE Transactions on Computers 57(10), 1423–1434 (2008)
[17] Ammari, H.M., Das, S.K.: Promoting heterogeneity, mobility and energy-aware Voronoi diagram in wireless sensor networks. IEEE Transactions on Parallel and Distributed Systems 19(7), 995–1008 (2008)
[18] Ammari, H.M., Das, S.K.: A trade-off between energy and delay in data dissemination for wireless sensor networks using transmission range slicing. Computer Communications 31(9), 1687–1704 (2008)
[19] Ammari, H.M.: Geographic forwarding in wireless sensor networks: A multiobjective optimization problem for energy-delay trade-off. Journal of Parallel and Distributed Computing (accepted for publication) (to appear, 2010)
[20] Ammari, H.M., Das, S.K.: Forwarding via Checkpoints: Geographic Routing on Always-On Sensors. Journal of Parallel and Distributed Computing (accepted for publication) (to appear, 2010)
[21] Ammari, H.M., Das, S.K.: Joint $k$-Coverage and Hybrid Forwarding in Duty-Cycled Three-Dimensional Wireless Sensor Networks. In: Proc. IEEE SECON (2008)
[22] Ammari, H.M., Das, S.K.: Clustering-based minimum energy m-connected $k$-covered wireless sensor networks. In: Verdone, R. (ed.) EWSN 2008. LNCS, vol. 4913, pp. 1–16. Springer, Heidelberg (2008) (best paper award)
[23] Ammari, H.M., Das, S.K.: Geographic forwarding in duty-cycled connected $k$-covered wireless sensor networks. In: Proc. IEEE PerCom 2008: Google Ph.D. Forum - A Ph.D. Forum on Pervasive Computing and Communications (2008) (best contribution paper)
[24] Ammari, H.M., Das, S.K.: Joint $k$-Coverage, Duty-Cycling, and Geographic Forwarding in Wireless Sensor Networks. In: Proc. IEEE ISCC (2009)
[25] Ammari, H.M.: Stochastic $k$-coverage in wireless sensor networks. In: Proc. WASA (2009) (invited paper)
[26] Ammari, H.M., Das, S.K.: On the Design of k-Covered Wireless Sensor Networks: Self- versus Triggered Sensor Scheduling. In: Proc. IEEE WoWMoM (2009)
[27] Ammari, H.M.: Mission-Oriented $k$-Coverage in Mobile Wireless Sensor Networks. In: Proc. ICDCN (2009)
[28] Ammari, H.M., Giudici, J.: On the connected $k$-coverage problem in heterogeneous sensor nets: The curse of randomness and heterogeneity. In: Proc. IEEE ICDCS (2009)
[29] Ammari, H.M., Das, S.K.: On computing conditional fault-tolerance measures for $k$-covered wireless sensor networks. In: Proc. ACM/IEEE MSWiM, pp. 309–316 (2006)
[30] Ammari, H.M., Das, S.K.: Coverage, connectivity, and fault tolerance measures of wireless sensor networks. In: Datta, A.K., Gradinariu, M. (eds.) SSS 2006. LNCS, vol. 4280, pp. 35–49. Springer, Heidelberg (2006)
[31] Ammari, H.M., Das, S.K.: An energy-efficient data dissemination protocol for wireless sensor networks. In: Proc. IEEE PerSenS, in conjunction with PerCom, pp. 357–361 (2006)

[32] Ammari, H.M., Das, S.K.: Data dissemination to mobile sinks in wireless sensor networks: An information theoretic approach. In: Proc. IEEE MASS, pp. 1–10 (2005)
[33] Ammari, H.M., Das, S.K.: Trade-off between energy savings and source-to-sink delay in data dissemination for wireless sensor networks. In: Proc. ACM/IEEE MSWiM, pp. 126–133 (2005)
[34] Ammari, H.M., Das, S.K.: Data dissemination in sensor networks: Trading off energy savings and source-to-sink delay. In: Poster, ACM Student Research Competition (ACM SRC), ACM MobiCom (2005)
[35] Antoniou, T., Chatzigiannakis, I., Mylonas, G., Nikoletseas, S., Boukerche, A.: A new energy efficient and fault-tolerant protocol for data propagation in smart dust networks using varying transmission range. In: Proc. ANSS, pp. 43–52 (2004)
[36] Aurenhammer, F.: Voronoi diagrams – A survey of a fundamental data structure. ACM Computing Surveys 23(3), 345–405 (1991)
[37] Bai, X., Kumar, S., Xuan, D., Yun, Z., Lai, T.H.: Deploying wireless sensors to achieve both coverage and connectivity. In: Proc. ACM MobiHoc, pp. 131–142 (2006)
[38] Balister, P., Bollobas, B., Sarkar, A., Kumar, S.: Reliable density estimates for achieving coverage and connectivity in thin strips of finite length. In: Proc. ACM MobiCom, pp. 75–86 (2007)
[39] Bandyopadhyay, S., Coyle, E.: Spatio-temporal sampling rates and energy efficiency in wireless sensor networks. In: Proc. IEEE Infocom, pp. 1729–1740 (2004)
[40] Baxter, R.J.: Ornstein-Zernike relation for a disordered fluid. Australian Journal of Physics 21, 563–569 (1968)
[41] Baxter, R.J.: Percus-Yevick equation for hard spheres with surface adhesion. Journal of Chemical Physics 49(6), 2770–2774 (1968)
[42] Bertin, E., Billot, J.-M., Drouilhet, R.: Continuum percolation in the Gabriel graph. Advances in Applied Probability 34(4), 689–701 (2002)
[43] Biswas, S., Morris, R.: ExOR: Opportunistic multi-hop routing for wireless networks. In: Proc. ACM SIGCOMM 2005 Conf. on Applications, Technologies, Architectures, and Protocols for Computer Communication, pp. 133–143 (2005)
[44] Bollobás, B.: The Art of Mathematics: Coffee Time in Memphis. Cambridge University Press, Cambridge (2006)
[45] Booth, L., Bruck, J., Franceschetti, M., Meester, R.: Covering algorithms, continuum percolation and the geometry of wireless networks. The Annals of Applied Probability 13(2), 722–741 (2003)
[46] Bose, P., Morin, P.: Online Routing in Triangulations. SIAM Journal on Computing 33(4), 937–951 (2004)
[47] Bose, P., Morin, P., Stojmenovic, I., Urrutia, J.: Routing with guaranteed delivery in ad hoc wireless networks. Wireless Networks 7(6), 609–616 (2001)
[48] Boukerche, A., Chatzigiannakis, I., Nikoletseas, S.: A new energy efficient and fault-tolerant protocol for data propagation in smart dust networks using varying transmission range. Computer Communications 4(29), 477–489 (2006)
[49] Boukerche, A., Chatzigiannakis, I., Nikoletseas, S.: Power-efficient data propagation protocols for wireless sensor networks. Simulation 81(6), 399–411 (2005)
[50] Boukerche, A., Cheng, X., Linus, J.: Energy-aware data-centric routing in microsensor networks. In: Proc. ACM MSWiM, pp. 42–49 (2003)
[51] Broadbent, S.R., Hammersley, J.M.: Percolation processes, I. Crystals and mazes. Proc. Cambridge Philosophical Society 53, 629–641 (1957)

[52] Bulusu, N., Heidemann, J., Estrin, D.: GPS-less low cost outdoor localization for very small devices. IEEE Personal Communications Magazine 7(5), 28–34 (2000)
[53] Cao, G., Kesidis, G., La Porta, T., Yao, B., Phoha, S.: Purposeful Mobility in Tactical Sensor Networks. In: Sensor Network Operations. Wiley-IEEE Press (2006)
[54] Cao, Q., Yan, T., Stankovic, J., Abdelzaher, T.: Analysis of target detection performance for wireless sensor network. In: Prasanna, V.K., Iyengar, S.S., Spirakis, P.G., Welsh, M. (eds.) DCOSS 2005. LNCS, vol. 3560, pp. 276–292. Springer, Heidelberg (2005)
[55] Cardei, M., Wu, J.: Energy-efficient coverage problems in wireless ad-hoc sensor networks. Computer Communications 29(4), 413–420 (2006)
[56] Casari, P., Maruccci, A., Nati, M., Petrioli, C., Zorzi, M.: A detailed simulation study of geographic random forwarding (GeRaF) in wireless sensor networks. In: Proc. IEEE VTC, pp. 59–68 (2004)
[57] Chang, J.-H., Tassiulas, L.: Energy conserving routing in wireless ad-hoc networks. In: Proc. IEEE INFOCOM, pp. 22–31 (2000)
[58] Chankong, V., Haimes, Y.: Multiobjective decision making: Theory and methodology. North Holland Series in System Science and Engineering (1983)
[59] Chen, B., Jameson, K., Balakrishnan, H., Morris, R.: Span: An energy-efficient coordination algorithm for topology maintenance in ad hoc wireless networks. Wireless Networks 8(5), 481–494 (2002)
[60] Chen, J., Kanj, I., Wang, G.: Hypercube network fault tolerance: A probabilistic approach. In: Proc. Int. Conf. on Parallel Processing (ICPP), pp. 65–72 (August 2002)
[61] Chen, A., Kumar, S., Lai, T.H.: Designing localized algorithms for barrier coverage. In: Proc. ACM MobiCom, Montreal, Quebec, Canada, pp. 75–86 (September 2007)
[62] Choi, W., Das, S.: A novel framework for energy-conserving data gathering in wireless sensor networks. In: Proc. IEEE INFOCOM, pp. 1985–1996 (2005)
[63] Choi, W., Das, S.: Design and performance analysis of a proxy-based indirect routing scheme in ad hoc wireless networks. Mobile Networks and Applications 8(5), 499–515 (2003)
[64] Coniglio, A., Angelis, U.D., Forlani, A., Lauro, G.: Distribution of physical clusters. Journal of Physics A: Mathematical and General 10(2), 219–228 (1977)
[65] Coniglio, A., Angelis, U., Forlani, A.: Pair connectedness and cluster size. Journal of Physics A: Mathematical and General 10(7), 1123–1139 (1977)
[66] Conway, J., Torquato, S.: Tiling, packing, and covering with tetrahedra. Proc. National Academy of Sciences of the United States of America (PNAS) 103(28), 10612–10617 (2006)
[67] Cortes, J., Martinez, S., Karatas, T., Bullo, F.: Coverage control for mobile sensing networks. IEEE Transactions on Robotics and Automation 20(2), 243–255 (2004)
[68] Cressie, N.: Statistics for spatial data. John Wiley & Sons, Inc., Chichester (1991)
[69] Dall, J., Christensen, M.: Random geometric graphs. Physical Review E 66(1), 016121, 1–9 (2002)
[70] Das, S.K., Ammari, H.M.: Wireless Sensor Networks: A Networking Perspective. In: Zheng, J., Jamalipour, A. (eds.) Routing and data dissemination in wireless sensor networks. Wiley/ IEEE Press (to appear) (2008)
[71] Datta, A.K., Gradinariu, M., Linga, P., Raipin-Parvedy, P.: Self-* distributed query region covering in sensor networks. In: Proc. IEEE SRDS (2005)
[72] DeSimone, T., Demoulini, S., Stratt, R.M.: A theory of percolation in liquids. Journal of Chemical Physics 85(1), 391–400 (1986)

[73] DeSimone, T., Stratt, R.M., Demoulini, S.: Continuum percolation in an interacting system: Exact solution of the Percus-Yevick equation for connectivity in liquids. Physical Rev. Letters 56(11), 1140–1143 (1986)
[74] Du, X., Lin, F.: Maintaining differentiated coverage in heterogeneous sensor networks. EURASIP Journal on Wireless Comm. and Networking 5(4), 565–572 (2005)
[75] Du, X., Lin, F.: Improving routing in sensor networks with heterogeneous sensors. In: Proc. IEEE VTC, pp. 2528–2532 (2005)
[76] Du, X., Lin, F.: Improving Sensor Network Performance by Deploying Mobile Sensors. In: Proc. IEEE IPCC, pp. 67–71 (2005)
[77] Duarte, M., Hu, Y.: Distance based decision fusion in a distributed wireless sensor network. In: Proc. IPSN, pp. 392–404 (2003)
[78] Duarte-Melo, E.J., Liu, M.: Analysis of energy consumption and lifetime of heterogeneous wireless sensor networks. In: Proc. Globecom, pp. 21–25 (2002)
[79] Efthymiou, E., Nikoletseas, S., Rolim, J.: Energy balanced data propagation in wireless sensor networks. Wireless Networks 12(6), 691–707 (2006)
[80] Elfes, A.: Using occupancy grids for mobile robot perception and navigation. IEEE Computer 22(6), 46–57 (1989)
[81] Esfahanian, A.: Generalized measures of fault tolerance with application to n-cube networks. IEEE Transactions on Computers 38(11), 1586–1591 (1989)
[82] Esfahanian, A., Hakimi, S.: On computing a conditional edge-connectivity of a graph. Information Processing Letters 27(4), 195–199 (1988)
[83] Essam, J.W.: Percolation theory. Reports on Progress in Physics 43(7), 833–912 (1980)
[84] Fonseca, C., Fleming, P.: Genetic algorithms for multiobjective optimization: Formulation, discussion and generalization. In: Forrest, S. (ed.) Proc. Int. Conf. on Genetic Algorithms, pp. 416–423 (1993)
[85] Franceschetti, M., Cook, M., Bruck, J.: A geometric theorem for wireless network design optimization. Tech. Report (2002),
[86] `http://paradise.caltech.edu/papers/`
[87] Fremlin, D.H.: The clustering problem: some Monte-Carlo results. Journal de Physique 37(7-8), 813–817 (1976)
[88] Gabriel, K.R., Sokal, R.R.: A new statistical approach to geographic variation analysis. Systematic Zoology 18(3), 259–278 (1969)
[89] Ganjali, Y., Keshavarzian, A.: Load balancing in ad hoc networks: Single-path routing vs. multi-path routing. In: Proc. IEEE INFOCOM, pp. 1120–1125 (2004)
[90] Gao, J., Zhang, L.: Load balanced short path routing in wireless networks. In: Proc. IEEE INFOCOM, pp. 1099–1108 (2004)
[91] Ghosh, A., Das, S.K.: Coverage and connectivity issues in wireless sensor networks: A survey. Pervasive and Mobile Computing 4(3), 303–334 (2008)
[92] Gilbert, E.N.: Random plane networks. Journal of the Society for Industrial and Applied Mathematics 9(4), 533–543 (1961)
[93] Glauche, I., Krause, W., Sollacher, R., Greiner, M.: Continuum percolation of wireless ad hoc communication networks. Physica A 325(3), 577–600 (2003)
[94] Gradshteyn, I., Ryzhik, I.: Tables of integrals, series, and product, 6th edn. Academic Press, San Diego (2000)
[95] Graham, R.L., Knuth, D.E., Patashnik, O.: Concrete Mathematics: A Foundation for Computer Science. Addison-Wesley, Reading (1994)
[96] Grimmett, G.: Percolation. Springer, Heidelberg (1989)

[97] Gupta, H., Zhou, Z., Das, S.R., Gu, Q.: Connected sensor cover: Self-organization of sensor networks for efficient query execution. IEEE/ACM Transactions on Networking 14(1), 55–67 (2006)

[98] Guo, W., Liu, Z., Wu, G.: An energy-balanced transmission scheme for sensor networks. In: Poster, Proc. ACM SenSys, pp. 300–301 (2003)

[99] Haenggi, M.: Twelve reasons not to route over many short hops. In: IEEE VTC, pp. 3130–3134 (2004)

[100] Haenggi, M., Puccinielli, D.: Routing in ad hoc networks: A case for long hops. IEEE Communication Magazine 43(10), 93–101 (2005)

[101] Hall, P.: Introduction to the theory of coverage processes. John Wiley & Sons Inc., New York (1988)

[102] Harary, F.: Conditional connectivity. Networks 13, 347–357 (1983)

[103] Heinzelman, W., Chandrakasan, A., Balakrishnan, H.: An application-specific protocol architecture for wireless microsensor networks. IEEE Transactions on Wireless Communications 1(4), 660–670 (2002)

[104] Helms, L.L.V.: Introduction to Potential Theory. Wiley-Interscience, New York (1969)

[105] Heo, N., Varshney, P.K.: Energy-Efficient Deployment of Intelligent Mobile Sensor Networks. IEEE TSMC – Part A: Systems and Humans 35(1), 78–92 (2005)

[106] Hill, T.L.: Statistical mechanics. Dover Publications Inc., New York (1956)

[107] Hill, T.: Molecular clusters in imperfect gases. Journal of Chemical Physics 23(4), 617–622 (1955)

[108] Horton, M., Culler, D., Pister, K., Hill, J., Szewczyk, R., Woo, A.: MICA: The commercialization of microsensor motes. Sensors Magazine, 40-48 (April 2002)

[109] Huang, C., Tseng, Y.: The coverage problem in a wireless sensor network. In: Proc. ACM WSNA, pp. 115–121 (2003)

[110] Huang, C., Tseng, Y., Lo, L.: The coverage problem in three-dimensional wireless sensor networks. In: Proc. IEEE Globecom, pp. 3182–3186 (2004)

[111] Huang, C., Tseng, Y., Wu, H.: Distributed protocols for ensuring both coverage and connectivity of a wireless sensor network. ACM Transactions on Sensor Networks 3(1), 1–24 (2007)

[112] Huynh, T.T., Hong, C.S.: An energy*Delay efficient routing scheme for wireless sensor networks. In: Dalmau Royo, J., Hasegawa, G. (eds.) MMNS 2005. LNCS, vol. 3754, pp. 11–22. Springer, Heidelberg (2005)

[113] Intanagonwiwat, C., Govindan, R., Estrin, D.: Directed diffusion: A scalable and robust communication paradigm for sensor networks. In: Proc. ACM MobiCom, pp. 56–67 (2000)

[114] Jarry, A., Leone, P., Powell, O., Rolim, J.: An optimal data propagation algorithm for maximizing the lifespan of sensor networks. In: Gibbons, P.B., Abdelzaher, T., Aspnes, J., Rao, R. (eds.) DCOSS 2006. LNCS, vol. 4026, pp. 405–421. Springer, Heidelberg (2006)

[115] Ji, X., Zha, H.: Sensor positioning in wireless ad-hoc sensor networks using multidimensional scaling. In: Proc. IEEE Infocom, pp. 2652–2661 (2004)

[116] Jiang, A., Bruck, J.: Monotone percolation and the topology control of wireless networks. In: Proc. IEEE INFOCOM, pp. 327–338 (2005)

[117] Johnson, D., Maltz, D.: Mobile Computing. In: Dynamic source routing in ad hoc wireless networks. Kluwer Academic Publishers, Dordrecht (1996)

[118] Kao, G., Fevens, T., Opatrny, J.: Position-based routing on 3-D geometric graphs in mobile ad hoc networks. In: Proc. CCCG, pp. 1–4 (2005)

[119] Karp, B., Kung, H.: GPSR: Greedy perimeter stateless routing for wireless networks. In: Proc. ACM/IEEE MobiCom, pp. 243–254 (2000)

[120] Keeney, R.L., Raiffa, H.: Decision Making with Multiple Objectives. Cambridge University Press, Cambridge (1993)

[121] Kim, H., Abdelzaher, T., Kwon, W.: Minimum-energy asynchronous dissemination to mobile sinks in wireless sensor networks. In: Proc. ACM SenSys, pp. 193–204 (2003)

[122] Klein, L.: A boolean algebra approach to multiple sensor voting fusion. IEEE Transactions on Aerospace and Electronic Systems 29(2), 317–327 (1993)

[123] Kong, Z., Yeh, E.: Analytical lower bounds on the critical density in continuum percolation. In: Proc. Workshop on Spatial Stochastic Models in Wireless Networks (SpaSWiN), pp. 1–6 (2007)

[124] Koushanfar, F., Meguerdichian, S., Potkonjak, M., Srivastava, M.: Coverage problems in wireless ad-hoc sensor networks. In: Proc. IEEE INFOCOM, pp. 1380–1387 (2001)

[125] Kranakis, E., Singh, H., Urrutia, J.: Compass routing on geometric networks. In: Proc. Canadian Conf. on Computational Geometry (CCCG), pp. 51–54 (1999)

[126] Krishnamachari, B., Mourtada, Y., Wicker, S.: The energy-robustness tradeoff for routing in wireless sensor networks. In: Proc. IEEE ICC, pp. 1833–1837 (2003)

[127] Krishnamurthy, L.: Sensor networks: Promise and reality. In: Sensors Expo and Conference (invited presentation) (2004)

[128] Kumar, S., Lai, T.H., Arora, A.: Barrier coverage with wireless sensors. In: Proc. ACM MobiCom, pp. 284–298 (2005)

[129] Kumar, S., Lai, T.H., Balogh, J.: On $k$-coverage in a mostly sleeping sensor network. In: Proc. ACM MobiCom, pp. 144–158 (2004)

[130] Kumar, S., Lai, T.H., Posner, M.E., Sinha, P.: Optimal sleep-wakeup algorithms for barriers of wireless sensors. In: Proc. IEEE BROADNETS (2007)

[131] Latifi, S., Hegde, M., Naraghi-Pour, M.: Conditional connectivity measures for large multiprocessor systems. IEEE Transactions on Computers 43(2), 218–222 (1994)

[132] Lazos, L., Poovendran, R.: Coverage in heterogeneous sensor networks. In: Proc. WiOpt (2006)

[133] Lazos, L., Poovendran, R.: Stochastic coverage in heterogeneous sensor networks. ACM Transactions on Sensor Networks 2(3), 325–358 (2006)

[134] Lazos, L., Poovendran, R., Ritcey, J.: On the deployment of heterogeneous sensor networks for detection of mobile targets. In: Proc. WiOpt, pp. 1–10 (2007)

[135] Leone, P., Nikoletseas, S., Rolim, J.: An adaptive blind algorithm for energy balanced data propagation in wireless networks. In: Prasanna, V.K., Iyengar, S.S., Spirakis, P.G., Welsh, M. (eds.) DCOSS 2005. LNCS, vol. 3560, pp. 35–48. Springer, Heidelberg (2005)

[136] Li, X., Calinescu, G., Wan, P., Wang, Y.: Localized Delaunay triangulation with application in ad hoc wireless networks. IEEE Transactions on Parallel and Distributed Systems 14(10), 1035–1047 (2003)

[137] Li, L., Halpern, J.Y.: Minimum-energy mobile wireless networks revisited. In: Proc. IEEE ICC, pp. 278–283 (2001)

[138] Li, J., Mohapatra, P.: Analytical modeling and mitigation techniques for the energy hole problem in sensor networks. Pervasive and Mobile Computing 3(3), 233–254 (2007)

[139] Li, J., Mohapatra, P.: An analytical model for the energy hole problem in many-to-one sensor networks. In: Proc. IEEE VTC, pp. 2721–2725 (2005)

[140] Li, X.-Y., Wan, P.-J., Frieder, O.: Coverage in wireless ad-hoc sensor networks. IEEE Transactions on Computers 52(6), 753–763 (2003)

[141] Lian, J., Naik, K., Agnew, G.: Data capacity improvement of wireless sensor networks using non-uniform sensor distribution. Intl. Journal of Distributed Sensor Networks 2(2), 121–145 (2006)

[142] Lindsey, S., Raghavendra, C.S.: PEGASIS: Power-efficient gathering in sensor information systems. In: Proc. IEEE Aerospace Conference, vol. 3 (2002)

[143] Lindsey, S., Raghavendra, C., Sivalingam, K.: Data gathering algorithms in sensor networks using energy metrics. IEEE Transactions on Parallel and Distributed Systems 13(9), 924–935 (2002)

[144] Lindsey, S., Raghavendra, C.S., Sivalingam, K.M.: Data gathering in sensor networks using the energy*delay metric. In: Proc. IPDPS, pp. 2001–2008 (2001)

[145] Liu, B., Brass, P., Dousse, O.: Mobility improves coverage of sensor networks. In: Proc. ACM MobiHoc, Urbana-Champaign, Illinois, USA, pp. 300–308 (May 2005)

[146] Liu, D., Hu, X., Jia, X.: Energy efficient information dissemination protocols by negotiation for wireless sensor networks. Computer Communications 29(11), 2136–2149 (2006)

[147] Liu, B., Towsley, D.: A study of the coverage of large-scale sensor networks. In: Proc. IEEE MASS, pp. 475–483 (2004)

[148] Liu, C., Wu, K., Xiao, Y., Sun, B.: Random coverage with guaranteed connectivity: Joint scheduling for wireless sensor networks. IEEE Transactions on Parallel and Distributed Systems 17(6), 562–575 (2006)

[149] Luo, J., Hubaux, J.-P.: Joint mobility and routing for lifetime elongation in wireless sensor networks. In: Proc. IEEE INFOCOM, pp. 1735–1746 (2005)

[150] Luo, H., Ye, F., Cheng, J., Lu, S., Zhang, L.: TTDD: Two-tier data dissemination in large-scale wireless sensor networks. Wireless Networks 11(1-2), 165–175 (2005)

[151] Ma, Y., Dalal, S., Alwan, M., Aylor, J.: ROP: A resource oriented protocol for heterogeneous sensor networks. In: Proc. Virginia Tech's Symposium on Wireless Personal Communication, pp. 59–70 (2003)

[152] Malde, P., Oellermann, O.: The F-connectivity of a graph. Scientia, Series A: Mathematical Sciences 1, 65–71 (1988)

[153] Manjeshwar, A., Agrawal, D.P.: APTEEN: A hybrid protocol for efficient routing and comprehensive information retrieval in wireless sensor networks. In: Proc. IPDPS, pp. 195–202 (2002)

[154] Manjeshwar, A., Agrawal, D.P.: TEEN: A routing protocol for enhanced efficiency in wireless sensor networks. In: Proc. IPDPS, pp. 2009–2015 (2001)

[155] Marler, R.T., Arora, J.S.: Survey of multi-objective optimization methods for engineering. Structural and Multidisciplinary Optimization 26(6), 369–395 (2004)

[156] Meester, R., Roy, R.: Continuum percolation. Cambridge University Press, Cambridge (1996)

[157] Megerian, S., Koushanfar, F., Potkonjak, M., Srivastava, M.: Worst and best-case coverage in sensor networks. IEEE Transactions on Mobile Computing 4(1), 84–92 (2005)

[158] Megerian, S., Koushanfar, F., Potkonjak, M., Srivastava, M.: Exposure in wireless ad-hoc sensor networks. In: Proc. ACM MobiCom, pp. 139–150 (2001)

[159] Megerian, S., Koushanfar, F., Potkonjak, M., Srivastava, M.: Coverage problems in wireless ad-hoc sensor networks. In: Proc. IEEE INFOCOM, pp. 1380–1387 (2001)

[160] Mhatre, V.P., Rosenberg, C., Kofman, D., Mazumdar, R., Shroff, N.: A minimum cost heterogeneous sensor network with a lifetime constraint. IEEE TMC 4(1), 4–15 (2005)

[161] Miller, M., Sengul, C., Gupta, I.: Exploring the energy-latency trade-off for broadcasts in energy-saving sensor networks. In: Proc. IEEE ICDCS, pp. 17–26 (2005)

[162] Nath, S., Gibbons, P.B.: Communicating via fireflies: Geographic routing on duty-cycled sensors. In: Proc. IPSN, pp. 440–449 (2007)

[163] Nicules, D., Nath, B.: Ad-hoc positioning system (APS) using AoA. In: Proc. IEEE INFOCOM, pp. 1734–1743 (2003)

[164] Oellermann, O.: Conditional graph connectivity relative to hereditary properties. Networks 21(2), 245–255 (1991)

[165] Olariu, S., Stojmenovic, I.: Design guidelines for maximizing lifetime and avoiding energy holes in sensor networks with uniform distribution and uniform reporting. In: Proc. IEEE INFOCOM, pp. 1–12 (2006)

[166] Ornstein, L.S., Zernike, F.: Accidental deviations of density and opalescence at the critical point of a single substance. Proc. Akad. Sci. 17, 793–806 (1914)

[167] Pike, G.E., Seager, C.H.: Percolation and conductivity: A computer study. Physical Review B 10(4), 1421–1434 (1974)

[168] Poduri, S., Pattem, S., Krishnamachari, B., Sukhatme, G.S.: Sensor network configuration and the curse of dimensionality. In: Proc. IEEE EmNets (2006)

[169] Pompili, D., Melodia, T., Akyildiz, I.F.: Routing algorithms for delay-insensitive and delay-sensitive applications in underwater sensor networks. In: Proc. ACM MobiCom, pp. 298–309 (2006)

[170] Pompili, D., Melodia, T., Akyildiz, I.F.: Deployment analysis in underwater acoustic wireless sensor networks. In: Proc. ACM WUWNet, pp. 48–55 (2006)

[171] Powell, O., Leone, P., Rolim, J.: Energy optimal data propagation in wireless sensor networks. Journal of Parallel and Distributed Computing 3(67), 302–317 (2007)

[172] Rao, R., Kesidis, G.: Purposeful Mobility for Relaying and Surveillance in Mobile Ad Hoc Sensor Networks. IEEE TMC 3(3), 225–232 (2004)

[173] Ravelomanana, V.: Extremal properties of three-dimensional sensor networks with applications. IEEE Transactions on Mobile Computing 3(3), 246–257 (2004)

[174] Roberts, F.D.K.: A Monte Carlo solution of a two-dimensional unstructured cluster problem. Biometrika 54(3-4), 625–628 (1967)

[175] Rodoplu, V., Meng, T.H.: Minimum energy mobile wireless networks. IEEE Journal on Selected Areas in Communications 17(8), 1333–1344 (1999)

[176] Sankar, A., Liu, Z.: Maximum lifetime routing in wireless ad-hoc networks. In: Proc. IEEE INFOCOM, pp. 1090–1098 (2004)

[177] Shah, R., Roy, R., Jain, S., Brunette, W.: Data mules: Modeling a three-tier architecture for sparse sensor networks. In: Proc. IEEE SNPA, pp. 1058–1068 (2003)

[178] Shakkottai, S., Srikant, R., Shroff, N.: Unreliable sensor grids: Coverage, connectivity and diameter. Ad Hoc Networks 3(6), 702–716 (2005)

[179] Shakkottai, S., Srikant, R., Shroff, N.: Unreliable sensor grids: coverage, connectivity and diameter. In: Proc. IEEE INFOCOM, pp. 1073–1083 (2003)

[180] Shih, E., Cho, S., Ickes, N., Min, R., Sinha, A., Wang, A., Chandrakasan, A.: Physical layer driven protocol and algorithm design for energy-efficient wireless Sensor Networks. In: Proc. ACM MobiCom, pp. 272–287 (2001)

[181] Sohrabi, K., Gao, J., Ailawadhi, V., Pottie, G.: Protocols for self-organization of a wireless sensor network. IEEE Personal Communications 7(5), 16–27 (2000)

[182] Sohrabi, K., Merill, W., Elson, J., Girod, L., Newberg, F., Kaiser, W.: Methods for scalable self-assembly of ad hoc wireless sensor networks. IEEE TMC 3(4), 317–331 (2004)
[183] Soltan, M., Maleki, M., Pedram, M.: Lifetime-aware hierarchical wireless sensor network architecture with mobile overlays. In: Proc. IEEE Radio and Wireless Symposium, pp. 325–328 (2007)
[184] Stojmenovic, I., Lin, X.: Loop-free hybrid single-path/flooding routing algorithms with guaranteed delivery for wireless networks. IEEE Transactions on Parallel and Distributed Systems 12(10), 1023–1032 (2001)
[185] Sun, T., Chen, L., Han, C., Gerla, M.: Reliable sensor networks for planet exploration. In: Proc. IEEE Int'l Conf. Networking, Sensing and Control, pp. 816–821 (2005)
[186] Tian, D., Georganas, N.: Connectivity maintenance and coverage preservation in wireless sensor networks. Ad Hoc Networks 3(6), 744–761 (2005)
[187] Tipler, P.: Physics for Scientists and Engineers: Electricity, Magnetism, Light, and Elementary Modern Physics, 5th edn. W. H. Freeman, New York (2004)
[188] Torquato, S.: Bulk properties of two-phase disordered media. I. Cluster expansion for the effective dielectric constant of dispersions of penetrable spheres. Journal of Chemical Physics 81(11), 5079–5088 (1984)
[189] Vicsek, T., Kertesz, J.: Monte Carlo renormalization-group approach to percolation on a continuum: test of universality. Journal of Physics A: Mathematical and General 14, L31–L37 (1981)
[190] Wan, P.-J., Yi, C.-W.: Coverage by randomly deployed wireless sensor networks. IEEE Transactions on Information Theory 52(6), 2658–2669 (2006)
[191] Wang, G., Cao, G., La Porta, T.: Proxy-based sensor deployment for mobile sensor networks. In: Proc. IEEE MASS, pp. 493–502 (2004)
[192] Wang, G., Cao, G., La Porta, T.: Movement-assisted sensor deployment. In: Proc. IEEE INFOCOM, pp. 2469–2479 (2004)
[193] Wang, G., Cao, G., La Porta, T., Zhang, W.: Sensor Relocation in Mobile Sensor Networks. In: Proc. IEEE Infocom, pp. 2302–2312 (2005)
[194] Wang, Y.-C., Hu, C.-C., Tseng, Y.-C.: Efficient Placement and Dispatch of Sensors in a Wireless Sensor Network. IEEE TMC 7(2), 262–274 (2008)
[195] Wang, Y., Wang, X., Agrawal, D.P., Minai, A.A.: Impact of heterogeneity on coverage and broadcast reachability in wireless sensor networks. In: Proc. IEEE ICCCN, pp. 63–67 (2006)
[196] Wang, W., Srinivasan, V., Chua, K.-C.: Using mobile relays to prolong the lifetime of wireless sensor networks. In: Proc. ACM MobiCom, pp. 270–283 (2005)
[197] Wang, Y.-C., Tseng, Y.-C.: Distributed deployment schemes for mobile wireless sensor networks to ensure multi-level coverage. IEEE Transactions on Parallel and Distributed Systems 19(9), 1280–1294 (2008)
[198] Wang, X., Xing, G., Zhang, Y., Lu, C., Pless, R., Gill, C.: Integrated coverage and connectivity configuration in wireless sensor networks. In: Proc. ACM SenSys, pp. 28–39 (2003)
[199] Woo, A., Tony, T., Culler, D.: Taming the underlying challenges of reliable multihop routing in sensor networks. In: Proc. ACM SenSys, pp. 14–27 (2003)
[200] Wu, X., Chen, G., Das, S.K.: Avoiding Energy Holes in Wireless Sensor Networks with Nonuniform Node Distribution. IEEE Transactions on Parallel and Distributed Systems 19(5), 710–720 (2008)

[201] Wu, X., Chen, G., Das, S.K.: On the energy hole problem of nonuniform node distribution in wireless sensor networks. In: Proc. IEEE MASS, pp. 180–187 (2006)
[202] Wu, J., Guo, G.: Fault tolerance measures for m-ary n-dimensional hypercubes based on forbidden faulty sets. IEEE Transactions on Computers 47(8), 888–893 (1998)
[203] Wu, J., Yang, S.: Optimal Movement-Assisted Sensor Deployment and Its Extensions in Wireless Sensor Networks. Simulation Modeling Practice and Theory 15, 383–399 (2007)
[204] Wu, J., Yang, S.: SMART: A Scan-Based Movement-Assisted Sensor Deployment Method in Wireless Sensor Networks. In: Proc. IEEE Infocom, pp. 2313–2324 (2005)
[205] Xing, G., Lu, C., Pless, R., Huang, Q.: On greedy geographic routing algorithms in sensing-covered networks. In: Proc. ACM MobiHoc, pp. 31–42 (2004)
[206] Xing, G., Wang, X., Zhang, Y., Lu, C., Pless, R., Gill, C.: Integrated coverage and connectivity configuration for energy conservation in sensor networks. ACM Transactions on Sensor Networks 1(1), 36–72 (2005)
[207] Yang, S., Cardei, M.: Movement-Assisted Sensor Redeployment Scheme for Network Lifetime Increase. In: Proc ACM MSWiM, pp. 13–20 (2007)
[208] Yang, S., Dai, F., Cardei, M., Wu, J.: On connected multiple point coverage in wireless sensor networks. Int. Journal of Wireless Information Networks 13(4), 289–301 (2006)
[209] Yang, X., Vaidya, N.: A wakeup scheme for sensor networks: Achieving balance between energy saving and end-to-end delay. In: Proc. IEEE RTAS, pp. 19–26 (2004)
[210] Yarvis, M., Kushalnagar, N., Singh, H., Rangarajan, A., Liu, Y., Singh, S.: Exploiting heterogeneity in sensor networks. In: Proc. IEEE INFOCOM, pp. 878–890 (2005)
[211] Ye, F., Luo, H., Cheng, J., Lu, S., Zhang, L.: A two-tier data dissemination model for large-scale wireless sensor networks. In: Proc. ACM MobiCom, pp. 148–159 (2002)
[212] Ye, F., Zhong, G., Cheng, J., Lu, S., Zhang, L.: PEAS: A robust energy conserving protocol for long-lived sensor networks. In: Proc. ICDCS, pp. 1–10 (2003)
[213] Yener, B., Magdon-Ismail, M., Sivrikaya, F.: Joint problem of power optimal connectivity and coverage in wireless sensor networks. Wireless Networks 13(4), 537–550 (2007)
[214] Younis, M., Akkaya, K.: Strategies and Techniques for Node Placement in Wireless Sensor Networks: A Survey. Ad Hoc Networks 6(4), 621–655 (2008)
[215] Yuce, M., Ng, P., Lee, C., Khan, J., Liu, W.: A wireless medical monitoring over a heterogeneous sensor network. In: Proc. IEEE EMBS, pp. 5894–5898 (2007)
[216] Zhang, W., Cao, G., La Porta, T.: Dynamic proxy tree-based data dissemination schemes for wireless sensor networks. In: Proc. IEEE MASS, pp. 21–30 (2004)
[217] Zhang, H., Hou, J.: Is deterministic deployment worse than random deployment for wireless sensor networks? In: Proc. IEEE INFOCOM, pp. 1–13 (2006)
[218] Zhang, H., Hou, J.: On the upper bound of a-lifetime for large sensor networks. ACM Transactions on Sensor Networks 1(2), 272–300 (2005)
[219] Zhang, H., Hou, J.: Maintaining sensing coverage and connectivity in large sensor networks. Ad Hoc & Sensor Wireless Networks 1(1-2), 89–124 (2005)
[220] Zhang, H., Hou, J.: On Deriving the Upper Bound of a-Lifetime for Large Sensor Networks. In: Proc. ACM MobiHoc, pp. 121–132 (2004)

[221] Zhang, H., Shen, H., Tan, Y.: Optimal energy balanced data gathering in wireless sensor networks. In: Proc. IEEE IPDPS, pp. 1–10 (2007)
[222] Zhao, J., Govindan, R.: Understanding packet delivery performance in dense wireless sensor networks. In: Proc. ACM SenSys, pp. 1–13 (2003)
[223] Zhou, Z., Das, S., Gupta, H.: Fault tolerant connected sensor cover with variable sensing and transmission ranges. In: Proc. IEEE SECON, pp. 594–604 (2005)
[224] Zhou, Z., Das, S., Gupta, H.: Connected *k*-coverage problem in sensor networks. In: Proc. ICCCN, pp. 373–378 (2004)
[225] Zhou, G., He, T., Krishnamurthy, S., Stankovic, J.: Impact of radio irregularity on wireless sensor networks. In: Proc. MobiSys, pp. 125–138 (2004)
[226] Zhou, Y., Lyu, M.R., Liu, J.: On setting up energy-efficient paths with transmitter power control in wireless sensor networks. In: Proc. MASS, pp. 1–9 (2005)
[227] Zorzi, M., Rao, R.: Geographic random forwarding (GeRaF) for ad hoc and sensor networks: Multihop performance. IEEE Transactions on Mobile Computing 2(4), 337–348 (2003)
[228] Zorzi, M., Rao, R.: Geographic random forwarding (GeRaF) for ad hoc and sensor networks: Energy and latency performance. IEEE Transactions on Mobile Computing 2(4), 349–365 (2003)
[229] Zou, Y., Chakrabarty, K.: A distributed coverage- and connectivity-centric technique for selecting active nodes in wireless sensor networks. IEEE Transactions on Computers 54(8), 978–991 (2005)
[230] Zou, Y., Chakrabarty, K.: Sensor deployment and target localization in distributed sensor networks. ACM Transactions on Embedded Computing Systems 3(1), 61–91 (2004)
[231] Motes, smart dust sensors, wireless sensor networks, `http://www.xbow.com/wireless_home.aspx`
[232] `http://mathworld.wolfram.com/BetaFunction.html`
[233] `http://mathworld.wolfram.com/Erf.html`
[234] `http://mathworld.wolfram.com/ReuleauxTriangle.html`
[235] `http://mathworld.wolfram.com/ReuleauxTetrahedron.html`

# Subject Index

MIX
Papier aus verantwortungsvollen Quellen
Paper from responsible sources
FSC® C105338

If you have any concerns about our products,
you can contact us on
**ProductSafety@springernature.com**

In case Publisher is established outside the EU,
the EU authorized representative is:
**Springer Nature Customer Service Center GmbH**
**Europaplatz 3, 69115 Heidelberg, Germany**

Printed by Libri Plureos GmbH
in Hamburg, Germany